Lychee Disease Management

Manoj Kumar • Vivek Kumar
Neera Bhalla-Sarin • Ajit Varma
Editors

Lychee Disease Management

Springer

Editors
Manoj Kumar
Amity Institute of Microbial Technology
Amity University
Noida, Uttar Pradesh
India

Vivek Kumar
Amity Institute of Microbial Technology
Amity University
Noida, Uttar Pradesh
India

Neera Bhalla-Sarin
School of Life Sciences
Jawaharlal Nehru University
New Delhi, Delhi
India

Ajit Varma
Amity Institute of Microbial Technology
Amity University
Noida, Uttar Pradesh
India

ISBN 978-981-10-4246-1 ISBN 978-981-10-4247-8 (eBook)
DOI 10.1007/978-981-10-4247-8

Library of Congress Control Number: 2017940394

Printed on acid-free paper

This Springer imprint is published by Springer Nature
The registered company is Springer Nature Singapore Pte Ltd.
The registered company address is: 152 Beach Road, #21-01/04 Gateway East, Singapore 189721, Singapore

Preface

"Lychee Disease Management" is a stand-alone book dedicated to the global agriculture community facing problems in production and commercialization worldwide.

This book is a comprehensive compilation of biotic and abiotic factors seemingly affecting lychee production and its commercialization. The book comprises disease management for its causal agents conferring leaf mite (*Aceria litchii* Keifer), leaf miner (*Conopomorpha cramerella*), fruit borers (*Conopomorpha cramerella*, *Platypeplus aprobola* Meyer, and *Dichocrocis* sp.), leaf webber/roller (*Platypeplus aprobola* Meyer), lychee bug (*Tessaratoma javanica* Thunberg), bark-eating caterpillar (*Indarbela quadrinotata*), shoot borer (*Chlumetia transversa*), etc.

Specialized chapters of the book uncover the statistical data at international level and recommend potential ways for lychee export, further illustrating the scope to increase the quantum of export; more so because the harvesting season is quite different in other parts of the world.

It also sheds light on systematic research for identification of additional potential areas and development and refinement of technologies for enhancing the productivity and quality of lychee. This book comprises the managerial understanding on post-harvest handling, processing and value addition, development-tolerant varieties, high yield, and processing. It includes explicit insights through a comprehensive visionary documentation addressing scientific and economical aspects for all the neglected fruit enhancements.

Noida, Uttar Pradesh, India — Manoj Kumar
Noida, Uttar Pradesh, India — Vivek Kumar
Delhi, India — Neera Bhalla-Sarin
Noida, Uttar Pradesh, India — Ajit Varma

Contents

1 **Lychee (*Litchi chinensis* Sonn.): Pre- and Post-harvest Disease Management** 1
Bhupendra Koul and Pooja Taak

2 **Microbial Interactions in Litchi Rhizosphere** 27
Kanchan Vishwakarma, Neha Upadhyay, Nitin Kumar, Rishi Verma, Jaspreet Singh, Pankaj Verma, Mitali Mishra, Shruti Jain, Durgesh Kumar Tripathi, Rohit Kumar Mishra, Vivek Kumar, and Shivesh Sharma

3 **Pre- and Postharvest Management Practices for Litchi Production in India** 45
Ravi Rajwanshi, Manoj Kumar, and Beche Lal

4 **Determination of Substrate Medium for Litchi Marcot Establishment in Nursery** 67
Prity Sagar, Anfal Arshi, and Awadh Kishore Roy

5 **Impact Assessment of Bio-inoculants on Growth of Litchi [*Litchi chinensis* (Gaertn.) Sonn.] Plants** 75
Prity Sagar, Anfal Arshi, and Awadh Kishore Roy

6 **Important Diseases of Litchi and Their Management** 89
Ajit Kumar Dubedi Anal, Vinod Kumar, Mritunjay Tripathi, and Vishal Nath

7 **Tissue Culture: Present Scenario and Future Prospects in Litchi Management** 99
Vishal Nath, Manvendra Singh, Alok Kumar Gupta, Evening Stone Marboh, and Karmabeer Yadav

8 **Systemic Information for Future Perspectives in Litchi Crop Improvement** 109
Alok Kumar Gupta, Vishal Nath, Awtar Singh, Manvendra Singh, Evening Stone Marboh, Swapnil Pandey, and Ashish Pathak

9 **Problems and Prospects of Lychee Cultivation in India** 139
Amar Singh Kashyap and Nitika Thakur

10 **Lychee Diseases and its Viable Control** 167
Shanker Kumar Pandey, Awadh Kishore Roy, Shyam Chandra Roy Chandel, Anuj Rani, Rashmi Singh, and Ravi Shankar Mishra

11 **Lychee Juice and Pulp Extracts as Potential Components for Production of Extracellular Phosphate-Binding Biopolymer from *Acinetobacter haemolyticus*** 181
Moushumi Ghosh

12 ***Litchi chinensis*: Taxonomy, Botany and Its Cultivars** 191
Renu Soni and Shachi Agrawal

13 **Procedural Insights on In Vitro Propagation of *Litchi chinensis* (Sonn.)** 217
Vivek Kumar, Manoj Kumar, Shivesh Sharma, Ajit Varma, and Neera Bhalla-Sarin

14 **Lychee (*Litchi chinensis*): Biochemistry, Panacea, and Nutritional Value** 237
Jasra Anjum, Rafiq Lone, and Khursheed Ahmad Wani

15 **Litchi Deterioration in India** 257
Shiwani Chaudhary, Manoj Kumar, and M.L. Aggarwal

16 **Integrated Approach for the Management of Differential Patterns of Diseases and Pest Incidence in Lychee** 265
Nitika Thakur

About the Editors

Manoj Kumar, Ph.D. Dr. Manoj Kumar is a scientist with sanguine behavior who is adoring about research and development, with a commitment to lifelong learning. He is determined on high-quality science that contributes broadly to increasing intellectual knowledge of both plant development and the ecological niche. He has a high level of professional desire for intellectual pursuits and the potential to fulfill his dream of high-impact publications and future recognition of these by academic peers. Dr. Kumar has pursued his Ph.D. in plant biotechnology at the prestigious Jawaharlal Nehru University and then been awarded two postdoctoral fellowships consecutively: DBT-PDF from IISc Bangalore in 2005 and then NRF-PDF from the University of Pretoria. Dr. Manoj Kumar is a researcher of plant biotechnology in the Division of Microbial Technology at Amity University, Uttar Pradesh, India. Recently he has accepted the affiliation from Ton Duc Thang University, Vietnam. Until recently, he was a coordinator of the Bio-resource Chapter (Northern India) and served on editorial boards of five international journals. Dr. Kumar has published several research papers, books, and review articles of international repute. During a decade of academic acquaintance, he has guided several research projects and dissertations and collaborated internationally. His diverse research background attracts global readers and researchers.

Vivek Kumar, Ph.D. Dr. Vivek Kumar is a scientist who is involved in teaching, research, and guidance, with a pledge to enduring knowledge. Dr. Kumar is working in the Division of Microbial Technology at Amity University, Uttar Pradesh, Noida, India. He is serving in the editorial board of reputed international journals, viz., *EnvironmentAsia*, the *International Journal of Biological and Chemical Sciences*, the *Journal of Advanced Botany and Zoology*, and the *Journal of Ecobiotechnology*. He is also reviewer of the *Journal of Hazardous Materials, Science International, Acta*

Physiologiae Plantarum, the *International Research Journal of Plant Sciences*, the *International Journal of Microbiology*, the *African Journal of Microbiology Research*, the *Journal of Microbiology and Antimicrobials, Environmental Science and Pollution Research*, and *Rhizosphere*. He has published 61 research papers, 19 book chapters, 6 review articles, and 2 books. Dr. Kumar has also served as a microbiologist for 8 years in the Department of Soil and Water Research, Public Authority for Agricultural Affairs and Fish Resources, Kuwait. Dr. Kumar's research areas are plant-microbe interactions, environmental microbiology, and bioremediation. He has been credited with first-time reporting and identification of pink rot inflorescence disease of date palm in Kuwait caused by *Serratia marcescens*. He has been awarded the "Young Scientist Award" for the year 2002 in "agricultural microbiology" by the Association of Microbiologists of India (AMI). Dr. Kumar is establishing an "unearthing and deliverance system," where a balance is being strived between the development of drought- and salinity-resistant microbiome for better crop production in rain-fed and saline areas. In the bioremediation research program, the isolation and characterization of autochthonous microbiome from textile dye effluent and soil performed very well in the remediation of dyes under laboratory conditions. The selected microbiome will be further employed in the bioremediation of textile dyes at a larger level.

Neera Bhalla-Sarin, F.N.A.Sc. Prof. (Dr.) Neera Bhalla-Sarin is working as professor and group leader at the School of Life Sciences, JNU, New Delhi. She has served the organization as professor and dean. She has set academic milestones as a chairperson of many academic councils at university level. Prof. Sarin has a proven record in plant developmental biology and accomplished numerous research projects sponsored by the government of India and international research funding (Indo-Switzerland, Indo-Korea, Indo-Australia, Indo-USA). She has guided more than 40 doctoral and postdoctoral researchers with her sanguine research capability. Her remarkable research and academic contributions in science and technology have been acknowledged nationally and internationally.

Ajit Varma, Fellow, AvH and NAASC Prof. Ajit Varma is distinguished scientist and professor of eminence at the Amity Institute of Microbial Technology (Amity University, Uttar Pradesh). He has been leading an international research group on microbial technology in collaboration with several prestigious institutions worldwide. He is also holding several other responsibilities in Amity University, like vice-chairman of the Amity Science, Technology and Innovation Foundation and chairman of the Faculty Research Council at university level. He has pursued his doctorate at Allahabad University in 1964 and then started his academic and scientific journey in the Indian Agricultural Research Institute, New Delhi, and then retired as an eminent professor at the prestigious Jawaharlal Nehru University in 2004. Since then, his leading role incepted in Amity University to harness the Amity Research at international level. Prof. Varma has numerous national and international research and academic awards to his credit and headed several councils in the plant-microbial world. He has visited several countries as a visiting scientist, professor, and academician for his world novel discovery *Piriformospora indica* – a magic fungus which has been popularized as ROOTONIC. Apart from the abovementioned facts, Prof. Varma has achieved academic height with several national and international accreditations.

Lychee (*Litchi chinensis* Sonn.): Pre- and Post-harvest Disease Management

1

Bhupendra Koul and Pooja Taak

Abstract

Lychee (*Litchi chinensis* Sonn.) belongs to the family Sapindaceae and is an esteemed member amongst the commercially important fruit crops. This delicious fruit is widely grown in tropical and subtropical regions of the world and is famous for its sweet fragrance, pleasurable taste and attractive colour. China is the largest producer of lychee in the world followed by India and Taiwan. Lychee tree is susceptible to various biotic stresses which include algae, fungi, insect pests, etc. These factors often become a hindrance for profitable fruit production. Moreover, post-harvest damage such as pericarp browning and desiccation ultimately declines the commercial value and shelf life of the fruit. Effective and timely approaches and post-harvest management practices are prerequisites in order to sustain the premium fruit quality, fruit yield and shelf life and also to control the pre-harvest diseases. Development in biotechnological techniques also provides an alternative approach for the crop improvement by introducing exotic genes with desirable characters. These genetically modified varieties hold a promising potential to upgrade the fruit quality standard in terms of disease resistance, increased shelf life and seedless character. The post-harvest practices in lychee production and their impact on fruit quality need more improvement. This chapter encompasses various diseases that downturn the yield, quality and market value of the lychee fruit and the strategies to check the pre- and post-harvest crop losses.

Keywords

Lychee • Fruit crop • Pre-harvest • Post-harvest • Shelf life

B. Koul (✉) • P. Taak
School of Bioengineering and Biosciences, Lovely Professional University, Phagwara 144411, Punjab, India
e-mail: bhupendra.18673@lpu.co.in

M. Kumar et al. (eds.), *Lychee Disease Management*,
DOI 10.1007/978-981-10-4247-8_1

Abbreviations

AA	Ascorbic acid
AFLP	Amplified fragment length polymorphism
FAO	Food and Agriculture Organization
GFP	Green fluorescent protein
GMO	Genetically modified organism
Ha	Hectare
HCl	Hydrochloric acid
IPM	Integrated pest management
PCR	Polymerase chain reaction
RAPD	Random amplified polymorphic DNA
RT-PCR	Reverse transcriptase PCR

1.1 Introduction

The lychee fruit (*Litchi chinensis* Sonn.) is a popular export commodity because of the attractive colour of its pericarp and exotic flavour. It is grown as an important commercial crop in China, South Africa, Madagascar, the USA, Australia, Mauritius, India, Pakistan, Thailand, Indonesia, Vietnam and the Philippines (Menzel 2001). China is the leading lychee-producing country in the world, having more than 584,000 ha area under cultivation with an annual production of 958,700 MT (Sarin et al. 2009). India is the second largest producer of lychee after China with an annual production of 585,300 MT, from 84,170 ha area (FAO 2014). There has been a year-wise expansion in the area under lychee cultivation. It has increased from 58,100 thousand ha to 84,200 ha in 1991–1992 to 2013–2014 with a similar trend in the production from 3,55,900 MT to 5,85,300 MT in the last decades. Several varieties of lychee are available throughout the world which can be distinguished from each other with respect to their morphological characteristics including fruit shape, size, shape of skin segment, colour, taste, aroma, etc. (Sarin et al. 2009). The economically most important lychee varieties cultivated in India are Dehra, Purbi, Kashba, Early and Late Bedana, China, Shahi, Deshi, etc. (Das and Rahman 2012).

Lychee is an average sized (10–15 m), evergreen, round-topped tree with smooth, grey-coloured trunk. Its leaves are pinnately divided, having leather texture and acuminate, glabrous and slightly reddish in colour when young and bright green at maturity. Flowers are small yellow in colour and are apetalous. Fruits are oval, rounded or heart shaped, dark pink to red in colour. The edible part of the fruit is called aril, which is creamy white, translucent, sweet fragrant and juicy. The seeds are oblong with smooth and glossy surface and are reddish brown in colour (Nacif et al. 2001; Menzel 2002). The lychee fruit production is affected by several biotic and abiotic stresses. Lychee tree and fruits are susceptible to several algae, fungi

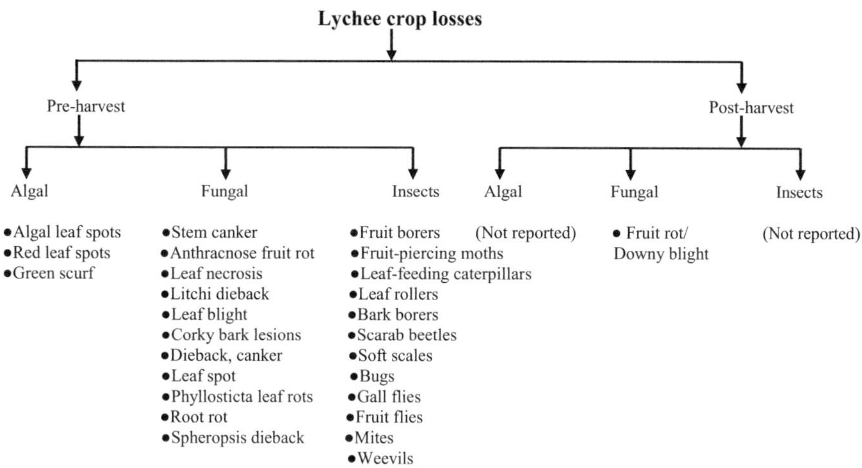

Chart 1.1 Diseases/damages (pre-harvest and post-harvest) caused by various biotic factors

and insect pests which damage the fruits and other vegetative parts. Amongst these, fungal diseases cause extensive pre- and post-harvest loss to the fruit quality and yield of the crop and hence considered as the most serious obstacle in lychee fruit production. These fungal pathogens belong to class *Dothideomycetes*, *Sordariomycetes*, *Leotiomycetes*, *Ascomycota* and *Oomycetes* (Mcmillan 1994; Menzel 2002; Wu et al. 2011). The major portion of the insects which damage the crop belongs to the order Lepidoptera. It has also been estimated that about 70% of the yield has been reported to retard as a result of the various abiotic stresses (Acquaah 2007). The lychee trees are also sensitive to salt-enriched soil. Salt stress in lychee also hinders the fruit production (Sinha and Das 2013). Hence, the cumulative effect of biotic as well as abiotic factors ultimately declines the agronomically importance of this cash crop. In this chapter, we shall solely deal with the effect of biotic stresses on lychee production as summarized in Chart 1.1.

1.2 Algal Leaf Spots, Red Leaf Spots and Green Scurf

Causative Organism *Cephaleuros virescens* (Class: Ulvophyceae)

Symptoms The algal leaf spots, red leaf spots and green scurf caused by *Cephaleuros virescens* are infrequent leaf spots of lychee (Alfieria et al. 1994). On lychee leaves, reddish brown to orange, velvety and cushion-like patches are formed. The spots are not observed on branches. The algal sporangia formed on fine hairs, germinate in moisture and produce zoospores which find their way through stomata and form mycelium-like chains of algal cells in the leaf tissues. The disease spread mostly during the rainy season from June to October. As the size of the leaf increases, the velvety growth becomes more dense and prominent. Older leaves show different types of malformations. The velvety growth turns dark brown to brick reddish in colour (Fig. 1.1a, Mcmillan 1994).

Control Measures Copper oxychloride spray (0.3%) can be done in the month of July and October. Bordeaux spray can also be done during autumn and spring season at 15-day interval. Ziram spray (0.25%) also reduces the risk of disease reoccurrence.

1.3 Fungal Diseases

1.3.1 Stem Canker

Causative Organism *Botryosphaeria* spp. (Class: *Dothideomycetes*)

Symptoms *Botryosphaeria* spp. normally attack the terminal branches of lychee tree (Alfieria et al. 1994). This fungus enters through the wounded surface on the tree and on dead and dying twigs. The disease is characterised by the presence of sunken, shrinking, irregular and dying tissues on the stem (Fig. 1.1b, Mcmillan 1994).

Control Measures Wound paint should be applied on the cut surfaces of the tree (Mcmillan 1994).

1.3.2 Anthracnose Fruit Rot

Causative Organism *Colletotrichum gloeosporioides* (Class: *Sordariomycetes*)

Symptoms Anthracnose fruit rot is the most important lychee disease (Alfieria et al. 1994). The fruit is highly susceptible to infection from flowering time. The small patches of infections coalesce and develop into large brown spots at fruit maturity. The infected fruits often develop a white mycelial layer over the fruit skin during refrigerated storage (Mcmillan 1994). Initially, the acervuli remain sub-epidermal but soon rupture the epidermis to expose the conidial mass. The fungus mostly attacks the leaves, flowers and fruits. Grey-coloured lesions appear on the leaf surface (Fig. 1.1c, Menzel 2002).

Control Measures Avoid overcrowding of trees and branches in orchard. Fungicides can be used during an initial stage but are not always effective (Menzel 2002). Application of chlorine dioxide (ClO_2) can reduce infection spreading by inhibiting the germination of fungal spores (Wu et al. 2011) Storing of the crop at lower temperature can also reduce the risk of fruit damage. Dipping of the fruits in hot benomyl at 0.05% at 52 °C for 2 min can also retard the rate of fruit deterioration.

1.3.3 Leaf Necrosis

Causative Organism *Colletotrichum gloeosporioides* (Class: *Sordariomycetes*)

Symptoms *Colletotrichum gloeosporioides* has been associated with leaf lesions caused by insect feeding and due to other mechanical injuries. In leaf necrosis, cylindrical pink-coloured conidia of *C. gloeosporioides* are produced in acervuli (Fig. 1.1d, Menzel 2002).

Control Measures Avoid overhead irrigation because the spores are spread by water splashes. Remove the highly infected plants from the orchard. In case of severe infestation, chlorothalonil, mancozeb and copper-based fungicides can be used to control the spreading of disease.

1.3.4 Litchi Dieback

Causative Organism *Diplodia* spp. (Class: *Dothideomycetes*)

Symptoms The dieback symptoms often appear after drought and other physiological stresses. The stem starts dying from tip downwards, often starting in the flowering stem. Infected wood turns brown or black and often somewhat shrivelled (Fig. 1.1e). Fungal pycnidia are produced in dead shoots (Mcmillan 1994).

Control Measures Pruning of the trees can reduce the spread of disease to some extent.

1.3.5 Leaf Blight

Causative Organism *Gloeosporium* spp. (Class: *Leotiomycetes*)

Symptoms Leaf blight is one of the important lychee diseases. The symptoms of this disease start from the tip of the leaf and then spread to the leaf margins. The disease generally starts appearing from late May to August. The leaf spots are light brown in colour and often appear to be that of scorched leaves (Fig. 1.1f). Defoliation can also occur during the rainy season (Mcmillan 1994).

Fig. 1.1 Algal and fungal diseases of lychee. (**a**) A leaf infected with *Cephaleuros virescens* showing reddish orange patches. (**b**) Stem canker caused by the fungus *Botryosphaeria* sp. (**c**) Anthracnose fruit rot and pericarp browning caused by the fungus *Colletotrichum gloeosporioides*. (**d**) Leaf necrosis caused by *C. gloeosporioides*. (**e**) A lychee tree showing symptoms of dieback disease caused by the fungus *Diplodia* spp. (**f**) A lychee twig showing symptoms of leaf blight caused by the fungus *Gloeosporium* spp.

Control Measures Spray of fungicides including thiophanate methyl (0.15%), chlorothalonil (0.15%), difenoconazole (0.05%) and copper oxychloride (0.25%) can be done during severe infestation.

1.3.6 Corky Bark Lesions

Causative Organism *Phoma* spp. (Class: *Dothideomycetes*)

Symptoms Symptoms of the corky bark lesions appear as small to large irregular patches of dark-coloured raised bark on the main trunk and then on the lateral branches of the tree. With the growth of the tree, the lumps and lesions also increase in size and become more corky and rough in appearance (Fig. 1.2a). The branches get covered with brown-coloured rough lesions which are of ¼ to ¾ inches in diameter (Mcmillan 1994).

Control Measures No fungicides are approved to completely stop the occurrence of corky bark lesions. Pruning of the severely damaged trees is useful to control the spreading of disease.

1.3.7 Dieback, Canker and Leaf Spot

Causative Organism *Phomopsis* spp. (Class: *Sordariomycetes*)

Symptoms *Phomopsis* spp. cause the tips of the branches to turn to black (Alfieria et al. 1994). Older trees are more prone to this disease. Leaf spots are small and reddish brown to dark black in colour with an average diameter of 1/8 in., while the cankers have rough, cracked outer cork and are greyish brown in colour (Fig. 1.2b, Mcmillan 1994).

Control Measures Pruning of the tree should be done carefully in order to ensure the complete eradication of the pathogen.

1.3.8 Phyllosticta Leaf Rots

Causative Organism *Phyllosticta* spp. (Class: *Dothideomycetes*)

Symptoms The leaf spots are rounded, large and brownish to black in colour, having concentric ridges (Fig. 1.2c). The younger leaves may become curled, while the older ones become shrivelled and hang down from the stem (Mcmillan 1994).

Control Measures Remove the infected leaves from the plant and dispose them in order to prevent the spreading of disease. Spray of fungicides such as ferbam, dithane M45, mancozeb and captan will help to control the infection. But it is not possible to completely eliminate the fungus from the plant.

1.3.9 Root Rot

(a) *Causative Organism* *Pythium* spp. (Class: *Oomycota*)

Symptoms The young roots of the infected tree become flabby, with rounded root tips, and dehydrated (Mcmillan 1994). The leaves of the infected plant are generally small and pale yellow in colour. Under severe damage, defoliation may also occur (Fig. 1.2d).

Control Measures Eliminate the infected trees and destroy them. No fungicide is approved for the treatment of this fungus.

(b) *Causative Organism* *Rhizoctonia solani* (Class: *Agaricomycetes*)

Symptoms The roots of the infected plant turn dark brown to black in colour which eventually die (Fig. 1.2e). A slow decline and sudden death of the lychee tree has also been reported. It can affect the whole tree or just few branches (Mcmillan 1994).

Control Measures Eliminate the infected tree from nursery and destroy them. No fungicide is approved (Mcmillan 1994). Addition of organic manure to the soil proves useful to improve nutrient balancing in the soil. A well drainage system should be applied in order to ensure that there is no standing water in the orchard.

1.3.10 Spheropsis Dieback

Causative Organism *Spheropsis* spp. (Class: *Ascomycota*)

Symptoms The disease starts spreading along the branches and forms cankers which produces a witch's broom (Fig. 1.2f). The twigs which develop from the brooms eventually dieback (Mcmillan 1994).

Control Measures Remove (pruning) the witch's broom as soon as they appear. No fungicide is approved for the control of this fungus.

Fig. 1.2 Fungal diseases of lychee. (**a**) Corky bark lesions caused by *Phoma* species. (**b**) Dieback, canker and leaf-spot symptoms caused by fungus *Phomopsis* spp. (**c**) Leaf rot caused *Phyllosticta* spp. (**d**) Root rot symptoms caused *Pythium* spp. (**e**) Root rot caused by *Rhizoctonia solani*. (**f**) Dieback symptoms caused by *Spheropsis* spp.

1.3.11 Fruit Rot/Downy Blight

Causative Organism *Peronophythora litchi* (Class: Oomycetes)

Symptoms Fruit rot or downy blight is one of the major diseases of lychee. The fungus mostly attacks the leaves, panicles and fruits (Fig. 1.3a,b). Immature fruits turn brownish in colour and some develop white-coloured mildew on their skin (Menzel 2002). Continuous rain leads to more spreading of the disease. The infection declines the yield and shelf life of the fruit.

Control Measures Remove the infected and dead branches from time to time so as to control the spreading of infection. The crude extract sprays of *Bacillus subtilis* can be done (Jiang et al. 2001a, b). Moreover, copper oxychloride spray in winter and copper sulphate in spring season can reduce the spread of the disease (Menzel 2002). Application of fungicide mandipropamid can also significantly protect the plant from downy blight disease (Tang et al. 2011). Spray of metalaxyl during flowering and fruiting can also reduce the risk of disease occurrence.

1.4 Insect Pests

1.4.1 Fruit Borers

(a) *Causative Organism* *Conopomorpha sinensis* (Order: Lepidoptera)

Symptoms *C. sinensis* is known as the lychee fruit borer in Thailand and stem-end borer in China. *C. sinensis* lays scale-like yellow-coloured eggs (size 0.4×0.2 mm long) on the fruit, young leaves and shoots. The eggs hatch within 3–5 days and the larva immediately starts penetrating into the fruit, leaf or shoot (1. 3C). They make a tunnel through the aril of the fruit which leads to fruit fall (Menzel 2002).

Control Measures Fruits must be weekly inspected for the eggs of *C. sinensis*. Infected fruit must be removed and destroyed. In case of severe infestation, permethrin must be sprayed weekly, for 2 weeks before harvest (Menzel 2002).

(b) *Causative Organism* *Conopomorpha litchiella* (Order: Lepidoptera)

Symptoms *C. litchiella* lays their small light yellow-coloured eggs on the new shoots of the plant which hatch within 3–5 days. The newly hatched creamy white-coloured larva penetrates into shoots, midrib veins and leaf blades (Fig. 1.3d). Infected shoots lead to wilting (Menzel 2002).

Control Measures Insecticides should be sprayed on the plants.

(c) *Causative Organism* *Argyroploce illepida* (Order: Lepidoptera)

Symptoms *A. illepida* lays eggs in groups of 15 or single on the fruit surface. The size of the egg is 1.0 × 0.8 mm. Eggs are creamy white in colour and oval to flat having reticulate surface. Larvae feed upon the skin of the fruit and then tunnel towards the seed (Fig. 1.3e). When the fruits are ripe, the larva penetrates directly into the seed, which is completely eaten. Hence, a single larva can damage the entire fruit (Menzel 2002).

Control Measures In regions of South Africa, triflumuron and teflubenzuron sprays are recommended when the fruits are immature. Covering of panicles with paper bags also improves fruit colour and quality. In Queensland, azinphos-methyl and carbaryl are also used to prevent insect damage.

1.4.2 Fruit-Piercing Moths

Causative Organism Eudocima fullonia, E. salaminia, E. jordani (Order: Lepidoptera)

Symptoms These moths drill a hole in the fruit skin and suck the fruit juice. Contamination of the hole with bacteria and yeasts ultimately destroys the fruit (Fig. 1.3f). After few days of the infestation, a frothy discharge comes out from the fruit (Menzel 2002).

Control Measures The Australian farmers make traps by putting a dark shade cloth on a framed wire and bait it with fermented bananas and citrus fruits. The moths get captured in this cloth.

1.4.3 Leaf-Feeding Caterpillars

Causative Organism Oxyodes scrobiculata, O. tricolor (Order: Lepidoptera)

Symptoms The caterpillars feed on the leaves and cause severe defoliation (Fig. 1.4a).

Control Measures Shaking the tree dislodges the larvae onto the ground. Application of carbaryl is recommended in Thailand. In Australia, endosulfan and methomyl are also used as a remedy (Menzel 2002).

1.4.4 Leaf Rollers

Causative Organism Olethreutes perdulata, Platypeplus aprobola, Adoxophyes cyrtosema Meyr., *Homona coffearia, Isotenes miseran* (Order: Lepidoptera)

Symptoms These pests are more common in China and India and damage the leaves and flowers. The young leaves are more prone to pest throughout the lychee

Fig. 1.3 Fungal and insect damage of lychee. (**a**, **b**) Fruit rot, leaf and panicle damage caused by *Peronophythora litchi*. Damage caused by insect pest. (**c**) Fruit damage by larval penetration of *Conopomorpha sinensis*. (**d**) Shoot and leaf damage caused by *C. litchiella*. (**e**) Fruit damage caused by *Aegyroploce illepida*. (**f**) Fruit damage caused by *Eudocima fullonia*, *E. salaminia* and *E. jordani*

growing area. The green-coloured caterpillar rolls the leaf and feeds on lamina within the roll (Fig. 1.4b).

Control Measures Methomyl or carbaryl can be used during the initial damage. In India, rolled leaves containing larvae are destroyed manually, but phosphamidon, fenitrothion and endosulfan are sprayed in case of heavy infestation (Menzel 2002). Spray of thiodan and dursban retards the rate of damage caused by pests.

1.4.5 Bark Borers

Causative Organism Aristobia testudo, Anoplophora maculate (Order: Coleoptera)

Symptoms The female damages the branches by chewing 10 mm strip of bark and lays eggs on the wound. The larva hatches in the last week of August and lives beneath the bark until January. Larvae bore through the xylem and make tunnels of up to 60 cm long. This larval stage lasts for about 10 months and the tunnels start damaging the branches (Fig. 1.4c).

Control Measures Manual removing of the beetles, eggs and young larvae can be performed. Established larvae can be located from the appearance of their frass and can then be removed with knives and wire hooks. Alternatively, dichlorvos is injected and the tunnels sealed with clay (Zhang 1997).

1.4.6 Scarab Beetles

Causative Organism Xylotrupes gideon (Order: Coleoptera)

Symptoms The larvae develop in the soil where they feed on humus and roots of the plant. Large and sexually dimorphic adults emerge during spring season. They are attracted to the ripened fruits and damage them (Fig. 1.4d).

Control Measures Use of chemical sprays is not satisfactory. Manual removing of beetles can be done in smaller trees but is quite difficult for larger trees (Menzel 2002).

1.4.7 Soft Scales

Causative Organism Pulvinaria psidii, Coccus hesperidum, Parasaissetia nigra, Saissetia coffeae (Order: Hemiptera)

Symptoms Causative organisms form scale-like lesions on the fruit surface and also produce honeydew which enhanced the growth of moulds on the infected fruit and panicle (Fig. 1.4e). These discoloured fruits reduced the market value of the crop (Menzel 2002).

Control Measures Severe infections can be controlled with the application of methidathion.

1.4.8 Bugs

Causative Organism Tessaratoma papillosa, *Tessaratoma javanica* Thunberg, *Tessaratoma quadrata* (Order: Hemiptera)

Symptoms During the spring season, the female lays their eggs on the backside of the leaves and the nymphs mature in June. Adults and nymphs both feed upon the terminal branches, flowers and fruits, causing them to fall (Fig.1.4f). The bugs also feed upon the developing seeds. Infected seeds have lesions on the testa (Menzel 2002).

Control Measures Endosulfan and trichlorfon should be applied at different concentrations because the bugs vary in their susceptibility to different concentrations. A maximum of two sprays applied (2 weeks' gap) during the first 6 weeks after fruit set is sufficient (Menzel 2002).

1.4.9 Gall Flies

Causative Organism Dasineura spp. (Order: Diptera)

Symptoms Galls are formed over the leaf surfaces. Female lays eggs on the younger leaves, and the larvae form water dots over the surface which later on become galls. Galls turn brown and then fall off, leaving a hole on the leaf (Fig. 1.5a).

Control Measures The infected leaves can be removed manually and burnt. Methyl parathion (2.5%) can be sprayed on the trees. Isofenphos (0.001%) can also be sprayed (Menzel 2002).

1.4.10 Fruit Flies

Causative Organism Bactrocera tryoni (Order: Diptera)

Symptoms The female flies lay eggs over the cracks and wound on the fruit skin (Fig. 1.5b). The level of the crop damage is quite low (Menzel 2002).

Fig. 1.4 Damage caused by insects. (**a**) Defoliation caused by *Oxyodes scrobiculata* and *O. tricolor*. (**b**) Leaf rolling and defoliation caused by *Olethreutes perdulata*, *Platypeplus aprobola*, *Adoxophyes cyrtosema*, *Homona coffearia* and *Isotenes miserana*. (**c**) Stem damage (tunnel formation) caused by *Aristobia testudo* and *Anoplophora maculata*. (**d**) Fruit damage caused by Scarab beetle (*Xylotrupes gideon*). (**e**) Discolouring of fruits and damage of leaves, flower and twigs caused by *Pulvinaria psidii*, *Coccus hesperidum*, *Parasaissetia nigra* and *Saissetia coffeae*. (**f**) Fruit damage and fruit fall caused by bugs: *Tessaratoma papillosa*, *Tessaratoma javanica* Thunberg and *Tessaratoma quadrata*

Control Measures In South Africa, pheromone-baited nets are used to capture the flies. The nets are sprayed with a combination of trichlorfon, mercaptothion and protein hydrolysate. The panicles can also be covered with paper bags (Menzel 2002).

1.4.11 Mites

Causative Organism *Aceria litchii* (Orders: Trombidiformes)

Symptoms *Aceria litchii* is also known as hairy mite, erinose mite, hairy spider and dog ear mite.
The female lays small and translucent white-coloured eggs on leaf surface. Mites are small and pinkish white in colour. All the stages of the life cycle are easily covered in moving from old leaves to young leaves. Their feeding stimulates the formation of hairy growth (erineum) over the leaf surface. Erenium covers the entire leaf and leads to leaf curling (Fig. 1.5c). In case of severe infestation, leaves are fully damaged. Mites can also damage the fruits and cause fruit disruption (Menzel 2002).

Control Measures Infected branches should be removed and burnt. Some insecticides can also be used. Three sprays of dimethoate can be applied during the emergence of leaves. In China dimethoate, chlorpyrifos, isocarbophos, dichlorvos, omethoate and dicofol are recommended to control the mites (Zhang 1997). Spray of wettable sulphur (4 g/l of water) after the fruit harvest can provide sufficient control over the mites. Rogor spray should be applied at the rate of 2 ml/l before and after the flowering season.

1.4.12 Weevils

Causative Organism *Apoderus* sp., *Conopomorpha cramerella* (Order: Coleoptera, Lepidoptera)

Symptoms The insects feed upon the leaf surface, and as a result of it, the leaves dry up giving a blighted appearance to the twigs (Fig. 1.5d). Young trees are totally devastated. Tan spots appear over the leaf surface giving the appearance of leaf burn (Kumar et al. 2011). The weevils remain active throughout the flushing period and the larvae of the weevils feed upon the roots of the plant.

Control Measures Pheromone traps and various biochemical agents including prophylactic spray of neem-based (*Azadirachta indica*) insecticide can be used to control the insects (Kumar et al. 2011).
Sevin spray at the rate of 3 g/l and dursban and thiodan at the rate of 2 ml/l can control the pest.

Fig. 1.5 Damage caused by insects. (a) Formation of galls on the leaf surface caused by *Dasineura* spp. (**b**) Fruit pericarp damage caused by *Bactrocera tryoni* (Froggatt). (**c**) **Leaf** bristles, leaf galls and leaf curling caused by *Aceria litchii* (Keifer). (**d**) Tan spots on leaf surface caused by Weevils: *Apoderus* sp. and *Conopomorpha cramerella*

1.5 Transgenic Approaches for Lychee Improvement

The economic importance of this fruit crops has led to selection and breeding over two decades. But, this practice resulted in relatively fewer genotypes containing restricted germplasm. Such kind of genetic uniformity expands the susceptibility of the crop towards various insects and pest's diseases. This ultimately leads to the immoderate use of the chemical pesticides (Norelli et al. 1994). Moreover, the biotechnological approaches can provide an alternative pathway to introduce a foreign gene which encodes for desirable traits (Hammerschlag and Litz 1992). Much attention has been focused on crop improvement, by incorporating genes for bacterial, fungal and insect resistance. It has now become possible to attain crop modification by genetic transformation to overcome the time-consuming conventional strategies. However, proper methodology is required for development of new cultivars with desirable characteristics (Puchooa 2004).

Puchooa (2004) introduced and successfully expressed a *gfp* gene (green-fluorescent protein gene) in lychee leaf tissue through *Agrobacterium*-mediated transformation. Screening for *gfp* gene expression may prove useful to improve transformation efficiency and to facilitate detection of transformed lychee plants. Ouyang and Zheng (1985) reported the transfer of T-DNA and formation of tumour induced by *Agrobacterium tumefaciens* in lychee. Hence, the use of *Agrobacterium* in transformation may provide better approach to produce genetically modified lychee plants with all the desirable traits. Nuclear technologies which include in vitro mutagenesis also have the potential to genetically transform the crop (Jain 2005). Both physical and chemical mutagens are used to induce the mutations, and this approach may provide mutant varieties within a short time period (Sarin et al. 2009).

In recent years, RAPD and AFLP markers have been successfully exploited for assessing genetic diversity in lychee (Ding et al. 2000; Tongpamnak et al. 2002; Kumar et al. 2006). There are two reports available on the production of transgenic lychee varieties. Das and Rahman (2012) transferred a gene named 'rice chitinase' through *Agrobacterium tumefaciens*-mediated transformation. PCR, RT-PCR and Western and Southern blotting techniques were used to confirm gene integration. Polyacrylamide in-gel assay was used to analyse the chitinase activity. It was found that transformed plants exhibited significant chitinase activity as compared to the non-transformed ones.

Sinha and Das (2013) transformed lychee with salt-tolerant Gly I and II genes in order to alter the glyoxylate pathway to enhance the salt tolerance. The integration of the genes was analysed by PCR and Southern and Western immunoblotting. The transformed plant showed significant tolerance towards salt as compared to the wild plants. The above findings suggest that transgenic approaches have the promising potential to produce lychee cultivars with fungal- and insect-resistant traits also. However, much work is still needed in order to understand the molecular basis of the transgene integrity and stability and to enhance the transformation efficiency as the level of the gene expression must be high in the transformed plant.

1.6 Post-harvest Strategies to Cope with Crop Loss

The lychee fruits are harvested selectively so as to ensure that only the mature ones are plucked and marketed (Lemmer and Kruger 2002). Hence, fruit picking is carried out repeatedly at regular intervals. During the peak of the harvest season, the fruits are harvested in clusters with the panicle at uniform maturity. Fruits are generally harvested manually by using ladders. They are mostly harvested in early morning in order to decrease the loss of moisture and weight. After that individual fruit is plucked from the panicles with the help of a cutter. The fruit separation process must be performed under shade, and the fruits are collected in clean plastic crates. The transfer of the harvested fruits from the orchard to packhouse is done rapidly in order to maintain the fruit quality.

1.6.1 Grading

Grading involves the separation of fruits into different grades according to fruit size, colour and quality. Grading system depends greatly upon the market requirement (Menzel 2002). Grading process should be carried out in well-ventilated, shady and temperature-controlled packing houses (Holcroft et al. 2005). FAO of the United Nations has established CODEX quality standard for lychee. According to CODEX standard, the mature fruit must have dominant red pericarp. The diameter for the superior class fruit must be 33 mm, while for standard class, it must be 20 mm (CODEX standards 2005). After grading of the fruits, post-harvest treatment is done to retain the fruit quality of the superior as well as standard fruits. Several researches are available on the post-harvest physiology of lychee. Moreover, there are numerous post-harvest technologies or strategies which can maintain the fruit standard for 3–4 weeks (Sivakumar et al. 2011). The following are the strategies which are performed after harvest in order to retain the overall fruit quality.

1.6.2 Sulphur Dioxide Fumigation

The lychee industry has been using SO_2 fumigation method commercially in order to prevent the post-harvest browning and infection by several post-harvest pathogens (Swarts 1983). Fumigation can be performed by burning 100 g of 90% sulphur powder per m^3 of fruit at a temperature of 25–28 °C for 20 min (Holcroft and Mitcham 1996). If excessive sulphur is used, the fruits changes to pale green or light yellow in colour (Timberlake and Bridle 1967). Treatment of HCl (hydrochloric acid) after the fumigation can also help to retain the red colour of the fruit (Zauberman et al. 1990). Some undesirable results of SO_2 fumigation have also been reported (Kremer-Köhne 1993).

1.6.3 HCl Dips

The hydrochloric dip treatment is performed to retain the original bright red colour of fruit fumigated with SO_2. The dip treatment should be given for 4–8 min. Moreover, the treatment time can be increased to increase the red colour intensity (Sivakumar et al. 2011).

1.6.4 Use of Metabisulphite Salts

Sheets impregnated with sodium metabisulphite are also helpful to control the browning and desiccation of fruit (Schutte et al. 1990). Combination of the above HCl dip and sodium metabisulphite treatment can effectively reduce the pericarp browning for 28 days (Sivakumar et al. 2011).

1.6.5 Gamma Irradiation

Irradiation of the gamma rays at low temperature can be used as a possible alternative to SO_2 fumigation treatment. Response of different cultivars can vary according to radiation dose (Ilangantileke et al. 1993). However, the irradiation process is not applied commercially in some countries because of their safety issues (Jiang et al. 2003a, b).

1.6.6 Dip Treatments

Various chemicals can be applied to perform the dip treatment. Polyamines also have the potential to maintain the red colour of the fruit pericarp (Jiang and Chen 1995). Spermine and putrescine are the polyamines which can be used in combination with fungicides reported to delay the fruit decay. However, certain countries do not recommend the use of fungicides because of their side effects on human and environment. A combination of citric acid and glutathione was reported to inhibit the pericarp browning, but this treatment is effective only for 4 days' storage (Zhang et al. 2001). Fruit treatment can also be performed with 'chitosan'. Chitosan inhibits the fruit decay and induces a defence mechanism in fungi (Zhang and Quantick 1997a, b). Effects of chitosan coating and ascorbic acid (AA) on lychee fruit storage were also investigated, and it was reported that AA significantly increases the anti-oxidation capacity and chitosan inhibits the dehydration and microbial attack. This strategy has the significant potential to store the lychee for longer duration (Sun et al. 2010). However, none of the dip treatment has the potential to retain the fruit quality for more than 21 days.

1.6.7 Heat Treatments

Heat treatment can retain the colour of the fruit pericarp to some extent, but this treatment fails at commercial level because the steam affects the pulp quality of fruit (Kaser et al. 1995). Some cultivars like "Tai So" and "Wai Chee" respond well to vapour heat treatment (45°C for 42 minutes) and retain the fruit quality (Jacobi et al. 1993a, b). But, in some of the cultivars this treatment leads to the loss of membrane integrity and browning of the pericarp (Wong et al. 1991). Brushing of hot water on fruits does not provide sufficient anti-fungal protection (Lichter et al. 2000). Hot water spray is an alternative to hot water brushing (Olesen et al. 2004). It has been reported that hot water treatment and fruit-dip in HCL eliminates the risk of fungal pathogens including *Penicillium* (Lichter et al. 2004a, b). Treatment of lychee fruit with pure O_2 enhances the fruit membrane integrity which reduces the decompartmentation of the enzymes responsible for pericarp damage (Duan et al. 2004). Benomyl which is a common fungicide used for the treatment of lychee fruit is no longer recommended due to its certain carcinogenic effects (National Research Council, 1987).

Detailed research on the ontogeny and morphology of the pericarp took tremendous attention of the scientists as the fruit quality is directly related to the structure of the pericarp and other physiological processes that occur during maturation (Cronje 2008). A large number of strategies have been framed to solve the problem of post-harvest, but one single approach cannot fix all the problems (Bhushan et al. 2015). Besides the above-mentioned techniques, careful harvesting and management of proper conditions during the storage are the prerequisites in order to prevent the fruit decay and desiccation and also to retain the attractive colour of the fruit pericarp for longer duration. It is estimated that 25% of the crop produced is spoiled due to post-harvest complications that means one-fourth of the fruit produced never reaches to the consumer for whom it was grown (FAO 2014).

1.7 Packing and Marketing

Packaging of fruit crop is one of the important steps in the extensive and complicated pathway from farmer to consumer. Type or selection of packaging depends upon the preferences and availability of market. The ideal packaging must protect the fruit from damage and retain the moisture content of the fruit (Menzel 2002). Level of the sanitation should be maintained in the packinghouses which includes the cleanliness of collection point, drying tunnels, off-loading areas, packing conveyer belts, etc. (De Jager et al. 2000). The infrastructure of the packing houses and processing plants should also be taken into consideration (De Roever 1999; Brackett 1999; Adams and Moss 2000). Packing of the fruits into moisture- and contamination-free bags can significantly reduce the risk of pericarp browning (Sivakumar et al. 2011). An efficient marketing significantly affects the income level of the growers as well as the satisfaction level of the consumers. Depending upon the market requirements, transportation of lychee can be done by land, sea and air. The workers included in whole transport operation also play an important role to maintain the correct temperature during transportation (Sivakumar et al. 2011). Transportation facilities with good infrastructure, updated market information and certain government policies play a vital role in the marketing of lychee crop.

1.8 Conclusions

Sustainable plant productivity (fruit and crop yield) in the coming years is the major constraint for food and nutritional security for the human population in developing countries where the arable land per capita is shrinking while human and livestock population is steadily increasing. Plant and crop productivity are the result of interaction of several physiological, biochemical and metabolic processes over a defined period of time, reflected in gain of total biomass or converted harvestable commodity like seeds, fruits or edible plant parts under a set of environmental conditions that consist of several physical, geochemical and biological components. Therefore, besides genetic potential of plant species, the phenotypic performance of plants in the field profoundly depends and is influenced by several physical, abiotic and biotic parameters and is highly variable. Hence, plant yield or harvest index is dependent on several factors and several of them beyond human control and part of climate change and environment. Amongst biotic components that influence plant/crop yield, perhaps infestation of plant pathogens and infestation of insect pest are major issues after the agronomic inputs and practices. The infestation of fungal and insect pests alone during field and storage condition may cause 24–65 ± 5% loss in fruit/grain yield of major crops (Ronald 2011). Control of agricultural insect pests under field and storage conditions largely depends on the widespread use of synthetic insecticides and pesticides which are harmful to ecosystem and human population (Hilder and Boulter 1999; Wahab 2009). In the recent past, with the development of diverse biotechnological tools and techniques of recombinant DNA and genetic engineering, it is now possible to transfer and express a desired gene in its native or modified form into the identified organism including plants, animals and microbes. Amongst the battery of genetically modified organisms (GMOs), the transgenic plants, expressing genes from either trans or cis-origin, are the latest introduction to sustainable crop and plant yield (Park et al. 2011).

Lychee is one of the delicious, juicy fruits and is known as the queen of the fruits because of its tastiness as well as abundant of nutrients. Since it is a rich source of minerals and other nutrients, it is also termed as a superfood. The production of this crop for human consumption is at risk due to the emergence of pests, pathogens (including algal and fungal) and certain animal pests. Reduction in yield of the crop can be controlled or reduced by applying various crop protection measures. The lychee fruit is susceptible to diseases before as well as after harvest. The availability of the lychee fruit has been limited because of yield variability, season differences and, the most importantly, post-harvest problems. Abiotic stresses can also limit the yield production but at less extent as compared to the biotic stresses. Furthermore, a large portion of the crop production is spoiled by the post-harvest damages which include pericarp decay and desiccation.

Breeding strategies for the selection of improved cultivars have the significant potential to solve the various post-harvest complications. Application of the new advance technologies such as RAPD and AFLP will prove to be quite useful for evaluation of genetic diversity of the crop and thus beneficial for the researchers to select and combine the traits of two different cultivars. Control of agricultural insect pests under field and storage conditions largely depends on the widespread use of synthetic insecticides and pesticides which are harmful to ecosystem and human population. The concept of integrated pest management (IPM) should be applied for pest control and to maintain the level of pesticides at ecologically and economically acceptable level. The use of IPM minimizes the application of pesticides as well as fungicides. However, the implementation of IPM is a challenging task because it requires in-depth understanding of the pests. Food and Agriculture Organization (FAO) regards the IPM as a pillar of both sustainable intensification of crop production and pesticide risk reduction.

Furthermore, the logical sequencing of the enzymes responsible for pericarp browning can also prove beneficial to manage the post-harvest damage.

Furthermore, research is required in order to thoroughly understand the mechanism of these enzyme-catalysed reactions. In vitro techniques including micropropagation have the immense potential for the production of elite cultivars.

There are a variety of fungicides and pesticides available in order to prevent the damage caused by the biotic factors. The consumers are now much aware about the adverse effects of these chemicals on human health and environment. Hence, the management of critical dose level of the chemical should be taken into consideration, so that these chemicals do no enter into the food chain. To conclude:

- IPM approaches should be taken into consideration in order to retard the risks associated with the use of fungicides and pesticides.
- More focus should be given to the production of genetically modified crops (GMO), which have inbuilt disease resistance genes.
- In-depth study of the physiological processes which occurs during fruit ripening is required in order to formulate such kind of post-harvest strategy which covers all the aspects related to post-harvest damage, without having any side effect.

References

Acquaah G (2007) Principles of plant genetics and breeding. Blackwell, Oxford

Adams MR, Moss MO (2000) Food microbiology. Royal Society of Chemistry, Cambridge

Ann PJ, Tsai JN, Yang HR (2011) Control of Peronophythora fruit downy blight of lychee by neutralized phosphoric acid. Plant Pathol Bull 20:90–97

Bhushan B, Pal A, Narwal R, Meena VS, Sharma PC, Singh J (2015) Combinatorial approaches for controlling pericarp browning in Litchi (*Litchi chinensis*) fruit. J Food Sci Technol 52 (9):5418–5426

Brackett RE (1999) Incidence, contributing factors, and control of bacterial pathogens in produce. Postharvest Biol Technol 15:305–311

Codex standard for litchi (1995, amended 2005) Available from http://www.codexalimentarius.net/download/standards/320/CXS_196e.pdf. Accessed 13 Sept 2005

Das DK, Rahman A (2012) Expression of a rice chitinase gene enhances antifungal response in transgenic litchi (cv. Bedana). Plant Cell Tiss Organ Cult 109:315–325

De Jager ES, Opperman M, Korsten L (2000) Quality standards in litchi packing houses: packing house sanitation. SA Litchi Growers Assoc Yearb 11:26–29

De Roever C (1999) Microbiological safety evaluations and recommendations on fresh produce. Food Control 10:117–143

Degani C, Deng J, Beiles A, EI-Batsri R, Goren M, Gazit S (2003) Identifying lychee (*Litchi chinensis* Sonn.) cultivars and their genetic relationships using intersimple sequence repeat (ISSR) markers. J Am Soc Hortic Sci 128(6):838–845

Hilder VA, Boulter D (1999) Genetic engineering of crop plants for insect resistance-a critical review. Crop Prot 18:177–191

Holcroft DM, Mitcham E (1996) Postharvest physiology and handing of litchi (*Litchi chinensis* Sonn). Postharvest Biol Technol 9:265–281

Holcroft DM, Lin H, Ketas S (2005) Harvesting and storage. In: Menzel CM, White GK (eds) Litchi and longan: botany, production and uses. CABI Publishing, Rockhampton, pp 273–295

Hwang JY, Lin JT, Liu SC, Hu CC, Shyu YS, Yang DJ (2013) Protective role of litchi (*Litchi chinensis* Sonn.) flower extract against cadmium-and lead-induced cytotoxicity and transforming growth factor ß1-stimulated expression of smooth muscle α-actin estimated with rat liver cell lines. J Funct Foods 5:698–705

Ilangantileke SG, Noomhorn A, Upadhyay IP, Srinivas Rao M (1993) Effect of irradiation and storage temperature on the shelf life and quality of Thai lychee. In: Champ BR, Highley E, Johnson GI (eds) Postharvest handling of tropical fruits, vol 50. ACIAR, Canberra, pp 352–354

Jacobi KK, Wong LS, Giles JE (1993a) Lychee (*Litchi chinensis* Sonn.) fruit quality following vapour heat treatment and cool storage. Postharvest Biol Technol 3:111–119

Jacobi KK, Wong LS, Janet EG (1993b) Lychee (*Litchi chinensis* Sonn.) fruit quality following vapour heat treatment and cold storage. Postharvest Bio Technol 3:111–119

Jain SM (2005) Major mutation-assisted plant breeding programs supported by FAO/IAEA. Plant Cell Tissue Organ Cult 82:113–123

Janisiewicz WJ, Korsten L (2002) Biological control of postharvest diseases of fruits. Annu Rev Phytopathol 40:411–441

Jiang YM, Chen F (1995) A study on polyamine change and browning of fruit during cold storage of litchi (*Litchi chinensis* sonn.) Postharvest Biol Technol 5:245–250

Jiang YM, Fu JR (1999) Postharvest browning of litchi fruit by water loss and its prevention by controlled atmosphere storage at high relative humidity. LWT-Food Sci Technol 32:278–283

Jiang YM, Yao L, Lichter A, Li J (2003) Postharvest biology and technology of litchi fruit. Food Agric Environ 2:76–81

Jiang Y, Duan X, Joyce D, Zhang Z, Li J (2004a) Advances in understanding of enzymatic browning in harvested litchi fruit. Food Chem 88:443–446

Jiang Y, Li Y, Li J (2004b) Browning control, shelf life extension and quality maintenance of frozen litchi fruit by hydrochloric acid. J Food Eng 63:147–151

Jiang YM, Zhu XR, Li YB (2001a) Post harvest control of Lychee fruit-rot by *Bacillus subtilis*. LWT Food Sci Technol 34:430–436

Jiang YM, Zhu XR, Li YB (2001b) Postharvest control of Litchi fruit rot by *Bacillus subtilis*. Lebensm-Wiss u-Technol 34:430–436

Joas J, Caro Y, Ducamp MN, Reynes M (2005) Postharvest control of pericarp browning of litchi fruit (*Litchi chinensis* Sonn cv Kwaï Mi) by treatment with chitosan and organic acids: I. Effect of pH and pericarp dehydration. Postharvest Biol Technol 38(2):128–136

Kao CW, Le LS (1980) Sporangium germination of *Peronophythora litchii*, the causal organism of litchi downy blight. Mycologia 72(4):737–748

Kaser C, Levin J, Wolstenholme BN (1995) Vapour, heat and low pH dips improve litchi (*Litchi chinensis* Sonn.) pericarp colour retention. J S Afr Soc Hortic Sci 4:6–12

Korsten L (2006) Advances in control of postharvest diseases in tropical fresh produce. Int J Postharvest Technol Innov 1(1):48–61

Kremer-Köhne S (1993) SO_2 residues in litchi fruit after post-harvest SO_2 treatments. SA Litchi Growers Assoc Yearb 5:15–16

Kumar M, Prakash NS, Prasad US, Sarin NB (2006) A novel approach of regeneration from nodal explants of field grown lychee (*Litchi chinensis* Sonn.) fruit trees. J Plant Sci 5(3):321–327

Kumar V, Kumar A, Nath V (2011) Emerging pests and diseases of litchi (*Litchi chinensis* Sonn.) Pest Manage Hortic Ecosyst 17(1):11–13

Kumar D, Mishra DS, Chakraborty B, Kumar P (2013) Pericarp browning and quality management of litchi fruit by antioxidants and salicylic acid during ambient storage. J Food Sci Technol 50(4):797–802

Lemmer D, Kruger FJ (2000) Factors influencing SO^{2-} residues on commercially fumigated HLH Mauritius and McLean's red litchi fruit. SA Litchi Growers' Assoc Yearb 11:42–46

Lichter A, Dvir O, Rot I, Akerman M, Regev R, Wiesblumb A, Fallika E, Zaubermana G, Fuchs Y (2000) Hot water brushing: an alternative method to SO_2 fumigation for color retention of litchi fruits. Postharvest Biol Technol 18:235–244

Lichter A, Dvir O, Ackerman M, Feygenberg O, Pesis E (2004a) Acidified peel of litchi fruits selects for postharvest Penicillium decay. Phytoparasitica 32:226–236

Lichter A, Dvir O, Rot I, Akerman M, Regev R, Wiesblum A, Fallik E, Zauberman G, Fuchs Y (2004b) Hot water brushing: an alternative method to SO_2 fumigation for color retention of litchi fruits. Postharvest Biol Technol 18:235–244

Mahajan PV, Goswami TK (2004) Extended storage life of litchi fruit using controlled atmosphere and low temperature. J Food Process Preserv 28:388–403

McMillan RT Jr (1994) Diseases of *Litchi chinensis* in South Florida. Proc Flo State Hortic Soc 107:360–362

Menzel C (2001) The physiology of growth and cropping in lychee. Acta Hortic 558:175–184

Menzel C (2002) The lychee crop in Asia and the pacific. Food and agriculture organization of the united nations. Regional office for Asia and the Pacific Bangkok, Thailand, pp 1–102

Nacif SR, Paoli AAS, Salomão LCC (2001) Morphological and anatomical development of the litchi fruit (*Litchi chinensis* Sonn. cv. Brewster). Fruits 56:225–233

Neog M, Saikia L (2010a) Control of post-harvest pericarp browning of litchi (*Litchi chinensis* Sonn). J Food Sci Technol 47(1):100–104

Norelli JL, Aldwinckle HS, Destéfano-Beltran L, Jaynes JM (1994) Transgenic 'Malling 26' apple expressing the attacin E gene has increased resistance to *Erwinia amylovora*. Euphytica 77:123–128

Olesen T, Nacey L, Wiltshire N, O'Brien S (2004) Hot water treatments for the control of rots on harvested litchi (*Litchi chinensis* Sonn.) fruit. Postharvest Biol Technol 32:135–146

Papademetriou MK, Dent MJ (2002) Lychee production in the Asia-pacific region. Food and agriculture organization of the United Nations, Bangkok, Thailand

Park JR, McFarlane I, Phipps RH, Ceddia G (2011) The role of transgenic crops in sustainable development. Plant Biotechnol J 9:2–21

Paul RE, Reyes MEQ, Reyes MU (1995) Litchi and rambutan insect infestation: treatments to minimize induced pericarp browning. Postharvest Biol Technol 6:139–148

Paull RE, Chen NJ (1987) Effect of storage temperature and wrapping on quality characteristics of Litchi fruit. Sci Hortic 33:223–236

Pesis E, Dvir O, Feygenberg O, Arie RB, Ackerman M, Lichter A (2002) Production of acetaldehyde and ethanol during maturation and modified atmosphere storage of litchi fruit. Postharvest Biol Technol 26:157–165

Ramma I (2003) Postharvest sulphur dioxide fumigation and low acid dip for pericarp colour retention and decay prevention on litchi. AMAS pp 41–48

Ronald P (2011) Plant genetics, sustainable agriculture and global food security. Genetics 188:11–20

Sarin NB, Prasad US, Kumar M, Jain SM (2009) Litchi breeding for genetic improvement. In: Jain SM, Priyadarshan PM (eds) Breeding plantation tree crops: tropical species. Springer, Heidelberg

Schutte GC, Botha T, Kotze JM (1990) Post-harvest control of decay and browning of litchi fruit by fungicide dips and paper sheets impregnated with sodium metabisulphite. SA Litchi Growers Assoc Yearb 3:10–14

Scott KJ, Brown BI, Chaplin GR, Wilcox ME, Bain JM (1982) The control of rotting and browning of litchi fruit by hot benomyl and plastic film. Sci Hortic 16:253–262

Sinha S, Das DK (2013) Transformation of litchi (*Litchi chinensis* sonn.) with Gly I+II gene leads to enhanced salt tolerance. IJBAF 1(11):483–495

Sivakumar D, Korsten L (2006) Influence of modified atmosphere packaging and postharvest treatments on quality retention of litchi cv. Mauritius. Postharvest Biol Technol 41(2):135–142

Sivakumar D, Regnier T, Demoz B, Korst L (2005) Effect of different post-harvest treatments on overall quality retention in litchi fruit during low temperature storage. J Hortic Sci Biotechnol 80(1):32–38

Sivakumar D, Korsten L, Zeeman K (2007a) Postharvest management on quality retention of Litchi during storage. Fresh Produce 1(1):66–75

Sivakumar D, Zeeman K, Korsten L (2007b) Effect of a biocontrol agent (*Bacillus subtilis*) and modified atmosphere packaging on postharvest decay control and quality retention of litchi during storage. Phytoparasitica 35(5):507–518

Sivakumar D, Arrebola E, Korsten L (2008) Postharvest decay control and quality retention in litchi (cv. McLean's red) by combined application of modified atmosphere packaging and antimicrobial agents. Crop Prot 27(8):1208–1214
Song L, Jiang Y, Gao H, Li C, Liu H, You Y, Sun J (2006) Effects of adenosine triphosphate on browning and quality of harvested litchi fruit. Am J Food Technol 1:173–178
Sun D, Liang G, Jianghui X, Lei X, Mo Y (2010) Improved preservation effects of litchi fruit by combining chitosan coating with ascorbic acid treatment during postharvest storage. Afr J Biotechnol 9(22):3272–3279
Swarts DH (1985) 'Litchis'. Postharvest handling of litchis farming in South Africa. Department of Agriculture, Pretoria
Tang ZH, Wang HC, Hou YP, Zhang SP, Wang JX, Zhou MG (2011) Baseline and differential sensitivity to mandipropamid among isolates of *Peronophythora litchii*, the causal agent of downy blight on Lychee. Crop Prot 30:354–359
Tian SP, Li BQ, Xu Y (2005) Effects of O_2 and CO_2 concentrations on physiology and quality of litchi fruit in storage. Food Chem 91:659–663
Timberlake CF, Bridle P (1967) Flavylium salts, anthocyanins. 1. Reactions with Sulphur dioxide. J Food Sci Agric 18:479–485
Underhill SJR, Simons DH (1993) Lychee (*Litchi chinensis* Sonn.) pericarp desiccation and the importance of postharvest micro-cracking. Sci Hortic 54:287–294
Wahab S (2009) Biotechnological approaches in the management of plant pests, diseases and weeds for sustainable agriculture. J Biopesticides 2:115–134
Wong LS, Jacobi KK, Giles JE (1991) The influence of hot benomyl dips on the appearance of cool stored lychee (*Litchi chinensis* Sonn.) Sci Hortic 46:245–251
Wu B, Li X, Hu H, Liu A, Chen W (2011) Effect of chlorine dioxide on the control of postharvest diseases and quality of litchi fruit. Afr J Biotechnol 10(32):6030–6039
Xuewu Duan, Yueming Jiang, Xinguo Su, Hai Liu, Yuebiao Li, Zhaoqi Zhang, Yonghua Zheng, Weibo Jiang (2004) Role of pure oxygen treatment in browning of litchi fruit after harvest. Plant Sci 167:665–668
Yang DJ, Chang YZ, Chen YC, Liu SC, Hsu CH, Lin JT (2012) Antioxidant effect and active components of litchi (*Litchi chinensis* Sonn.) flower. Food Chem Toxicol 50(9):3056–3061
Yi-Hsieng Samuel Wu, Chih-Hsien Chiu, Deng-Jye Yang, Yi-Ling Lin, Jung-Kai Tseng, Yi-Chen Chen (2013) Inhibitory effects of litchi (*Litchi chinensis* Sonn.) flower-water extracts on lipase activity and diet-induced obesity. J Funct Foods 5(2):923–929
Yueming Jiang, Lihu Yao, Amon Lichter, Jianrong Li (2003b) Postharvest biology and technology of litchi fruit. Food, Agric Environ 1(2):76–81
Zauberman G, Ronen R, Akerman M, Fuchs Y (1990) Low pH treatment protects litchi fruit color. Acta Hortic 269:309–314
Zauberman G, Ronen R, Akerman M, Weksler A, Rot I, Fuchs Y (1991) Post-harvest retention of the red colour of litchi fruit pericarp. Sci Hortic 47:89–97
Zhang DL, Quantick PC (1997) Effects of chitosan coating on enzymatic browning and decay during postharvest storage of litchi (*Litchi chinensis* Sonn.) fruit. Postharvest Biol Technol 12:195–202
Zhang Z, Pang X, Xuewu D, Ji Z, Jiang Y (2001) Role of peroxidase in anthocyanin degradation in litchi fruit pericarp. Food Chem 75:217–221
Zhang Z, Yuan P, Wang B, Qiu Y, Li J (1997) Litchi pictorial narration of cultivation. Pomology Research Institute, Guangdong Academy of Agricultural Science, Guangdong
Zhanga Z, Panga X, Jia Z, Jiang Y (2001) Role of anthocyanin degradation in litchi pericarp browning. Food Chem 75:217–221

Microbial Interactions in Litchi Rhizosphere

2

Kanchan Vishwakarma, Neha Upadhyay, Nitin Kumar, Rishi Verma, Jaspreet Singh, Pankaj Verma, Mitali Mishra, Shruti Jain, Durgesh Kumar Tripathi, Rohit Kumar Mishra, Vivek Kumar, and Shivesh Sharma

Abstract

The litchi is one of the chief fruits of the Asian continent. It is grown almost half of the year and is enjoyed by people for its sweet juice as well as its soft pulp. Also, it is very rich in vitamin C and important minerals and antioxidants. Modern farming does not use chemical fertilizers; thus, bio-fertilizers are used, exploiting the relationship of microbial interactions to plant growth. Mycorrhizal fungi associations are found in the litchi rhizosphere, and these fungi acquire their nutrition from the associated host plants and consequently enhance access to phosphorus and nitrogen. The association between fungi and trees such as the litchi is recognized as vesicular arbuscular mycorrhizae (VAM). In soil habitat, arbuscular mycorrhizal fungi (AMF) are present in symbiosis with litchi roots and cause increase in root length by secreting plant hormones. Availability of appropriate temperature and nutrients will strongly influence the association of arbuscular mycorrhizal fungi with plant growth. Thus, the present

K. Vishwakarma • N. Upadhyay • N. Kumar • R. Verma • J. Singh • P. Verma
Department of Biotechnology, Motilal Nehru National Institute of Technology (MNNIT), Allahabad 211004, Uttar Pradesh, India

M. Mishra • S. Jain • S. Sharma (✉)
Department of Biotechnology, Motilal Nehru National Institute of Technology (MNNIT), Allahabad 211004, Uttar Pradesh, India

Centre for Medical Diagnostic and Research, Motilal Nehru National Institute of Technology (MNNIT), Allahabad 211004, Uttar Pradesh, India
e-mail: shiveshs@mnnit.ac.in; ssnvsharma@gmail.com

D.K. Tripathi • R.K. Mishra
Centre for Medical Diagnostic and Research, Motilal Nehru National Institute of Technology (MNNIT), Allahabad 211004, Uttar Pradesh, India

V. Kumar
Department of Biotechnology, Himalayan Institute of Biosciences, Swami Rama Himalayan University, Jolly Grant, Dehradun 248016, Uttarakhand, India

M. Kumar et al. (eds.), *Lychee Disease Management*,
DOI 10.1007/978-981-10-4247-8_2

chapter focuses on understanding how microbial flora interaction with the litchi rhizosphere helps to enhance the potential productivity of the litchi.

Keywords
Litchi • Arbuscular mycorrhizal fungi • Microbial flora • Plant growth promotion • Rhizosphere

2.1 Introduction

The litchi (*Litchi chinensis* Sonn.) is classified as an evergreen tree that is native to subtropical China. Its height is usually between 33 and 49 feet. The bark of the litchi is greyish black in colour with branches brown red in colour. The leaves of the litchi generally range from 10 to 25 cm or more in length, and the leaflets occur in pairs of two to four. Depending on weather, locality, and cultivar, maturity of fruit will occur within 80 to 112 days.

The litchi fruit is approximately 5 cm in length, with a width of 4 cm with round, heart or ovoid shape. The skin of the fruit is tough, thin, inedible and green in colour during the immature stage. It turns to red or pinkish red once ripening starts and consists of slight and sharp protrusions. During harvest, the colour of the skin becomes brown and dry. The aril is the edible plump part of the fruit, surrounding a inedible dark brown seed of 1 to 3.5 cm long and 6 to 1.2 cm wide.

The litchi is a distinctive fruit of which only the aril is suitable for consumption. The shelf life of litchi is approximately 36 h. Aridness and browning is the major challenge for the storage of litchi, which may result from the increased levels of phenols (Jaiswal et al. 1986); this further makes selling of the fruit difficult.

Lychee is a source of vitamin C (72 g/100 g). It is believed that the daily need of vitamin C in an adult can be met by consuming just nine litchis. Other constituents of litchis are minerals: 14% daily value (DV) of Cu, 9% DV of P and 6% DV of K. Similar to most fruits, lychees have little nonsaturated fat and sodium and are also cholesterol free. The level of polyphenols in lychees, as observed in a French study, was reported to be greater when compared to numerous other fruits. In terms of phenols, monomers and dimers of flavan-3-ol were the main components, constituting 87% of the phenols. The levels of phenols decrease with storage. The major anthocyanins present are malvidin-3-acetylglucoside, cyanidin-3-rutinoside and cyanidin-3-glucoside (Anon 2003). Friedelin, stigmasterol and stigmasteryl acetate are supported in the plant (Rastogi and Mehrotra 1999) and are reduced with storage or browning. Traces of malvidin-3-glucoside are also present. Over-consumption of lychee may result in fainting spells or skin rashes. The Chinese used the seeds found in the Malay Peninsula as an anodyne, and these are strongly recommended to treat neurological diseases and orchitis (Kirtikar and Basu 1999). The lychee also has anticancer activity (Wang et al. 2011) and strong free radical scavenging activity (Takuya et al. 2008).

The considerable colour changes that occur during fruit maturation provide a significant indication of the maturity and quality of a number of fruit species (Anderson and Jordheim 2006). In the case of litchi, the outer red colour is of commercial importance, even though, because of slower chlorophyll degradation in the pericarp, some litchi cultivars do not have a significant colour formation. Chlorophylls present in the pericarp in high concentration not only camouflage the red colour of the fruit surface provided by anthocyanins but also decrease their rate of biosynthesis (Davenport 2000; Wang et al. 2005). This action can be improved to some extent by the practice of bagging during cultivation, which enhances the degradation of chlorophyll in the litchi pericarp and facilitates fruit pigmentation. The degradation of chlorophyll contributes to a deeper colour formation during ripening of the fruit as most of the incident red light is absorbed by photosynthetic pigments; thus, the extent of phytochrome control of anthocyanin production is reduced (Jose and Schafer 1978; Lee and Wicker 1991). Subsequently, high-level irradiation is needed for synthesising anthocyanin pigments in green tissues in comparison to etiolated tissues. Numerous stages in the chlorophyll deprivation pathway in the leaves experiencing ageing have been characterised recently by using mutants with disrupted chlorophyll breakdown machinery (Hörtensteiner 2013). In addition, chlorophyll catabolism enzymes (CCEs) and the stay-green (SGR) protein have been reported to be significant (Sakuraba et al. 2012). Also, studies relevant to kiwifruit have inferred that genes involved in chlorophyll breakdown such as stay-green protein 2 showed high expression in golden-coloured fresh cultivars compared to a green-coloured cultivar, thereby causing more and earlier persistent degradation of chlorophyll (Pilkington et al. 2012). However, degradation of chlorophyll in several fruits is still poorly known.

Molecular studies provide limited research results on litchi fruit as very little gene sequence information has been reported. Although extensive studies have been performed, only a few hundred gene sequences have been assembled. RNA-seq-mediated transcriptome profiling technologies based on next-generation sequencing (NGS) have provided data for gene expression patterns. This technique can also be used to study those species whose genome sequence is still not published, such as litchi, by providing data on molecular marker development, gene discovery, transcriptional analysis and pathway enrichment analysis.

2.2 *Litchi chinensis* and Its Varieties

An evergreen subtropical fruit, litchi is well known for its nutritive value and delicious taste. Of 2.11 million tonnes in worldwide production, more than 95% of cultivation occurs in Asian countries. In comparison, Mexico, the USA and Central and South America account for a relatively small amount (Evans et al. 2004). Ninety percent of the world's litchi production is contributed by India and China (Anon 2012); other top litchi-producing countries are Taiwan, Thailand and Vietnam (FAO 2002).

The second largest producer of the litchi is India, but China has high productivity. In 2010–2011 the acreage under litchi production was 80,400 ha with a production of 538,100 tonnes (Anon 2012). In India, litchi production spreads from north to east, that is, from Jammu and Kashmir to Manipur. Litchi production contributes to the economy of Bihar (Muzaffarpur, Vaishali, Samastipur, East Champaran, West Champaran, Sitamarhi, Bhagalpur), West Bengal (Murshidabad, 24-Parganas), Punjab (Gurdaspur, Ropar, Pathankot, Hoshiarpur), Uttarakhand (Dehradun, Udham Singh Nagar, Pithauragarh, Nainital, Haridwar), Uttar Pradesh (Saharanpur, Kushinagar, Meerut), Jharkhand (Hazaribagh, Ranchi, Lohadaya), Assam (Kamrup, Sonitpur, Bongaigaon), and Tripura (South and East Tripura).

Litchi chinensis Sonn. of the family Sapindaceae is a fruit of Southeast Asian origin. It is a non-climacteric fruit (Nakasone and Paull 1998; Rivera-López et al. 1999). Its white and translucent aril gives an artistically sweet citrus-flavoured taste. Its pericarp is red in colour from the presence of anthocyanins in the covering (Rivera-López et al. 1999; Zhang et al. 2004). Litchi can be taken fresh, or as a processed wine, or can be used for sorbets and ice creams (Salunke and Desai 1984; Hui 2008).

Of 40 known litchi cultivars in India, only 12 or 13 are available for commercial purposes. On the other hand, of 200 known cultivars in China, only about 20 are commercialised. The same cultivar or variety may be known by different names in various parts of the country (Karnick 2014).

Currently the major litchi-producing state in India, contributing 40% to the total country production, is Bihar. Seventy-five percent of the total litchi production in India occurs in Muzaffarpur, Bihar (Table 2.1). Muzaffarpur has the most appropriate climate for litchi production as it has plenty of rainfall and frost-free areas. Litchi is sensitive to hot winds and humidity which can easily damage it.

2.2.1 Economic Importance

Sugar content determines the food quality of litchi which varies among its different varieties. Litchi is abundant in vitamins such as B_1, C and riboflavin. Other than this, 0.7% proteins, 0.3% fats, 9.4% carbohydrates, 0.7% minerals, 2.25% fibrous matter, 0.21% calcium, 0.31% phosphorus and 0.03% iron and carotene are found. Litchi is one of the fruits which can be canned for the purpose of exporting across the globe. During summers a well-flavoured mash from litchi fruits is also prepared and used as a food in Asian countries; people also make many other products such as pickles, preserves and wine. A popular form called litchinut is also made in China by drying it. The litchi has also been incorporated in traditional Asian medicines.

In Taiwan, it is one of the most important commercialised crops; it grows between March and maturity in June (Nakasone and Paull 1998). Its flower is small and pale white to light yellow in colour, providing a main nectar source for bees that turn these nectars into honey (Baltrušaityte et al. 2007).

Table 2.1 Commercially grown varieties in India and China are listed by states

Varieties in India	
State:	
Bihar and Jharkhand	China, Deshi, Purbi, Early and Late Bedana, Mclean, Muzaffarpur, Rose Scented, Shahi, Kasba
Orissa	Muzaffarpur, Bombai, China
Punjab and Haryana	Saharanpur, Dehradun, Calcutta, Muzaffarpur, Seedless (Late) and Rose Scented
Uttaranchal	Rose Scented, Calcutta, Early and Late Seedless
Uttar Pradesh	Seedless Early, Seedless Late, Early Large Red, Late Large Red, Calcutta, Rose Scented, Dehradun
West Bengal	Muzaffarpur, China, Deshi, Purbi, Elachi Early, Elachi Late, Bombai, Goothi, Bedana, Potee, Kalyani Selection
Varieties in China	
Ripening period:	
Early	Sanyuehong, Baitangying
Mid	Dazao, Heiye, Baila, Feizixiao and Shuidongheiye
Late	Xiangli, Guiwei, Noumici, Huaizhi, Xuehuaizi, Lanzhu, Yuanhong, Xiafanzhi and Nanmuye

From database of National Horticulture Board

2.2.2 Properties

Shahi and China are the major varieties of litchi found in India. These varieties contain high amounts of vitamin C and flavonoids and are considered to be an important source of these factors. The complex texture of the litchi is thought to be the result of the fruit pericarp. Its pulp or juice is slightly acidic in nature, consisting mostly of reducing sugars.

In studies carried out by a number of researchers, it has been shown that litchi pulp, which is the part of the whole fruit most consumed, contains a great number of phenolic compounds (Duan et al. 2007; Sarni-Manchado et al. 2000; Rivera-López et al. 1999). Litchi pulp displays outstanding antioxidant activities when assayed with ferric reducing power and 2, 2-diphenyl-1-picrylhydrazyl (Mahattanatawee et al. 2006; Saxena et al. 2011). These compounds have the ability to quench reactive free radicals (Sorata et al. 1984) and act as antioxidants for other possible mechanisms contributing to anticarcinogenic, antiinflammatory and cardiac protection impacts as well as preventing degenerative disorders (Biglari et al. 2008). With the aid of sophisticated instruments such as high pressure liquid chromatography (HPLC) with UV detector and electron spray ionization mass spectrometry (ESI-MS), two phenolic compounds, viz. trans-cinnamic acid and pelargonidin-3-*O*-glucoside, were identified from litchi pulp from Thailand (Bhoopat et al. 2011). Furthermore, six independent phenolic compounds have been identified from litchi pulp in Southeast Asia, namely, gallic acid, (+)-catechin, caffeic acid, epicatechin, chlorogenic acid and rutin, with the help of HPLC (Zhang et al. 2013). In Florida, litchi pulp was shown to contain the flavones quercetin and kaempferol and glycosides (Mahattanatawee et al. 2006).

2.3 Growing Conditions

The cultivation of litchi originates from southern China, where it has been grown continuously for 2500 years. The seeds of the litchi have a short-term lifespan which with unusual soil and weather conditions results in somewhat slow geographic dissemination of the fruit. India displays the classic example of this where litchi cultivation has been known since about the middle of the 1700s. Other subtropical areas such as Asia, Israel, Hawaii, Australia, Mexico and South Africa also farm litchi crops commercially apart from India and China (Jiang et al. 2001). A fifth of global cultivation derives from India, which also contributes to good export potential.

The characteristic features of litchi involve the denseness of the fruit, its round top and slowly growing nature. Its leaves are evergreen with 69 elliptical oblong and lanceolate abruptly pointed leaves. Leaf colour ranges from light green to dark green. The greenish or yellowish white flowers occur in bunches. The colour of litchi and the flavour of the aril vary with the type of cultivar. Usually seeds are thick and bold, but they are only partially developed in a few cultivars because of failed pollination. Such seeds are called 'chicken tongue' seed (FAO 2002).

The requirements for a litchi plant to grow include a moist subtropical or tropical atmosphere and low altitude with an elevation of about 800 m. Soil ideal for the cultivation of the crop has the characteristics of deep properly drained loam soil having high organic properties with pH in the range of 5 to 7 (Karnick 2014).

Limiting factors for thriving production of litchi take into account frost in winters and dry heat in summers. For their proper and firm establishment, the plants at the young stage need to be safeguarded from frost and hot winds for a number of years. However, a slight disparity in temperature is mandatory for appropriate fruiting of plants (Menzel and Simpson 1995). The temperature in summer is maintained up to 40 °C, and in winter, it is not below the freezing point. At the time of flowering, special care is taken in case of prolonged rain because that obstructs pollination.

Traditional production of litchi in India is kerbed to the north in foothills of the Himalayas from Tripura to Jammu and Kashmir and the lower plain areas of Madhya Pradesh and Uttar Pradesh. Nevertheless, its cultivation for commercial purposes has been extended to many other states such as Bihar, Jharkhand, Madhya Pradesh, and Chhattisgarh with rising demand for the plant.

The appropriate shape of the litchi tree is attained by carrying out a certain amount of pruning after plantation (Table 2.2). Once the required shape is accomplished, there is no further requirement for pruning with the exception of removing dead or diseased branches or broken shoots. During harvest, a fraction of the shoot-bearing fruits is cut to favour further development. In case of excessive vegetative growth, both the roots and shoots are subjected to pruning. However, very heavy pruning may lead to copious vegetative growth at the cost of developing flower and

Table 2.2 Details of usual practices to grow litchi

Details	Usual practice
Planting time	August, September
	Trees are planted in spring and early summer if irrigation facility exists
Planting distance	10 m (between both the plants and the rows)
	8 m (during dry weather and not much fertile soil)
	Average number of plants: 200/ha
Size of pits	Generally 1 × 1 × 1 m (digging of pits 1 week before plantation)
Filling of pits	Pits are left as is during initial stage for about 15–20 days
	Filling with top soil + manure + fertilizer at the rate of 20 to 25 kg FYM, 2 kg bone meal, and 300 g muriate of potash per pit
	Soil from old litchi farm is added to every pit to ensure mycorrhizal interaction with roots of litchi
	Watering is done in pits for soil to settle down
Planting	Usually square system of plantation is followed
	Creation of small hole at the centre of pit for planting the seed followed by immediate watering

Singh and Babita FAO repository

fruits. Heavy pruning is suggested at the stage when trees become very old and produce small-sized fruits.

A sound development approach should be developed for litchi plants with respect to the particular area such as enhanced water efficacy in case of hot and dry circumstances. This provision enables relocation of organic compounds from source to sink and meristematic sites via moving and circulating water at the vital phase. The major interventions for behavioural and quality characteristics for litchi plants take into account the provision of irrigation at crucial phases of crop growth and preservation of soil moisture reservoirs. In research carried out in Israel, manipulation of autumn vegetative flushing and improvement of flowering and yield of fruit have been done by drought (Goren and Gazit 1996).

Shelf Life Litchi plants have a shelf life of about 72 h, or less at atmospheric temperature. The post-harvest losses for litchi are approximately 25–30% and possibly as much as 50% (Jiang et al. 2001). Just after plucking, fruits begin to degrade; thus, they are graded and packing is done in boxes embedded with green leaves which act as cushions during further transport for commercialisation (Shi et al. 2001).

Browning is a very common phenomenon in fruits. A number of techniques are introduced to avoid browning as well as to control post-harvest rot and to enhance the shelf life of litchi fruit, including fumigation with sulphur, dips against fungi, materials promoting growth of plants, microbial agents (antagonist; viz. *Bacillus subtilis*), coats of wax and chitosan, radiation and treatments with heat. Of all these techniques, application of sulphur-mediated disinfection and dips having antifungal activities are commercialised (Jiang et al. 2003). A few examples of fungicides are benomyl, thiabendazole, iprodione and prochloraz for controlling diseases in litchi

(Wong et al. 1990, 1991). Benomyl is considered to possess a very powerful and wide range of fungicidal properties and is thought to be efficient in controlling fruit decay. However, it is indexed as a post-harvest compound in several countries because of its oncogenic properties (National Research Council 1987).

Litchi chinensis of Indian origin is attacked by multiple pests, including mites such as *Aceria litchii* Kiefer and the shoot borer *Chlumetia transversa* (Heather and Hallman 2008). Several others include the bark-eating weevil *Amblyrrhinus poricolis*, caterpillars (*Indarbela tetraonis*, *I. quadrinotata*), the fruit stone borer *Agyroploce carpophaga*, and the butterfly *Virachola isocrates* (Butani 1977). The presence of such pests in fruits can be managed with the help of gamma radiation, heat/cold treatment, and fumigation as these methods are considered to be insect disinfestation procedures (IAEA 1992).

2.4 Role of Plant–Microbe Interactions in Promotion of Plant Growth

The rhizosphere facilitates the growth and establishment of enormous and widespread microbial communities, involving microbes that have the capacity for plant growth promotion. Such microbes are called plant growth-promoting rhizobacteria (PGPR), and they colonise the roots of monocotyledons as well as dicotyledons. These free-living bacteria have the potential of decreasing the adverse effects of continual use of chemical fertilizers, increasing the synthesis of phytohormones and vitamins, inhibiting ethylene production and enhancing stress resistant. In addition, they facilitate enhanced uptake of nutrients, solubilisation of inorganic phosphates and mineralisation of organic phosphates. Also, PGPR can cause modification of physiological and functional properties of plant tissue situated at a considerable distance from the affected areas such as the shoots. Stimulation of plant nutrition modifies the primary metabolism, thereby contributing to enhanced growth.

Plant root secretions and the associated microbial community are the most important element of the rhizosphere ecosystem. The microorganisms present in the plant root system have a crucial role in enhancing the bioavailability of mineral and nutrients. The root exudates secreted by plants act as a potential source of nutrient and energy for microorganisms, and in return microorganisms accelerate secretions from plant roots. In the co-evolutionary system, the microbial community associated with plant roots either competes or coexists for their survival in the continuous varying environment, and the interactions between the microbes and roots are either beneficial or detrimental. The mechanism behind the increase in mobility of nutrients and minerals involves (a) release of H^+ ions which causes acidification and by making acid–mineral complexes; (b) formation of phytochelatins and several amino acids; (c) secretion of enzymes which catalyse microbial transformations; and (d) enhancement of microbial activity in the plant rhizosphere which indirectly accelerates plant growth and survival and ultimately enhances plant growth promotion (Ström et al. 2002; Zaidi et al. 2009; Pérez-Esteban et al. 2013; Sessitsch et al. 2013).

Every single plant species has the potential to choose their own rhizospheric microflora from the nearby soil, and they are characterised as having specific microbial communities (Hartmann et al. 2009). On the basis of co-evolutionary pressure, plant–microbe interactions are extremely active and vigorous in nature (Chaparro et al. 2013). The rhizospheric ecosystem of plants efficiently communicates with the microbial community in the surrounding soil ecosystem by secreting chemicals which act as signalling molecules for these microbial species, whereas the plant-associated microbial community builds an effective association with plants by stimulating the functional signals of the host, for example, microbial chemotaxis and root colonisation (Doornbos et al. 2012; Bulgarelli et al. 2013; Drogue et al. 2012). There are two categories of mechanisms of enhancement of plant growth by root-accompanying microbes: (1) growth promotion directly through the association between beneficial microorganisms and their host plant, causing enhancement in growth-promoting substances, and (2) growth promotion indirectly by preventing the growth of the plant pathogens.

The adaptation of microbial inoculants for advances in sustainable agriculture proved to be a reliable tool for an integrated explanation of the current agro-environmental problems. The reason behind the utilisation of microbial inoculants is their potential to promote plant growth, improve nutrient availability, enhance nutrient uptake, and maintain plant health (Dobbelaere et al. 2001; Hodge et al. 2001; Bonfante 2003; Weller 2007; Adesemoye et al. 2008). Microbial inoculants are characterised into three categories: (1) arbuscular mycorrhizal fungi (AMF), (2) symbiotic nitrogen fixer (rhizobia), and (3) PGPR.

It has been stated previously that plant and PGPR interactions proved to be advantageous for plant growth and productivity as well as responsible for enhancing seed germination percentage, root growth and nutrient uptake, therefore increasing plant nutrient content and resistance to environmental stress with the potential to inhibit plant pathogens (Bashan et al. 2004; Mantelin and Touraine 2004; Raaijmakers et al. 1997; Bakker et al. 2007; Yang et al. 2009; Mahaffee and Kloepper 1994). Some of the important beneficial characteristics of PGPR inoculants include increasing potential for phosphate solubilisation (Rodriguez and Fraga 1999), free nitrogen fixation (Bashan et al. 2004), siderophore production which chelates iron from surroundings (Raaijmakers et al. 1997; Bakker et al. 2007), production of plant hormones such as indole-3-acetic acid (Gutierrez-Manero et al. 2001) and synthesising enzymes such as 1-amino cyclopropane-1-carboxylate (ACC) deaminase, which is responsible for reducing the plant ethylene concentration, consequently reducing stress on plants (Glick et al. 2007). The plant–PGPR association process involves a varied series of mechanisms, and a detailed overview of these mechanisms was reported in the reviews by Glick et al. in 2007. One of the basic mechanisms reported by which the plant growth-promoting bacteria influence plant nutrient intake, simultaneously increasing plant growth and expanding plant roots, is that plant roots with heavy surface area and high number of root hairs are capable of absorbing more nutrients (Biswas et al. 2000; Adesemoye et al. 2008).

The potential of AMF to stimulate plant growth by altering water transportation and nutrient concentration has been extensively described over the years (Ames et al. 1983; Giovannetti et al. 2006). The AMF have significantly excessive affinity for phosphorous uptake and therefore increase phosphorous nourishment to plants. This process is based on the mechanism that AMF scavenge the accessible phosphorus through their hyphae having large surface areas and these hyphae further acts as a transport channel between soil and plant roots (Liu et al. 2000; Bianciotto and Bonfante 2002). Many problems have also been reported with the utilisation of AMF including in vitro cultivation of AMF, as well as the molecular mechanism of P solubilisation, when rhizosphere competence is not clearly available (Amijee et al. 1989; Koide 1991).

To establish a healthy fruit orchard having good productivity, it is a precondition to have a nourishing nursery. The presence of AMF in the cortical cells of roots of litchi was first reported by Coville (Pandey and Misra 1971). The root infections by AMF consist of intercellular hyphae and vesicles along with sharply branched arbuscules that develop within the cortical cells of the host (Aradhana et al. 2013b). These AMF obtain carbon from the plants and improve the access of phosphorus and nitrogen in return. These mycorrhizal fungi are reported to increase the tolerance of the plants towards drought, seasonal temperature and pH imbalance, reducing the harmful effects of transplant shock. These fungi also help protect the plants from detrimental diseases and pathogens and thus minimise the reliance of plants on chemical fertilizers. The data showed that bio-inoculation of PGPR stains can improve nutrient uptake, growth and nutritional quality (Cakmakci Turan et al. 2014; Zaidi et al. 2009). Research carried out by Yao et al. (2005) inferred that the inoculation of the AMF *Gigaspora margarita* and *Glomus intraradices* onto seedlings of *Litchi chinensis* effectively enhanced the amounts of indole-3-acetic acid (IAA) and isopentenyl adenosine in shoots and roots. IAA is involved in controlling a number of processes in plant growth and development and several essential physiological mechanisms of plants such as cell broadening, cellular division, root initiation and vascular tissue differentiation (Aloni et al. 2006).

To establish a successful plant–microbe association, the ability of microorganisms to colonise plant habitats is crucial (Kamilova et al. 2006; Lugtenberg et al. 2002). The colonisation process takes place in several steps: recognition, adherence and invasion, and numerous strategies are involved to develop colonisation. Plant roots release signalling molecules for soil microorganisms which in response produce signals for root colonisation (Bais et al. 2006). Although the application of the strategies to develop a sustainable agriculture and ecosystem needs to be explored, advancements in molecular technology have provided considerable information in plant–AMF or plant–PGPR interactions.

2.5 Studies Related to Rhizospheric Interaction of Litchi

In the majority of plants, the root develops a symbiotic association with specific kinds of fungi, which are referred to as mycorrhizae. Two different types of associations are reported: the first is ectomycorrhizal association in which fungi intercellularly colonise the plant roots and the second is endomycorrhizal association wherein the colonisation of plant roots by fungi takes place intracellularly. The endomycorrhizal association is further categorised as vesicular, arbuscular, orchidaceous or ericaceous. The most common association between fungi and several types of trees and agricultural and horticultural crops is recognised as vesicular–arbuscular mycorrhizae (VAM). In this association, a special type of organ described as vesicles and arbuscules is developed by fungal hyphae within the cortical cells of the plant. The main function of the vesicles is acting as food storage organs for fungus, and arbuscules are responsible for the transport of nutrients. The mycorrhizal fungi enhance nutrient and mineral uptake to plants and also improve water transportation, therefore being proven to be beneficial for plant growth. The mechanism behind the uptake of nutrients is such that, in the case of phosphorus uptake, the phosphorus in the hyphae of the fungus is converted into polyphosphate granules and further translocated to the arbuscules and subsequently transferred to the host plant. The VAM association enhances the uptake of various minerals including zinc (Zn), copper (Cu), sulphur (S) and potassium (K) by improving nodule formation in leguminous plants; in addition, it also inhibits disease caused by fungal pathogens such as root rot disease. The advantageous results of mycorrhizal associations have previously reported in litchi (Pandey and Misra 1971; Singh and Prasad 2006; Singh et al. 2013; Kumar et al. 2016). Studies on VAM symbiosis with litchi signify that numerous species of AMF, living in red soil, naturally develop symbiosis with roots of litchi plants (Menzel 2002). The colonisation is highest in field conditions (15%) and sometimes decreases when grown in pot culture (3–15%).

It was reported that inoculation of the litchi plant with AMF significantly improves phosphorous uptake; however, Morton (1987) reported the enhancement in plant growth of litchi when it is grown with VAM inoculation, which was further confirmed by several researchers (Yao et al. 2005). Additionally, a strong effect of AMF inoculation was seen on the morphology of the plant root system (Kaldorf and Muller 2000). Study of litchi plants inoculated with two VAM species, *Gigaspora margarita* and *Glomus intraradices*, shows considerable increase in lateral root length (Yao et al. 2005), and Sharma et al.(2009) also reported the enhancement of root length by dual inoculation of litchi with *Glomus fasciculatum* and *Azotobacter* species (Sharma et al. 2009).

Reasons accounting for the alterations in morphology induced by using AM fungal inoculants are complicated. Enhanced phosphorous nutrition prompted this modification in mycorrhizal root structure of leek (Trotta et al. 1991); however, it could not entirely justify the morphological vicissitudes. Additional tests are required to support the finding that AM fungi association can alter the morphological characteristics of roots by plant hormone secretion. A study in maize inoculation with *Glomus intraradices* showed the increment in auxin, that is, indole-3-butyric acid (IBA), at various phases of colonisation, and additionally the enhancement in lateral roots of maize was also observed (Kaldorf and Muller 2000).

The symbiotic association of AMF is prevalent with popular terrestrial plants (Wang and Qiu 2006). AMF acquire their nutrition from the associated host plants and consequently enhance access of phosphorus and nitrogen (Fitter 2006; Kiers et al. 2011; Guether et al. 2009). It was reported previously that inoculation of litchi plants with AMF enhanced growth and development after propagation by air layering. Sharma et al. (2009) observed both qualitative and quantitative variations in AMF species in association with several cultivated varieties of litchi growing in the North-Western Himalayan Region (NWHR) of India. Aradhana et al. (2013a) reported advanced colonisation of litchi plants in NWHR regions. These studies clearly exhibit the value of AMF colonisation with litchi; nonetheless, information concerning the factors that determine AMF symbiotic functioning and community structure is restricted. Development of AMF association and colonisation could rely on edaphic or climatic conditions (Muthukumar and Udaiyan 2002; He et al. 2002).

It has been constantly discovered that when mycorrhizal fungi is inoculated in litchi, this resulted in enhancement in the growth of plants. The AM population showed noticeable differences in incidence and distribution in the litchi farm of Siwalik range of NWHR caused by differences in various farming techniques (Sharma et al. 2002). Litchi trees are very much dependent on AM fungi since immemorial times. The reason could be that the new plants are inoculated with the soil of older and mature litchi trees.

Singh and Prasad (2006) conducted a study in four different regions of Uttar Pradesh, namely, Faizabad, Sultanpur, Basti and Pratapgarh. They concluded that maximum vesicular–arbuscular mycorrhizal (VAM) colonisation and spore formation were observed in litchi trees (*Litchi chinensis*) in Basti and Pratapgarh. The VAM fungi that were dominant involve species of *Gigaspora*, *Acaulospora*, *Glomus*, *Rhizophagus* and *Endogone*. In a study by Sharma et al. (2009), the AMF which were found in the soil samples in a majority of the litchi growing areas of Himachal Pradesh include *Glomus fasciculatum*, *G. melanosporus*, *G. magnicaulis*, *G. radiatus*, *G. macrocarpum*, *G. clarum*, *G. mosseae*, *G. epigeaum*, *Gigaspora calospora*, *G. heterogamma*, *G. gigantea*, *G. pellucida*, *Scutellospora pellucida*, *Sclerocystis coremioides*, and *Acaulospora scrobiculata*.

In a study by David et al. (2001), it was observed that growth of shoots was enhanced in litchi air layering inoculated with native AM fungi. Therefore, when both AM fungi and *Azotobacter* were inoculated in litchi shoots during the period of air layering, an efficient and enhanced root proliferation was witnessed that further led to the improvement of availability of nitrogen and phosphorus to the plant; this will cause more synthesis of protein and result in improved morphological growth (Sharma et al. 2009).

A study was carried out in China where *Hemicriconemoides fujianensis* sp. nov. was identified from the rhizosphere of *Litchi chinensis*. This species closely resembled *H. mangiferae*, *H. litchi* and *H. birchfielai* (Sheng 1998). *Hemicriconemoides communis* is obtained from *Citrus sinensis* (including 15 other hosts from varied families) rhizosphere in Allahabad (UP) India. It is differentiated from a closely related species, *H. cocophilus*. *H. litchi* spp. (*H. litchi* subsp.) was described from *Litchi chinensis* rhizosphere in Gola Gokarannath UP (Edward and Misra 1964).

Latest findings recommended that AMF adaptation to external abiotic conditions, for example, nutrient availability and temperature, will affect the AMF symbiosis on plant growth and development (Johnson 2010; Antunes et al. 2011), However, it is yet not been reported on the litchi ecosystem. The study of soil nutrition influence on AMF colonisation apart from the biotic factors will likely be important to utilise the AMF colonisation for more suitable productiveness and high quality of litchi. The soils of litchi orchard of Bihar, also known as 'the litchi hub,' are characterised as saline with acidic pH, where P is unavailable to trees because of fixation by calcium. By AMF colonisation the phosphorous availability is improved and growth of trees and seedlings is also enhanced. A study reported by Kumar et al. 2016 examines the AMF diversity and effect of AMF colonisation of litchi root and soil nutrient conditions.

2.6 Conclusion

This chapter provides a detailed understanding of the interaction of microorganisms with the litchi rhizosphere. It showed the rife manifestation of arbuscular mycorrhizal fungi in connection with the litchi tree. The profusion of spores clearly gives a good indication of the AMF potential of soils. Although it was not plausible to differentiate between biotic and abiotic parameters which influence the copiousness of spores and AMF colonisation in roots under native circumstances, several studies postulated a clear-cut interpretation of how soil nutrients, management status, and other factors affect AMF synergy with litchi. There are fewer findings on the effects generated by soil micronutrients, but the outcomes listed exhibited their importance as a regulatory component of AMF colonisation. In the future, there is a need for further research addressing the chemistry of exogenous and endogenous parameters in AMF; specifically, how nutrients impinge on symbiotic signalling and on the consequent cellular programming in the rhizosphere of litchi.

Acknowledgments The authors are thankful to the Director of MNNIT Allahabad for providing the necessary research facilities for this work. The support provided by MHRD sponsored project DIC (Design and Innovation Centre) is also acknowledged.

References

Adesemoye AO, Torbert HA, Kloepper JW (2008) Enhanced plant nutrient use efficiency with PGPR and AMF in an integrated nutrient management system. Can J Microbiol 54:876–886

Aloni R et al (2006) Role of cytokinin and auxin in shaping root architecture: regulating vascular differentiation, lateral root initiation, root apical dominance and root gravitropism. Ann Bot 97:883–893

Ames RN, Reid CP, Porterf PLK, Cambardella C (1983) Hypha uptake and transport of nitrogen from two ^{15}N-labelled sources by *Glomus mosseae*, a vesicular-arbuscular mycorrhizal fungus. New Phytol 95:381–396

Amijee F, Tinker PB, Stribley DP (1989) The development of endomycorrhizal root systems. VII. A detailed study of effects of soil phosphorus on colonization. New Phytol 111:435–446

Andersen Ø, Jordheim M (2006) The anthocyanins. In: Andersen ØM, Markham KR (eds) Flavonoids. CRC Press, Boca Raton, pp 471–551

Anon (2003) The wealth of India: a dictionary of Indian raw materials and industrial products. In: First Supplement Series (Raw Materials), vol 4. J-Q. National Institute of Science Communication and Information Resources (NISCIR), CSIR, New Delhi, pp 47–48

Anon (2012) Final area and production estimates for horticulture crops for 2011–2012. National Horticulture Board, Government of India. http://nhb.gov.in/area%20_production.html

Antunes PM, Koch AM, Morton JB, Rillig MC, Klironomos JN (2011) Evidence for functional divergence in arbuscular mycorrhizal fungi from contrasting climatic origins. New Phytol 189:507–514

Aradhana D, Sharma DR, Dohroo NP (2013a) Biodiversity of arbuscular mycorrhizal fungi in agricultural crops of western Himalayas. Ann Plant Prot Sci 21:391–395

Aradhana D, Sharma DR, Dohroo NP (2013b) Occurrence of arbuscular mycorrhizae in rhizospheric soils of different crops and agroclimatic zones of Himachal Pradesh, India. Indian J Agric Res 47(4):373–376

Bais HP, Tiffony L, Weir LT, Perry LG, Gilroy S, Vivanco JM (2006) The role of root exudates in rhizosphere interactions with plant and other organisms. Annu Rev Plant Biol 57:233–266

Bakker PAHM, Pieterse CMJ, van Loon LC (2007) Induced systemic resistance by fluorescent *Pseudomonas* spp. Phytopathology 97:239–243

Baltrušaitytē V, Venskutonis PR, Cěksterytē V (2007) Radical scavenging activity of different floral origin honey and beebread phenolic extracts. Food Chem 101:502–514

Bashan Y, Holguin G, de-Bashan LE (2004) *Azospirillum*–plant relationships: physiological, molecular, agricultural, and environmental advances (1997–2003). Can J Microbiol 50:521–577

Bhoopat L, Srichairatanakool S, Kanjanapothi D, Taesotikul T, Thananchai H, Bhoopat T (2011) Hepatoprotective effects of lychee (*Litchi chinensis* Sonn.): a combination of antioxidant and anti-apoptotic activities. J Ethnopharmacol 136(1):55–66

Bianciotto V, Bonfante P (2002) Arbuscular mycorrhizal fungi: a specialized niche for rhizospheric and endocellular bacteria. Antonie van Leeuwenhoek 81:365–371

Biglari F, AlKarkhi AFM, Easa AM (2008) Antioxidant activity and phenolic content of various date palm (*Phoenix dactylifera*) fruits from Iran. Food Chem 107:1636–1641

Biswas JC, Ladha JK, Dazzo FB (2000) Rhizobia inoculation improves nutrient uptake and growth of lowland rice. Soil Sci Soc Am J 64:1644–1650

Bonfante P (2003) Plants, mycorrhizal fungi, and endobacteria: a dialog among cells and genomes. Biol Bull 204:215–220

Bulgarelli D, Schlaeppi K, Spaepen S, Ver Lorenvan Themaat E, Schulze-Lefert P (2013) Structure and functions of the bacterial microbiota of plants. Annu Rev Plant Biol 64:807–838

Butani DK (1977) Pests of litchi in India and their control. Fruits 32:269–273

Cakmakci Turan RM, Gelluce M, Sahin F (2014) Rhizobacteria for reduced fertilizer inputs in wheat (*Triticum aestivum* spp. *vulgare*) and barley (*Hordeum vulgare)* on aridisols in Turkey. Int J Plant Prod 8(2):163–182

Chaparro JM, Badri DV, Bakker MG, Sugiyama A, Manter DK, Vivanco JM (2013) Root exudation of phytochemicals in *Arabidopsis* follows specific patterns that are developmentally programmed and correlate with soil microbial functions. PLoS One 8:55731

Database of National Horticulture Board, Ministry of Agriculture, Govt. of India. http://nhb.gov.in/model-project-reports/Horticulture%20Crops/Litchi/Litchi1.htm

Davenport TL (2000) The process influencing floral initiation and bloom in litchi: the role of phytohormones in a conceptual flowering model. Hortic Technol 10:733–739

David PJ, Michelle SS, Bruce S, Jonatha HC (2001) Inoculation with arbuscular mycorrhizal fungi enhances growth of *Litchi chinensis* Sonn. Trees after propagation by air-layering. Plant Soil 233:85–94

Dobbelaere S, Croonenborghs A, Thys A, Ptacek D, Vanderleyden J, Dutto P, Labandera-Gonzalez C, Caballero-Mellado J, Anguirre JF, Kapulnik Y, Brener S, Burdman S, Kadouri D, Sarig S, Okon Y (2001) Response of agronomically important crops to inoculation with *Azospirillum*. Aust J Plant Physiol 28:871–879

Doornbos RF, van Loon LC, Bakker PAHM (2012) Impact of root exudates and plant defense signaling on bacterial communities in the rhizosphere: a review. Agron Sustain Dev 32:227–243

Drogue B, Dore H, Borland S, Wisniewski-Doré F, Prigent-Combaret C (2012) Which specificity in cooperation between phytostimulating rhizobacteria and plants? Res Microbiol 163:500–510

Duan X, Jiang Y, Su X, Zhang Z, Shi J (2007) Antioxidant properties of anthocyanins extracted from litchi (*Litchi chinensis* Sonn.) fruit pericarp tissues in relation to their role in the pericarp browning. Food Chem 101:1365–1371

Edward JC, Misra SL (1964) *Hemicriconemoides communis* n. sp. and *H. litchi* n. sp. (Nematoda: Criconematidae), from Uttar Pradesh, India. Nematologica 9(3):405–411

Evans E, Degner R, Crane J, Rafie R, Balerdi C (2004) Is it still profitable to grow lychee in South Florida? EDIS FE496, Dept. Food and Resource Econ., Univ. Fla., Gainesville. http://edis.ifas.ufl.edu/fe496. Accessed 25 Sep 2013

FAO (2002) Lychee production in the Asian-Pacific Region. RAP Publ., Bangkok, Thailand

Fitter AH (2006) What is the link between carbon and phosphorus fluxes in arbuscular mycorrhizas? A null hypothesis for symbiotic function. New Phytol 172:3–6

Giovannetti M, Avio L, Fortuna P, Pellegrino E, Sbrana C, Strani P (2006) At the root of the wood wide web. Self recognition and nonself incompatibility in mycorrhizal networks. Plant Signal Behav 1:1–5

Glick BR, Todorovic B, Czarny J, Cheng Z, Duan J, McConkey B (2007) Promotion of plant growth by bacterial ACC deaminase. Crit Rev Plant Sci 26:227–242

Goren M, Gazit S (1996) Management of lychee orchards in Israel. In: Proceedings of fourth National Lychee Seminar. Australian Lychee Growers Association, Yeppoon, pp 74–77

Guether M, Neuhäuser B, Balestrini R, Dynowski M, Ludewig U, Bonfante P (2009) A mycorrhizal-specific ammonium transporter from *Lotus japonicus* acquires nitrogen released by arbuscular mycorrhizal fungi. Plant Physiol 150:73–83

Gutierrez-Manero FJ, Ramos-Solano B, Probanza A, Mehouachi J, Tadeo FR, Talon M (2001) The plant-growth promoting rhizobacteria *Bacillus pumilus* and *Bacillus licheniformis* produce high amounts of physiologically active gibberellins. Physiol Plant 111:206–211

Hartmann A, Schmid M, van Tuinen D, Berg G (2009) Plant-driven selection of microbes. Plant Soil 321:235–257

He X, Mouratov S, Steinberger Y (2002) Temporal and spatial dynamics of vesicular-arbuscular mycorrhizal fungi under the canopy of *Zygophyllum dumosum* Boiss. in the Negev Desert. J Arid Environ 52:379–387

Heather N, Hallman G (2008) Pest management and phytosanitary trade barriers. CABI Publishing, Wallingford

Hodge A, Campbell CD, Fitter AH (2001) An arbuscular mycorrhizal fungus accelerates decomposition and acquires nitrogen directly from organic material. Nature (Lond) 413:297–299

Hörtensteiner S (2013) Update on the biochemistry of chlorophyll breakdown. Plant Mol Biol 82 (6):505–517

Hui YH (2008) Lychee. Handbook of fruits and fruit processing. Wiley, New Delhi, pp 606–611

IAEA (1992) Use of Irradiation as a quarantine treatment of food and agricultural commodities. In: Proceedings of a meeting, Malaysia 1990. International Atomic Energy Agency, Vienna

Jaiswal BP, Sah NL, Prasad US (1986) Studies on phenolics of rind and aril during ripening and senescence of litchi fruits. Plant Physiol Biochem 13(1):40–45

Jiang YM, Zhu XR, Li YB (2001) Post-harvest control of litchi fruit rot by *Bacillus subtilis*. LWT Food Sci Technol 34:430–436
Jiang Y, Yao L, Lichter A, Li J (2003) Post-harvest biology and technology of litchi fruit. J Food Agric Environ 1(2):76–81
Johnson NC (2010) Resource stoichiometry elucidates the structure and function of arbuscular mycorrhizas across scales. New Phytol 185:631–647
Jose AM, Schafer E (1978) Distorted phytochrome action spectra in green plants. Planta 138 (1):25–28
Kaldorf M, Muller LJ (2000) AM fungi might affect the root morphology of maize by increasing indole-3-butyric acid biosynthesis. Physiol Plant 109:58–64
Kamilova F, Kravchenko LV, Shaposhinkov AI, Azarova T, Makarova N, Lugtenberg B (2006) Organic acids, sugars and L-tryptophan in exudates of vegetables growing on stone wool and their effects on activities of rhizosphere bacteria. Mol Plant-Microbe Interact 19:250–256
Karnick A (2014) Cultivars and varieties of litchi. In: Agropedia. http://agropedia.iitk.ac.in/content/cultivars-and-varieties-litchi
Kiers ET, Duhamel M, Beesetty Y, Mensah JA, Franken O, Verbruggen E, Fellbaum CR, Kowalchuk GA, Hart MM, Bago A, Palmer TM, West SA, Vandenkoornhuyse P, Jansa J, Bücking H (2011) Reciprocal rewards stabilize cooperation in the mycorrhizal symbiosis. Science 333:880–882
Kirtikar KR, Basu BD (1999) Indian medicinal plants, vol I. International Book Distributors, Dehradun, pp 636–637
Koide RT (1991) Tansley review no. 29: nutrient supply, nutrient demand, and plant response to mycorrhizal infection. New Phytol 117:365–386
Kumar V, Kumar R, Kumar A, Anal D, Info P (2016) Spore population, colonization, species diversity and factors influencing the association of arbuscular mycorrhizal fungi with litchi trees in India. J Environ Biol 37:91–100
Lee HS, Wicker L (1991) Anthocyanin pigments in the skin of lychee fruit. J Food Sci 56 (466–468):483
Liu SH, Lin JT, Wang CK, Chen HY, Yang DJ (2009) Antioxidant properties of various solvent extracts from lychee (*Litchi chinensis* Sonn.) flowers. Food Chem 114:577–581
Lugtenberg BJ, Chin-A-Woeng TF, Bloemberg GV (2002) Microbe plant interactions: principle and mechanisms. Antonie van Leeuwenhoek 81:373–383
Mahaffee WF, Kloepper JW (1994) Applications of plant growth promoting rhizobacteria in sustainable agriculture. In: Pankhurst CE, Doube BM, Gupta VVSR, Grace PR (eds) Soil biota: management in sustainable farming systems. CSIRO, Melbourne, pp 23–31
Mahattanatawee K, Manthey JA, Luzio G, Talcott ST, Goodner K, Baldwin EA (2006) Total antioxidant activity and fiber content of select Florida-grown tropical fruits. J Agric Food Chem 54(19):7355–7363
Mantelin S, Touraine B (2004) Plant growth-promoting bacteria and nitrate availability: impacts on root development and nitrate uptake. J Exp Bot 55:27–34
Menzel C (2002) The lychee crop in Asia and the Pacific. FAO, Bangkok
Menzel CM, Simpson DR (1995) Temperatures above 2°C reduce flowering in lychee (*Litchi chinensis*-Sonn.) J Hortic Sci 70:981–987
Morton J (1987) Lychee. In: Morton JF (ed) Fruits of warm climates. Miami, pp 249–259
Muthukumar T, Udaiyan K (2002) Seasonality of vesicular-arbuscular mycorrhizae in sedges in a semiarid tropical grassland. Acta Oecol 23:337–347
Nakasone HY, Paull RE (1998) Tropical fruits, crop production science in horticulture series. CAB International, Wallingford, pp 173–207
National Research Council (1987) Regulating pesticides in food—the Delaney paradox. National Research Council, Board on Agriculture. National Academy Press, Washington, DC
Pandey SP, Misra AP (1971) *Rhizophagus* in mycorrhizal association with *Litchi chinensis* Sonn. Mycopathol Mycol Appl 45:337–354

Pérez-Esteban J, Escolástico C, Moliner A, Masaguer A (2013) Chemical speciation and mobilization of copper and zinc in naturally contaminated mine soils with citric and tartaric acids. Chemosphere 90:276–283

Pilkington SM, Montefiori M, Jameson PE, Allan AC (2012) The control of chlorophyll levels in maturing kiwifruit. Planta 236(5):1615–1628

Raaijmakers JM, Weller DM, Thomashow LS (1997) Frequency of antibiotic-producing *Pseudomonas* spp. in natural environments. Appl Environ Microbiol 63:881–887

Rastogi RP, Mehrotra BN (1999) Compendium of Indian medicinal plants, vol 2, 1970–1979. CDRI/NISC, Lucknow/New Delhi

Rivera-López J, Ordorica-Falomir C, Wesche-Ebeling P (1999) Changes in anthocyanin concentration in lychee (*Litchi chinensis* Sonn.) pericarp during maturation. Food Chem 65:195–200

Rodriguez H, Fraga R (1999) Phosphate solubilizing bacteria and their role in plant growth promotion. Biotechnol Adv 17:319–339

Sakuraba Y, Schelbert S, Park SY, Han SH, Lee BD, Andres CB et al (2012) STAY-GREEN and chlorophyll catabolic enzymes interact at light-harvesting complex II for chlorophyll detoxification during leaf senescence in *Arabidopsis*. Plant Cell 24(2):507–518

Salunke DK, Desai B (1984) Postharvest biotechnology of fruits, vol II. CRC Press, Boca Raton, pp 77–80

Sarni-Manchado P, Le Roux E, Le Guernevé C, Lozano Y, Cheynier V (2000) Phenolic composition of litchi fruit pericarp. J Agric Food Chem 48:5995–6002

Saxena S, Hajare SN, More V, Kumar S, Wadhawan S, Mishra BB et al (2011) Antioxidant and radioprotective properties of commercially grown litchi (*Litchi chinensis*) from India. Food Chem 126(1):39–45

Sessitsch A, Kuffner M, Kidd P, Vangronsveld J, Wenzel WW, Fallmann K et al (2013) The role of plant-associated bacteria in the mobilization and phytoextraction of trace elements in contaminated soils. Soil Biol Biochem 60:182–194

Sharma SD, Bhutani VP, Awasthi RP (2002) Effect of vesicular-arbuscular mycorrhizae and phosphorus on leaf and soil nutrient status of apple seedlings. Indian J Hortic 59:140–144

Sharma SD, Kumar P, Raj H, Bhardwaj SK (2009) Isolation of arbuscular mycorrhizal fungi and *Azotobacter chroococcum* from local litchi orchards and evaluation of their activity in the air-layers system. Sci Hortic 123(1):117–123

Sheng ZS (1998) Two new species of Hemicriconemoides (Nemata: Criconematidae). Acta Phytopathol Sin 28(4):367–373

Shi JX, Wang CS, An XZ, Li JH, Zhao M (2001) Postharvest physiology, storage and transportation of litchi fruits: a review. In: Huang H, Menzel C (eds) International symposium on Litchi and Longan Guangzhou. ISHS Acta Horticulturae, p 558

Singh HP, Babita S Lychee production in India. In: Lychee production in the Asia Pacific region. FAO corporate document repository

Singh RP, Prasad V (2006) Occurrence and population dynamics of vesicular arbuscular mycorrhizae in the Indian orchards of litchi (*Litchi chinensis* Sonn), aonla (*Phyllanthus emblica* L.) and banana (*Musa paradisiaca* L.) Asian J Biol Sci 1:154–156

Singh R, Roy AK, Rani A, Mishra RS, Pandey SK (2013) Diversity in rhizosphere fungi associated with common varieties of litchi (*Litchi chinensis*) plants of North Bihar. Asian J Microbiol Biotechnol Environ Sci 15:77–83

Sorata Y, Takahama U, Kimura M (1984) Protective effect of quercetin and rutin on photosensitized lysis of human erythrocytes in the presence of hematoporphyrin. Biochim Biophys Acta 799:313–317

Ström L, Owen AG, Godbold DL, Jones D (2002) Organic acid mediated P mobilization in the rhizosphere and uptake by maize roots. Soil Biol Biochem 34:703–710

Takuya S, Hiroshi N, Hajime F, Norihiko N, Takako K, Zsolt R, Tetsuya I, Shukoh H, Hideki O (2008) Antioxidative effects of a new lychee fruit-derived polyphenol mixture, oligonol, converted into a low-molecular form in adipocytes. Biosci Biotechnol Biochem 72(2):463–476

Trotta A, Carminati C, Schellenbaum L, Scannerini S, Fusconi A, Berta G (1991) Correlation between root morphogenesis VA mycorrhizal infection and phosphorus nutrition. In: McMichael BL, Persson H (eds) Plant roots and their environment. Elsevier, Amsterdam, pp 333–389

Wang B, Qiu YL (2006) Phylogenetic distribution and evolution of mycorrhizas in land plants. Mycorrhiza 16:299–363

Wang HC, Huang XM, Hu GB, Yang Z, Huang HB (2005) A comparative study of chlorophyll loss and its related mechanism during fruit maturation in the pericarp of fast- and slow-degreening litchi pericarp. Sci Hortic 106(2):247–257

Wang LJ, Lou GD, Ma ZJ, Liu XM (2011) Chemical constituents with antioxidant activities from litchi (*Litchi chinensis* Sonn.) seeds. Food Chem 126(3):1081–1087

Weller DM (2007) *Pseudomonas* biocontrol agents of soilborne pathogens: looking back over 30 years. Phytopathology 97:250–256

Wong LS, Jacobi KK, Peacock BB (1990) Effects of hot benomyl and cold prochloraz on the external appearance of lychee (*Litchi chinensis*) during storage. Singap J Prim Ind 18:15–17

Wong LS, Jacobi KK, Giles JE (1991) The influence of hot benomyl dips on the appearance of cool stored lychee (*Litchi chinensis*). Sci Hortic 46:245–251

Yang J, Kloepper JW, Ryu CM (2009) Rhizosphere bacteria help plants tolerate abiotic stress. Trends Plant Sci 14:1–4

Yao Q, Zhu HH, Chen JZ (2005) Growth responses and endogenous IAA and iPAs changes in litchi (*Litchi chinensis* Sonn) seedling induced by arbuscular mycorrhizal fungal inoculation. Sci Hortic 105:145–151

Zaidi A, Khan MS, Ahmed M, Ovis M (2009) Plant growth promotion by phosphate solubilising bacteria. Acta Microbiol Immunol Hung 56:263–284

Zhang ZQ, Pang XQ, Yang C, Ji ZL, Jiang YM (2004) Purification and structural analysis of anthocyanins from litchi pericarp. Food Chem 84(4):601–604

Zhang R, Zeng Q, Deng Y, Zhang M, Wei Z, Zhang Y et al (2013) Phenolic profiles and antioxidant activity of litchi pulp of different cultivars cultivated in southern China. Food Chem 136(3):1169–1176

Pre- and Postharvest Management Practices for Litchi Production in India

3

Ravi Rajwanshi, Manoj Kumar, and Beche Lal

Abstract

Litchi (*Litchi chinensis*) is an important woody mycorrhizal fruit tree originated in China. In India, the agroclimatic conditions of foothills of the Himalayas in northern states like Bihar, West Bengal, Uttarakhand, Jharkhand, Punjab, and northeastern states such as Assam and Tripura provide immense scope for litchi cultivation. Various abiotic and biotic factors affect the litchi cultivation and production. Optimum temperature, humidity, soil nutrition, and climatic conditions are the deciding factors to support litchi cultivation, but insect and pest infestation severely affect the overall production of litchi. Insects are the major limiting factor affecting litchi production compared to the diseases. Various agricultural practices such as propagation methods and girdling also have an influence on the litchi plantation and overall productivity of this delicious fruit. The present chapter focuses on economically important pests and diseases and their control measures to reduce the infestation altogether with pre- and postharvest management practices to increase the productivity and shelf life of mature litchi fruits during storage as well as transportation process, respectively.

R. Rajwanshi (✉)
Department of Biotechnology, School of Life Sciences, Assam University, Silchar 788011, Assam, India
e-mail: rrajwanshi@gmail.com

M. Kumar
Amity Institute of Microbial Technology, Amity University, Noida, Uttar Pradesh-201311, India

B. Lal
National Bureau of Plant Genetic Resources, Plant Quarantine Division, Indian Council of Agricultural Research (ICAR), New Delhi 110012, India

M. Kumar et al. (eds.), *Lychee Disease Management*,
DOI 10.1007/978-981-10-4247-8_3

Keywords
Erinose mite • *Hemicriconemoides litchi* • *Indarbela quadrinotata* • Powdery mildew • Sulfur dioxide • Girdling • Quality parameters • *Litchi chinensis*

3.1 Introduction

The litchi (*Litchi chinensis* Sonn.) is one of the most delicious fruits which China has given to the world. It belongs to the family *Sapindaceae*. A monograph on litchi was written by Tsai Hasing in 1059 AD, is considered to be the oldest publication on litchi fruit. No Chinese dinner was considered complete without dried litchi nuts. The spread of the litchi to other parts of the world has been limited and has taken place only comparatively in recent times. India is probably the second largest litchi producer in the world. It was introduced in Bengal at the end of the eighteenth century.

Litchi is a typical non-climacteric subtropical fruit and only commercially grown in two narrow strips from 17° to 26° altitude of southern and northern hemispheres (Menzel and Simpson 1986). Litchi cultivation has been reported since 1500 BC by the people of Malayan descent and has been growing for thousands of years in the southern Guangdong province of China. From China, it reached Myanmar by the end of seventeenth century and was introduced in India about 100 years later. It reached West Indies by 1775, followed by Madagascar and Mauritius around 1870. It was introduced in Hawaii in 1873 by a Chinese trader. From India it arrived in Florida between 1870 and 1880 and was introduced in California in 1897. Litchi was reported to be brought to Australia by Chinese migrants in 1954 and reached Israel sometime between 1930 and 1940. Presently, litchi is grown in Central and South Africa, throughout Asia (China and India), South Africa, Australia, Mauritius, Madagascar, Thailand, and Vietnam, and these are now the major litchi-producing countries in the world. Some of the major litchi-producing countries and their major cultivars are listed in Table 3.1. A survey was conducted in 2013–2014 and reported that the area under litchi production in India was 84.2 thousand Ha. and total production estimated 585.30 thousand MT. The major area under litchi cultivation falls in Bihar, West Bengal, Uttarakhand, Assam, Chhattisgarh, Jharkhand, Odisha, Tripura, and Punjab (Table 3.2). Among all litchi-growing states of India, Bihar contributes 40% of total litchi production in India. Other states like West Bengal, Jharkhand, Assam, Chhattisgarh, Uttarakhand, Punjab, Odisha, and Tripura contribute 16, 10, 8.2, 6.4, 5.2, 4.8, 3.5, and 3.4%, respectively (Saxena et al. 2014). There is a scope for the further extension of this fruit crop along the entire foothill region of Uttarakhand, submountain region of Uttar Pradesh, and Punjab states of India (Nijjar 1981).

Litchi is one of the most environmentally sensitive subtropical fruit crops. It is a non-climacteric fruit, and shelf life at room temperature (30 °C) is less than 72 h. However, maintaining cold chain with optimum temperature range of 1–2 °C and relative humidity between 90 and 95%, can enhance the shelf life of sulfur treated

Table 3.1 Litchi producing countries and major cultivars

S. No.	Country	Major cultivars
1.	China	Boh Lup, Baitong-Ying, Fay Zee Siu, Hoak Yip, Kwai May (Red), Long Lue, No-Mai Chee and Wai Chee, Sum Yee, Hong, Souey Tung, Tai So, and Brewster
2.	Australia	Fay Zee Siu, Tai So, Bengal,Wai Chee, Kwai May Pink, Salothiel
3.	India	Bedana, China, Culcuttia, Late Seedless, Late large Red, Elaichi, Shahi, Rose Scented, Purbi, Dehradun
4.	Indonesia	Local selections
5.	Israel	Mauritius
6.	Madagascar	Madaras, Mauritius, Tai So
7.	Philippines	Sinco, Tai So, ULPB Red
8.	South Africa	Mauritius, Mc Lean's red, Tai So, Madras, Bengal
9.	Thailand	Chacapat, Hoak Yip, Kom, Tai So, Wai Chee, Baidum
10.	USA	Brewster, Hoak Yip, Kwai Wai, No Mai Chee, Shanchi, Kaimana Brewster.
11.	Vietnam	Vaithieu
12.	Taiwan	Hoak Yip, Sah Keng

Source: The litchi crop in Asia and Pacific regional office for Asia and the Pacific, FAO corporate document repository. http://www.fao.org/documents/en/

Table 3.2 State-wise area production and productivity of litchi in India

	2011–2012		2012–2013		2013–2014	
State	Area (000 Ha.)	Production (000 MT)	Area (000 Ha.)	Production (000 MT)	Area (000 Ha.)	Production (000 MT)
Bihar	31.1	236.4	31.28	256.43	31.48	234.20
West Bengal	8.9	85.3	9.19	90.00	9.30	93.90
Jharkhand	4.8	57.5	5.27	58.24	5.27	58.24
Assam	5.3	41.5	5.63	49.64	5.38	48.08
Chhattisgarh	4.5	27.1	4.99	30.89	5.36	37.63
Uttrakhand	9.5	19.0	9.49	19.16	9.44	30.71
Punjab	1.7	24.5	1.75	26.52	1.85	28.00
Odisha	4.5	20.1	4.46	20.26	4.47	20.32
Tripura	3.2	16.6	3.46	17.97	3.88	20.18
Others	3.2	16.6	7.23	10.98	7.74	14.04
Total	**80.40**	**538.10**	**82.70**	**580.10**	**84.17**	**585.30**

Source: All Indian 2013–2014 (Final Estimates), Department of Agriculture & Cooperation

as well as fresh litchi by 3–5 weeks. Dehydration, brown discoloration of pericarp, and rotting greatly reduce commercial value of fresh litchi. In the production side, the major problem is low and irregular yield due to poor flowering and fruit set which may be due to different biotic and abiotic conditions.

3.2 Abiotic Factors Affecting Litchi Cultivation

3.2.1 Climate and Temperature

The litchi cultivation and productivity depends on appropriate climatic conditions which is the most limiting factor in the extension and spread of this delicious fruit. It requires a moist subtropical or tropical climate without heavy frost or hot dry winds. Low humidity can be compensated to some extent by liberal irrigation in the dry periods. Dry atmosphere and cold snaps between 1–4 °C in winters are beneficial. Seasonal variations in temperature are necessary for bumper fruiting. In Southern India, where the seasonal fluctuations in temperature are limited, litchi grows successfully only on hill slopes up to an elevation of 35,000 f. which are cooler than plains. In litchi-growing tracts of India, temperature varies from 21 to 37 °C during the flowering and fruiting seasons. In Southern China, the home of litchi, the annual rainfall is 60 in. but this high rainfall is by no means absolutely essential. It grows successfully with drip irrigation in the state of western Uttar Pradesh, India, where the annual rainfall is not more than 35 in. Prolonged rains are not desirable, especially at the time of flowering, when it interferes with pollination. Alternate spells of rain and dry heat in summers cause fruit splitting and drops. The split fruit rots quickly and is not fit for consumption and marketing. Splitting of fruits also takes place if the fruiting season is extremely dry. The relative humidity varies between 80% and 90% during day and night in many litchi-growing areas. This is, however, not essential if good irrigation facilities are available. The limiting factors in the extension of the litchi cultivation areas in India is the hot and dry wind prevalent during the months of April to June when fruit matures. At present, the entire area under litchi cultivation is confined to the foothills of the Himalayas, where the humidity in these months is comparatively high. It can be extended away from the Himalayas if sufficient irrigation facilities in the summer months and adequate wind breaks are provided.

The young plants during the first 3 or 4 years are very susceptible to the frost and need to be protected, but mature trees can withstand mild frost easily. Horticultural studies revealed that frost resistance varies at different stages of tree development. Trees are more susceptible to frost during their dormancy in the cold weather, whereas they are not susceptible to frost when the trees are in flush. Consequently, early frost is less dangerous than the late frost.

Sunburning and skin cracking of developing fruit could be a serious problem in litchi and are promoted by high temperature, low humidity, and low soil moisture in orchards. Temperature higher than 38 °C in combination with humidity lower than 60% promotes litchi fruit cracking. Inadequate moisture during the early period of fruit growth results in the hardening of skin which results to sun burn. It may crack when subjected to increased internal pressure as a result of rapid aril growth following irrigation. The different litchi cultivars show variation in sunburning and skin cracking symptoms which subsequently damages the pulp. The cultivars which have relatively thin skin, few tubercles per unit area, and rounded or flat shape are less prone to cracking. Frequent irrigation during the critical period of aril growth and spraying of zinc sulfate (1.5%) at weekly interval starting from pea stage of fruit growth to harvest and spray of gibberellic acetic acid (GAA) at 40 ppm and ethephon at 1.00–10.00 ppm application reduces the incidence of fruit cracking.

3.2.2 Soil

Litchi can grow in a variety of soil types provided they are well drained, but it grows best in a deep well-drained loam, rich in organic matter. The water table should be at least 1.5–2 m deep. The litchi roots can stand immersion for a considerable time, provided the water is flowing as stagnant water causes root decay. In West Bengal, it grows near the bank of Hooghly River. It has been reported that litchi grows well in slightly acidic soil compared to neutral or slightly alkaline soil. In the state of Bihar in India, litchi grows well in soil containing organic matter up to 30%. Thus, it is apparent that it grows well in acidic soil as it can also grow in calcareous soils. Addition of lime to maintain pH between 5.0 and 5.5 would be beneficial. The roots are covered with tubercles containing mycorrhizal fungi to a larger extent. These fungi grow on the roots of the litchi which mutually benefit both plant and fungi. Arbuscular mycorrhizal (AM) fungi supply mineral nutrients to the roots, influence plant development, and enable mycorrhizal plants to overcome biotic and abiotic stresses. The litchi tree provides carbohydrates to fungi which help it to complete its life cycle. Hence, it is recommended that the new saplings of litchi may be planted in the vicinity of the old litchi tree which helps in the introduction of mycorrhizal fungi to the roots of the new plants of litchi. High organic content and well-acraled soils which do not dry up are not suitable for the growth and development of the mycorrhizal fungi. Studies on the mycorrhizal fungi and litchi host interaction have shown that litchi plant could allocate up to 30% of their manufactured carbohydrate to AM fungi, whereas the fungi in return provide 80% of plant phosphate and nitrogen. A fairly high percentage of clay in the soil is beneficial, but in India and Florida (USA), litchi plants have flourished on light loamy soils as well. Deep plowing may be necessary if the soil is compact. Litchi trees tolerate poor drainage but do not grow well in standing water. Surface drainage should be installed if such conditions exist. If nematodes are expected to be a problem, fumigation should be carried out.

3.3 Nutritional Management (Fertilization)

Fertilization practices in commercial litchi orchards differ due to differences in climate, soil, and availability of different kinds of organic and inorganic matters. The practices for litchi cultivation on alluvial or red basaltic soils require application of 300–400 g urea, 100 g superphosphate, and 25–50 kg compost per tree just after harvest but not later than July. At the time of blossoming, 100 g urea, nitrogen–phosphorous–potassium (NPK) per tree needs to be applied. To prevent

Table 3.3 Manure and fertilizer schedule for litchi cultivation

			Per plant/year (in Kg)	
Age of plant	FYM (Kg)	C.A.N.	Super phosphate	Nitrate of Potash
1–3 years	10–20	0.3–1.00	0.2–0.6	0.05–0.15
4–6 years	25–40	1.00–2.00	0.75–1.25	0.20–0.30
7–10 years	40–50	2.00–3.00	1.50–2.00	0.30–0.50
10 years and above	60	3.5	2.25	0.60

Source: Nijjar (1981)
FYM Farm yard manure, *CAN* Calcium Ammonium Nitrate

fruit drop, trees are sprayed with 0.1–0.2% urea and sometimes with 0.2% magnesium sulfate. A light fertilization should be carried out with care immediately after field transplanting due to the sensitivity of litchi roots to fertilizer burns. Heavy application should be delayed after a year or after one or two growth flushes. One-year-old trees are provided with 30 g urea per plant per month. Highly analyzed mixed fertilizer such as NPK (15:4:11) is given every 3 months together with the urea. Fertilizers in subtropical areas are withheld during cold months to prevent flushing. Whereas in Australia general recommendations for 5-year-old litchi trees are 150 g urea, 300 g single superphosphate, and 150–200 g potassium sulfate application. These quantities are increased by 20–30% each year till trees attain the age of 15 years, when each application rate increases to 1200 g urea, 1200 g superphosphate, and 600–800 g potassium sulfate. No fertilizer is applied to flower-bearing trees in the spring to prevent vegetative growth in the autumn to enhance the flowering. The conditions for floral induction, besides tree condition, are temperature and water stress conditions for litchi. Manure and fertilizer schedule for litchi is listed in Table 3.3.

3.4 Biotic Factors Affecting Litchi Cultivation

3.4.1 Insect

The major loss faced by litchi cultivation is due to insects which are of greater concern in litchi production than diseases. These are described as below:

3.4.1.1 Erinose Mite/Eriophyid Mite, *Aceria* (*Eriophyes*) *litchii* Keifer

It is also known as hairy mite, hairy spider, or dog ear mite. It is a small whitish insect hardly visible to the naked eye. It lives at the base of the hairs on the undersurface of the leaves and causes a brown velvety growth. Pits are formed which may develop into galls and finally the leaves curl up. The adult sucks the sap from the leaves during winters. The breeding starts in March, and maximum activities are found in July. The life cycle is completed in about a fortnight. The mites are very small measuring 0.13 mm long and pinkish-white in color. All stages have only four legs (Waite and Gerson 1994; Waite and McAlpine 1992). The fruit setting is disrupted if the mites move from leaves to the fruits. The mites can be controlled by destroying the affected leaves on the trees manually as well as those which have fallen down should be burnt or put in the pits and buried in soil. Banding of the trees in the month of December with one and a half feet wide plastic band, the upper and lower end of band wrapped with cotton thread (sutli) and coated with coal tar, can prevent the adult and nymphs of mite and mealybug from climbing up the litchi tree. Spraying the trees with any contact miticide, namely, Kelthane (0.1%) and Phosphamidon (0.3%), is also effective. The leaves can be destroyed either by burning or by burying in deep soil. For banding, coal tar or cloth band soaked in weak crude oil emulsion can be used. If the crude oil is too strong, then it may also kill the litchi tree.

3.4.1.2 Bark-Eating Caterpillar (*Indarbela quadrinotata* Walker, *I. tetraonis* Moore)

The attack of bark-eating caterpillar has been observed on the older trees, but it is not a very serious pest of litchi. The branches, especially the old ones, have elongated zigzag ribbonlike messy web with the presence of brownish ribbonlike loose masses of excreta which remain attached with the main scaffold branches at the point of injury. The caterpillar is very harmful to the litchi trees as it feeds on the bark, and in this process it seriously injures the plant vessels through which nutritional plant sap is transported between the plant system, which adversely affect the growth and fruit-bearing capacity of the tree. At times, the infested branches can dry up, and during serious infestation, the whole tree may die. It is a peculiarity of this pest that it prefers older trees than the young ones.

The adult is a large-sized moth with a wing span of about 4 cm in female and 3 cm in male, light gray to light brick red in color with dark brown patches or dots on it. The female adult moth lays eggs in groups of 15–25 during May–June which hatch after 8–11 days. The larvae take shelter in a web of wood bark powder and excreta, and continue feeding on the surface of the bark up till September. Afterward, they bore into bark and underlying tissues in the branches. These become fully grown up by December and pupate in late April. Caterpillars bore inside the bark or main stems and branches about 150–250 mm deep during night. They come out during the night to feed on the bark which is protected by large silken webs that cover the entire affected portion. Later they bore through the bark into the wood. In case of severe infestation, sap movement is disturbed and flush is adversely affected. The larval habit is quite important. It remains in hiding practically the whole day and comes out for feeding on fresh areas during the night only. The pupation starts from April onward and takes place within the larval gallery. Pupal period is only 3–4 weeks.

The rational approach is to clean the web and fumigate with any ordinary fumigants such as carbon disulfide, petroleum, or formalin and check the caterpillars. This operation may be carried with any one of the above fumigants by putting the whole swap of cotton wool dipped in liquid fumigant and inserting the same into the hole which should thereafter be plucked with mud. Another somewhat simpler approach is to inject a strong persistent insecticidal liquid formulation into the web of the larvae and also smear the same up to some distance outside the hole so that when the larvae come out in the night, it may get poisoned. Spraying of either of the insecticide that is dichlorvos (0.03%), trichloroform or endosulfan (0.05%), parathion (0.05%), and methidathion or azinophosmethyl (0.05%) is recommended for its control.

3.4.1.3 Anar Butterfly (Fruit-Eating Caterpillar) *Virachola isocrates* Fab

This pest causes moderate damage to several fruits including litchi, which is distributed throughout India. The larval stage is the most important stage of this pest. The larvae bore into the fruit and feed on its content. It actually creates a lot of mess, and offensive smelling matter oozes out from the entrance hole. Obviously, the wound caused by this pest gets infected with a number of bacteria and fungi which results in the rotting of the fruits.

The female butterfly lays eggs singly on various parts of the shoot, but the young caterpillar which hatches out within a week or 10 days starts feeding on fruit and bores into the litchi fruit. They do not feed on leaves. The larval period may last from 2 weeks to one and a half month. The preparation takes place and pupal period varies from a week or so to more than a month. The breeding of the pest continues throughout the year since it is a polyphagous pest in nature which is influenced by the climatic or weather conditions. As the larvae bore directly into the fruit, there is no satisfactory control measure for this pest. There are few control measures such as vapor heat treatment (VHT) or hot water treatment (HWT) which may be applied for its control. Sometimes infestation is so serious that it may cause 90% damage to the fruit by this pest alone. The effective options may be searched to attract the adults for trapping. The infested orchards may be sprayed with phosphamidon (0.05%) or metasystox (0.3%) at monthly interval. The first spray should be carried out in the month of April in northern parts of India.

3.4.1.4 Mango Mealybug [*Drosicha mangiferae* (Monophlebus stebbingi)]

The orchard of litchi or any alternative hosts in surrounding areas such as mango, citrus, kinnow, papaya, guava, pomegranate, *Ficus* spp., etc. are prone to infestation with these large fleshy flat-bodied bugs that are about one and half centimeter long and little less than a centimeter in width, covered with waxy white mealy powder. These insects crawl downwards from the tree after fertilization to lay eggs and take shelter near the tree trunk, soil craveces, fallen leaves, and pack-houses etc. These are insects belonging to the bug group; most of these bugs suck the tender parts of the plant sap. More than 65 host plants are recorded for this pest. The sucking activities of these bugs results in the production of honey dew which encourages the growth of the sooty mould, giving a very unhealthy appearance to the plant. At times, they are found clustering in masses on young shoots like fungus outgrowth. The female is wingless, and males are winged form with one pair of wings. They have a very delicate reddish body which flies actively and fertilizes the females. The male adults have much shorter longevity than the female adults which live for a month. The adult gravid females after fertilization crawl down from the trees along the trunk to the ground where they lay eggs at depth of about 2–6 in. and in a cluster of 300–400 eggs each. After egg laying, the female dies and remains inside the soil plant debris till the emergence of the next season. The oviposition is generally confined to an area of a few feet in diameter around the base of the tree. The female adult migrates from the trees toward the ground and oviposition in the soil where the males die soon after mating and the female soon after oviposition.

The eggs laid in the soil take quite a few months before they hatch, and their hatching has been reported to be quite appreciably influenced by the temperature and moist conditions of the soil. In northern India, hatching starts from last week of December to the third week of January. Therefore, it is advisable to start plant protection measures by the third week of December. Late monsoon and winter rains have been reported to delay the hatching. The young hatched nymphs after hatching crawl to search some suitable host plant on which they spend some time. Afterward they start migrating on other host plant, and this upward migration lasts several weeks. On reaching the freshly grown shoots, the nymphs congregate there and begin to suck the plant sap. They molt thrice during their nymphal period which lasts about 3 months or more, depending upon the environmental conditions, especially temperature. Therefore, male nymphs undergo short pupation period and transform themselves into winged adult males, whereas female-producing nymphs do not undergo any pupation changes except increase in size. Thus, there is only one generation occurring in a year.

Raking of the soil surrounding the base of the tree trunk which has been infested helps the air masses get exposed to the sun heat and get killed or eaten up by the different predators. The newly hatched nymphs can be killed by the application of soil insecticide in the affected area. The application of sticky band around the tree trunk helps in checking the upward movement of the nymphs on the litchi tree. Incorporation of the slippery material like polythene seeds in the sticky band is likely to increase the effectiveness of this band. Strong organophosphorus insecticides are sprayed which can penetrate the waxy covering of the nymphal body and can also control the pest, but such applications are likely to be more effective if carried out when the nymphs are young. The strength of the spray has to be considerably increased if they have to be applied when the pest is in the advanced developmental stage. The farmers have to be careful while selecting the insecticides so that they should not be harmful to the pollinating insects which are the main source of pollination in litchi.

3.4.1.5 Soft Brown Scale (*Coccus hesperidum* Linnaeus)

It is a pest of minor economical importance in India. The scale causes no significant damage to the fruit, but their severe presence on the fruit reduces its commercial value. The females are sometimes mistaken with mango mealybug because the egg masses are covered with white waxy filaments at the end of the scale. The scales also produce honeydew which helps in the development of the sooty mould on the fruits and panicle. This also causes discoloration of the fruits which downgrades the fruit quality and reduces its market value.

3.4.1.6 The Cocoa Moth or Fruit Borer (*Conopomorpha cramerella* Snellen)

Earlier it was known as *Acrocercops cramerella* Snellen (Bradley 1986). The moth was found to start its breeding activities on the litchi tree from August to February, and during the offseason (March to July), its restricted breeding was observed on *Eugenia jambolana* (Jamun) and *Cassia tora* (Chota Amaltas). The highest peak population of the moth was observed during September and seems to be associated with high temperature followed by high relative humidity. Low temperature and high humidity seems to be associated with its retarded growth and development. The larvae of the moth were noticed to mine the leaves and also bore the shoots and fruits of litchi. The branches of the infested trees appear withered and drop the poor flowers and fruits. The duration of the life cycle of the moth range between 13 and 20 days during August to September. The "shahi" variety was found more susceptible, while "longia" variety of litchi is highly resistant. The larvae of *Acrocercops cramerella* after hatching from the eggs bore into fruits by tunneling the fruit pulp and start feeding on it. When the larvae are full grown, they come out and pupate on the leaf surface. The larvae of *A. cramerella* were found to mine the young leaves and buds of *Cinnamomum* sp. (Cachar plant) and cause a large blotch by mining the upper epidermis of the leaf. The larvae of *A. cramerella* were found to mine the young leaves, shoots, and also bore in the fruit. The larvae on hatching from eggs were found first to penetrate the basal part of the leaves and young shoots before damaging the epidermal cell. As a result of infestation, leaf mining causes leaves to dry and fall on the ground. The infested twigs present a dropping appearance. The leaf infestation varied between 7.1 and 72.5%, while pre-infestation ranged between 47.7% and 88.9% during August to February. The highest tree infestation (88%) was observed in the month of August and lowest (47.7%) in the month of January.

Egg incubation period ranges from 6–7 days, larval period 15–18 days, and pupal period 5–8 days. Eggs are laid by a female varied between 30 and 46 numbers. Larvae undergo 5-month hibernation before pupation. Pupation usually took place inside the oval cocoon on the leaf surface. The duration of the life cycle varied between 13 and 22 days during August to February.

To control the said pest, foliar spray with the methyl demeton (0.25%) or dimethoate (0.3%) in the month of September gave good control of the pest. The larvae of *A. cramerella* could be controlled effectively by spraying of carbaryl/Sevin (0.4%).

The application of 4 kg of Castor cake and 1 kg Neem cake per litchi tree in the soil of the basin of the trees after the first shower of monsoon showed higher population reduction (56.55%).

3.4.1.7 Leaf Rollers (*Platypeplus aprobola* Meyrick and *Isotenes miserana* Walker)

The insects roll or web leaves together to form a shelter in which they feed and subsequently pupate. Heavy infestation causes extensive leaf damage, especially to developing leaves/flushes. Due of their sheltered habit, it is difficult to kill the insect with a contact insecticide. Spraying of endosulfan or carbaryl at 2 g/L water during new leaf growth and before flowering time (at least once just before flowering) was found effective in controlling the pest.

3.4.1.8 Macadamia Nut Borer (*Cryptophlebia ombrodelta*)

The litchi berry shows light brown sawdust like frass symptoms at the fruit holes near the stalk. When peeled, an aril colored grub 1–1.5 cm in size may be seen at the pedicle end or in the hole or even in the seed (Butani 1977; Bradley 1953).

It can be controlled by the application of one spray of azinphos-methyl (350 g/kg or 140 g/L water) or Guthion to the whole tree for 2 weeks after fruit setting. One spray gives protection for about 2 weeks. Examine young fruits with hand lens for signs of damage, and if noticed, apply second and third spray as per need during fruit development.

3.4.1.9 Litchi Stink Bug (*Tessaratoma papillosa*)

This bug leaves brown stains of about 1/8–½ in. on the infested fruit. No hole(s) is seen and neither flesh found discolored on peeling. Damage is caused by the stink bug excreting on the fruit while feeding from the fruit petiole. These bugs appear on the trees during spring (February–April). They suck the sap and blemish young fruits with their excretions. The young at the nymphal stage are often seen on the trees during late spring.

Control measures include spraying of trees with endosulfan at 150 ml/100 L of water once or twice at the beginning of fruit setting (at the late fruit developmental stage).

3.4.1.10 Fruit Spotting Bug (*Amblypelta* sp.)

After infestation, young green fruits of 1–2 cm drop heavily (Waite and Huwer 1998). There is no apparent external fruit damage noticed. As soon as the skin is removed, numerous brown pinprick holes in flesh and/or deep lesions on the developing seeds can be noticed. Treatment is necessary at the first sign of heavy fruit shedding. Spraying of endosulfan at 150 ml/100 L of water provides good control.

3.4.1.11 *Rhynchaenus mangiferae*

It feeds on newly emerged leaves and flowers and ultimately destroys them. Pre-bloomed spraying of 2.5 ml metasystox in 10 L of water or any systemic insecticide shows good control measure of the pest.

3.4.1.12 Mango Hopper (*Idiocerus clypealis*, *I. niveosparsus*, and *I. atkinsoni*)

These are greenish wedge-shaped small-sized insects (also known as Jassids), which varies greatly according to the species. Both adults and nymphs suck the sap from the tender plant parts such as young shoots and panicles. The insect thus causes withering away of panicles, minimizes the fruit setting, and results in the premature fruit drop. Physical injury is caused to the flower buds by ovipositions.

Nymphs are more harmful. Hoppers are mostly active in the flowering and fruiting periods.

Spray the litchi trees with malathion at 0.15% during February–March before the emergence of flowering panicles. Two to three applications are needed to control the pest effectively.

3.4.1.13 Litchi Leaf Miner (*Conopomorpha litchiella* Bradley)

It produces the symptoms like fruit borer. The female lays light yellow eggs on the leaf surface (probably on the lower surface), and larvae bore inside the midrib of leaves. The eggs hatch within 3–4 days, and the creamy white newly hatched larvae start boring into the shoots as well as the leaf blades. The larvae tunnels through the midrib and may enter leaf veins subsequently. Since the vascular system is destroyed, the leaf lamina, in parts or in full, dries up and turns brown. It also bores into newly formed panicles and the entire panicle may dry up. Maximum damage has been observed in the month of September–October resulting in the destruction of entire autumn flush.

Control measures include spraying of the trees with monocrotophos/monocil or nuvacron at 100 ml/100 L of water. Contact insecticide to be sprayed only on the new flush.

3.4.1.14 Shoot Borer (*Chlumetia transversa* Walker)

The caterpillars make a hole/bore inside the newly developed shoots/twigs and start feeding on them which causes drying of the new flesh and twigs. In case of severe infestation, the cell sap of the plant parts gets interrupted, which adversely affects the growth of the plants/saplings.

For the effective management of the pest, it is suggested that pruning should be carried out when initial infestation/damage is noticed. The dried twigs/branches should be burnt or buried by putting them in deep soil pits. For the effective control, the chemical control measures may be applied. Carbosulfan or quinalphos (0.05%) provide good control measure for the pest.

3.4.1.15 Fruit Borer (*Platypeplus aprobola* Meyer, *Dichocrosis* sp.)

It is a serious pest of litchi and causes maximum damage when plants are at fruit developing stage. The newly hatched larvae bore through the fruit stalk at the end of the fruit and damage the nuts/skin of the fruits. The higher humidity and intermediate rains provided favorable conditions for the pest development and infestation. It results in the fruit drop which adversely minimizes the fruit yield per tree as well as the fruit quality.

For the pest control, the cultural practices are most effective in minimizing the pest population such as plowing, burning of wrapping materials, sanitation (weed control and their destruction), and burying of the damaged fruits in the soil. For insecticidal control measures, twice spraying of Neem-based materials or Kamdhenu Keet Niyantrak (4–5 ml/L) provides best results. The spraying of imidacloprid (0.05%) provides best results. The first application should be applied when fruits are pea size and second application should be followed after 15–20 days interval which helps in effective pest control.

3.4.1.16 Gall Flies (*Dasineura* sp.)

It is a major pest of litchi recorded in Muzaffarpur district of Bihar, India. It causes damage to litchi leaves especially during winters. The larvae produce galls on the leaves. The severe infestation is recorded in dense litchi plantation. The larvae pupate in the soil. After the adult emergence, they start infesting the leaves and about eight overlapping generations are recorded per annum. The adults lay eggs on the young leaves and flushes. The larvae start mining the leaves and cause watery dots and afterward they appear as a gall. These galls turn brown and ultimately drop. After the emergence of these galls, a shoot hole appears on the leaves. These galls can be controlled by dusting beneath the tree trunk with methyl parathion (2.5%) or spray of isofenphos (0.001%).

3.5 Disease Management

A number of disease-causing organisms infecting the litchi trees are listed in the literature, but none are considered serious. All are of the postharvest nature. Few are causing considerable losses in preharvest stage which are as discussed below:

3.5.1 Powdery Mildew (*Oidium* sp.)

It is a fungal disease caused by *Oidium* sp. specially observed in those litchi orchards where mango trees also exist. These diseases occur during the flowering period when the humidity is very high, accompanied by cool nights. The infected shoots/flowers/fruits show grayish whitish powdery appearance on the panicle, flower buds, fruitlets, and rachis of the panicles which later shows dark brown lesions on the litchi fruit. In severe cases, the whole panicle looks as if scorched. Within few days of its first visible symptoms, all the panicles get affected. Sometimes, growing tips of the shoots also get affected, while other parts of the plant remain quite free.

Two sprays of either 0.2% wet sulfur powder or kerathan 25% WP 0.06–0.09% during pre- and post-bloom stages at 15–20-day interval gave good control of the pathogen.

3.5.2 Anthracnose (*Botryodiplodia theobromae* Pat., *Colletotrichum gloeosporioides* Penz.)

The two types of leaf spot caused by fungi starts from the tip or the margin of laminae where deep chocolate colored spots appear. The limiting margin of the spots with irregular outline is Vandyke brown. Black pycnidia appear on both surface of the leaves but more often on the upper surface of the leaves.

3.5.3 *Colletotrichum gloeosporioides* Penz.

It has irregular spot (brick brown in color with a prominent marsh brown margin encircling them) usually start from the tip of the laminae and extend toward the base. Mummy brown, waxy, subepidermal acervuli appear on the surface, especially the upper surface, of the infected leaves. In severe cases, flowering panicles, flowers, and fruits are also affected.

For the control of anthracnose caused by *B. theobromae*, avoid overcrowding of the trees and branches in the orchard. Burning of affected plant parts along with other sanitary measures is the control measure of this pathogen for the control of *C. gloeosporioides*. Three sprays of 3:3:50 Bordeaux mixture in February–April and September–October or application of 1.8 kg Capton (30% WP) in 45 L of water along with spreader is more effective.

3.5.4 Red Rust (*Cephaleuros mycoides*)

It is caused by an algal parasite *C. mycoides*, on the infected young leaves; small lesions of velvety white growth appear on the lower surface of the leaves. On the upper surface of the leaves just opposite to the lesions, chlorotic patches occur. As the leaves unfold and increase in size, the velvety growth becomes more prominent and dense. Larger areas of the leaves are covered with such growth. Old and thick leaves show various types of malformation (depressions and curling). The velvety growth turns light brown and finally dark brown to brick red. The affected leaves become lathery and brittle. Disease results in considerable decline in tree vigor and fruit yield. Three sprays of lime sulfur in autumn (September–October) and three sprays during February–March at 15-day interval depending upon the severity of infection gave good control of the pathogen.

3.5.5 Nematodes

Nematodes have not been reported as a serious threat to litchi cultivation. Tree decline is associated with more than a dozen species of nematodes which are reported from different parts of India (Choudhary et al. 2004; Nath et al. 1996). Roots become stubby and brown in color, and secondary feeder root development is inhibited. Although these nematodes have no significant effect on the production and quality of the litchi fruit, however, they have an adverse role to play when litchi plants are less than 4–5 years old. Under favorable climatic conditions, they may adversely affect the plant growth which may result in the death of the plant at early stage. Some important nematode species associated with roots of litchi plant are listed below:

(a) *Rotylenchulus reniformis*
(b) *Helicotylenchus indicus*
(c) *Tylenchorhynchus leviterminalis*
(d) *Xiphinema* sp.
(e) *Hemicriconemoides litchi*
(f) *Meloidogyne incognita*
(g) *Helicotylenchus dihystera*
(h) *Hoplolaimus indicus*
(i) *Xiphinema brevicolle*

Under field conditions, these parasitic nematodes can be controlled by using Neem cake manure or mixture of carbofuran granules with organic manures. The flooding of the plantation also helps in reducing the nematode population under field condition.

3.5.6 Weed Control

Weed control is most important from the time of litchi plantation in field up till 3–4-year-old orchards. As a tree grows and expands horizontally, there is a decreasing trend of weed growth due to intercropping, shading, and underneath canopy. Use of polyethylene mulch around the plant at 1 m^2 area during the plantation time reduces the weed growth. Organic mulches are highly recommended around the base of litchi trees which are reported to be adversely affected, around the tree trunk base.

3.6 Other Factors Affecting Litchi Cultivation

3.6.1 Propagation

Litchi plant (saplings) grown from seed stage takes about 7–12 years to bear the fruits, in some areas even up to 20 years to attain fruit-bearing stage. Moreover, litchi seeds lose their viability within 4–5 days after removal from the fruits. Hence, the litchi is preferably propagated vegetatively. The most common method of propagation in India is air layering "gootee" and grafting. With the advancement of the plant tissue culture, micropropagation techniques can be opted to overcome the propagation problems.

The best time for air layering is June or beginning monsoon season. The air layers are thus removed late in August. They took the advantage of the most part of the moist monsoon season, and at the time of detachment of the young plants (branches with roots), the atmosphere is humid. These plants are shifted to nursery and may be transplanted in the field in late September. However, it would be preferable to do the planting in the beginning of next monsoon. The layering can also be started in August, and the plants (2–3 cm diameter and 30–60 cm long) can be removed from the trees in October and planted during the next monsoon.

The process can be simplified and watering can be eliminated, if the air layer is wrapped with moist sphagnum moss and covered with polythene wraps. Treatment of ringed portion with a hormone such as 50% aqueous solution of "rootone" or 200 ppm of a naphthalene acetic acid (NAA) in lanolin helps root formation. The roots are formed in about a 2-month duration. This method is much cheaper than former. Recently, plastic material for wrapping the air layer has become easily available and is very useful. The new plants are thus raised and planted in the field.

Inarching is another method of propagation followed in case of few varieties. If the litchi seeds are kept in distilled water, they will remain viable for 2–3 weeks. Another effective technique of storing the seeds is to place a single layer of seeds between two layers of moist sphagnum moss and finally rolling it up between moisture proof papers. These can be stored up to 8 weeks in summer. The seeds are likely to germinate in storage and must be planted very carefully without damaging the young sprouts. The seeds are sown half an inch deep in a partially shady place and the soil is kept moist. The foot hills of Nilgiri, in south India, up to 30% success has been obtained in the inarching of litchi.

Recent experiments have shown that litchi can be propagated successfully from cuttings. Two-year-old cuttings treated for 24 h in 0.02%, aqueous solution of Indole acetic acid (IAA), or 0.005–0.01% solution of indole butyric acid (IBA) before planting, give very good results. Higher concentration of these hormones is harmful. The cuttings should be taken from young plants. Cuttings from older plant do not give good results.

3.6.2 Cultural Practices

The young litchi plants are very delicate and if proper care is not taken, the mortality after planting is very high. Planting during the monsoon season and frequent shallow irrigation or water application through drip irrigation afterward help in reducing mortality. Since the litchi prefers to grow in association of the mycorrhizal fungi, the land should not be allowed to become completely dry; otherwise, the beneficial fungi working in the root nodules of the plants are damaged, and the young plant itself suffers. However, in the humid places, if liberal irrigation facilities are available, planting can be done in spring.

The plant should be planted at least 30–40 in. distance apart under very favorable conditions for growth. A distance up to 50 in. apart is recommended in South Africa. If the trees are planted too close (high-density plantation), lack of adequate sunlight and air decreases flowering and fruiting. Nevertheless, partial shading of trees to guard against the effect of desiccating winds is desirable. Under such conditions, the trees may be planted at 25 in. spacing.

Prior to planting, the pits must be adequately supplemented with manure as the litchi plants require heavy doses of manure. Inclusion of canal silt in the pit, if available, is also desirable. The manure dose, recommended for the pit, is 25 kg of farmyard manure and 1.75 kg of bone meal for each pit.

The young plants should be protected against frost from severe winters; this can be done by providing thatch shelter on three sides (south, west, and north) and the top. The eastern side is left open to avail sunshine. Frequent irrigation in the evening when frost is accepted is helpful in reducing cold injury. During the dry periods, the young plants should be watered frequently, i.e., twice a week, and the soil should be kept moist to prevent drying of the soil. Lack of irrigation in dry areas during fruit setting period causes fruit drop and splitting of fruits.

The irrigation of the young trees should be done by the basin system. As the trees grow, the basin size should be gradually enlarged. The older plantations are irrigated by flooding or by furrow irrigation. In Uttarakhand state of India, irrigation of old trees is done by drench method. The drenches are dug at the distance of 5–7 in. from the trunk of the litchi trees and 3–4 in. deep which is filled with water during litchi cropping season.

The cultural requirement of litchi is similar to that of the mango cultivation in India except that it is a shallow-rooted crop; hence, deep tillage or deep plowing is not recommended, which is harmful for litchi cultivation. The litchi orchards should be given tillage three or four times a year and must be kept free from weeds. Raising of cover crops or intercropping is very beneficial to the litchi crop and economical for the growers. Summer cover crops are especially useful for maintaining humidity. Intercropping of young orchards provides the much-needed income during the period when litchi plants are not in bearing stage. Leguminous crops like cowpea, beans, groundnut, and grams are to be prepared for this purpose. These young orchards can also be planted with filler trees of papaya and phalsa.

It is certain that manural requirement of litchi are high. In particular, it requires high doses of organic matter. In China about 227 kg of night soil per tree are applied every year. In Bihar state of India, little or no manure is applied to litchi since it is normally grown in naturally rich soils. Even so, the addition of manure is bound to improve the performance of trees and resulted good yield; 23–227 kg of farmyard manure or leaf mold per tree is given depending upon its age and is recommended for the Indian conditions.

It should be spread under the drip of the trees and forked into ground. The fallen leaves of the trees should not be removed. They form good mulch. The application of 2.5 kg Castor cake or 1.75 kg of Neem cake, 1.75 kg bone meal, and 3.75 kg of wood ash per tree has been recommended by authorities. If the soil is deficient in zinc, a foliar spray of zinc containing 3.62 kg of zinc sulfate and 1.81 kg of hydrated lime in 378.54 L of water is beneficial for litchi production and quantity produce.

After the initial training and building up a good framework of the trees, very little pruning is required. The litchi flowers are borne mostly on new shoots. The old branches rarely produce flowers. Snipping of the old branches to promote fresh growth is, therefore, desirable. The fruits are harvested in bunches along the shoots, and this also serves the purpose of pruning. If the trees are making too much vegetative growth both roots and shoots pruning is sometime applied. If the crown is too dense, branches should be thinned out. It should be remembered that heavy pruning causes profuse vegetative growth, which takes place at the expense of flowering and fruiting. When the trees become too old and produce fruits of small size, heavy pruning may be carried out. This has limited commercial utility, since the fruit yield is reduced and it is also not effective for more than few years.

3.6.3 Girdling

Trunk girdling increases carbohydrate content above the girdling site which promote flowering in cultivated litchi by inhibiting new shoot growth (Huang et al. 2003; Yuan and Huang 1993). Girdling reduces AM colonization due to low glucose, fructose, starch, sucrose and quebrachitol content (Shu et al. 2016; Kiers et al. 2011). Experiments conducted in the Uttarakhand state of India have shown that in some varieties girdling done by running a pruning saw around the branches or trunk greatly increases flowering and fruiting. Girdling is, however, ineffective if the trees are in poor health and lack of major soil nutrients may decrease new flush/tree growth, 6 months prior to bud initiation. It should not be done in windy areas as the tree trunks may break which eventually results in the death of the plant. In dry areas it should be done in alternate years or only half of the trunk should be girdled in 1 year. Before girdling, the trees should be given a complete fertilizer in July after harvesting, and for next 3 or 4 weeks, they should be given sufficient irrigation in order to induce vegetative flush.

3.6.4 Maturity of Fruits

The maturity of fruits for harvesting is judged by the flattening of the tubercles. When fruit matures, the pericarp become smooth and red in color. The fruits are harvested in bunches along with a portion of branch and a few leaves. If individual fruits are harvested, skin of the fruit and stems get damaged which results in the rotting of the fruit quickly. The color of the fruits undergoes rapid changes on ripening. In the colored varieties, the bark or skin is of bright reddish color. For local consumption, the fruit is harvested when it has attained this color. However, for distant markets, the fruit should be harvested when it has just started to turn reddish (Zauberman et al. 1991).

3.7 Deficiency Symptoms of Different Fertilizers in Litchi Orchards

Application of nutrients to the soil is essential for desired growth and production. Nitrogen is the major nutrient and occupies an important place in litchi cultivation. The other major fertilizers needed are phosphorous, potassium, calcium and magnesium (Ghosh and Mitra 1990). Micronutrient consists of iron, boron, copper, zinc, and manganese which are required in very small amount to maintain the tree health

Table 3.4 Litchi leaf nutrient standards

S.No.	Nutrients (% or ppm)	Optimum level
1.	Nitrogen (%)	1.3–1.4
2.	Phosphorous (%)	0.08–0.20
3.	Potassium (%)	0.8–1.2
4.	Calcium (%)	0.5–2.5
5.	Magnesium (%)	0.4–0.7
6.	Iron (ppm)	50–200
7.	Magnesium (ppm)	30–500
8.	Zinc (ppm)	15–150
9.	Copper (ppm)	5–15
10.	Boron (ppm)	25–100
11.	Sodium (ppm)	200
12.	Chloride (%)	2.5

Source: Menzel and Simpson (1986)

Table 3.5 Litchi leaf nutrient composition at different stages of growth

S.No.	Nutrients	Two months before flowering	Ten days before flowering	After harvest
1.	Nitrogen (%)	1.32–1.34	1.46–1.48	1.08–1.22
2.	Phosphorous (%)	0.15–0.16	0.16–0.18	0.14–0.15
3.	Potassium (%)	0.96–0.98	1.02–1.04	0.88–0.92
4.	K.N. ratio	1:1.4	1:1.4	1:1.3

Source: Ghosh and Mitra (1990)

and yield. Some of the litchi leaf nutrient standards and leaf nutrient composition at different stages of growth of litchi trees are mentioned in Tables 3.4 and 3.5, respectively. The deficiency of different mineral elements showed the following symptoms in litchi.

Nitrogen: Yellowing of old leaves, stunted growth, poor flowering and small fruit size.

Phosphorous: Tip and marginal necrosis of old leaves, leaf curl, desiccation and early falling of leaves.

Potassium: Yellowing of leaves, necrotic leaf, leaf tips and margins, poor fruit set and stunted growth.

Calcium: Death of growing points.

Magnesium: Small leaves, leaf necrosis, leaf drop, poor flowering.

Zinc: Bronzing of leaflets and reduced fruit size.

Iron: Yellowing of leaf and producing die-back symptoms.

Copper: Die-back, small fruits, reduced pulp recovery.

Boron: Small fruits.

3.8 Postharvest Treatment of Litchi with Sulfur Dioxide

There are two areas in the postharvest handling chain of litchi which deserve special attention. Firstly, pre-cooling should be applied to remove field heat and provide effective temperature management during transportation. This enables maintenance of fresh quality and flavors, reduces desiccation, and prevents browning of the rind (Huang and Scott 1985). The pericarp browning is mainly attributed to desiccation of pericarp and degradation of anthocyanin pigments along with oxidation of phenolic compounds by polyphenol oxidase (PPO) and/or peroxidase (POD) enzymes which is the most important postharvest problem associated with litchi (Jiang et al. 2004). Several approaches such as heat treatment, wax coating, sulfur fumigation, application of fungicide, acid dipping, irradiation, and modified atmosphere packaging have been tried to overcome the problem of pericarp browning and shelf-life extension of litchi (Kumar et al. 2012; Yueming et al. 2008). Secondly, effective postharvest fungicidal treatment is needed to prevent fruit decay.

Litchi is a non-climacteric fruit which at 25 °C exhibits a moderate rate of respiration (30-40 ml CO_2/kg/h) and a low rate of ethylene production (less than 0.1 μl/kg/h). However, it deteriorates rapidly after harvest (Tian et al. 2005). Shelf life at room temperature (30 °C) is less than 72 h. One of the major problems associated with high temperature and humidity is the growth of saprophytic fungi, mainly *Botryodiplodia* sp., on the fruit surface. At a lower pH, deterioration by decay is reduced, but the fruit loses its freshness. The rind turns brown, dry, and brittle, and the aril wilts and shrivels. The rot problem is reduced but not entirely eliminated by cold storage. Fruits stored at 5°–7 °C also suffer from chilling injury as indicated by browning of the rind, and upon removal to ambient temperatures, the injured fruits are more susceptible to fungal infection.

The process of "sulfitation" of litchi which was initially developed in South Africa has a fungicidal effect and also ensures fixing of red color of the rind and prevents brittleness during storage and transportation. The accepted level of SO_2 in litchi pulp (aril) was fixed at 10 ppm in imported litchi by French authorities in 1989.

In the fumigation system, SO_2 is added to an endorser in order to control or eliminate undesirable microorganisms. The most appropriate system where the transportation period is less than 2 weeks to importing countries is the high concentration, short duration fumigation system. For consistency and reproducibility, a standardized fumigation procedure is used where the ratio of fruit weight and free space volume of the fumigation chamber is maintained at 1:5 particularly where gaseous SO_2 is used (Underhill et al. 1992; Underhill and Simons 1993). A French expert, Dr. J. Marchal, reported that in Madagascar it was an established practice to use 550 g sulfur for fumigation of 1 MT of litchi for 45 min by burning pure sulfur in closed chamber, and it was good enough for storing litchi at 0°–2 °C ± 1 °C for a period of up to 6 weeks (Kudachikar et al. 2007; Kumar et al. 2011). Accordingly, for the storage of litchi for a period of 7 days (under transportation) at the temperature of 10°–15 °C, 140 g pure sulfur powder was burnt for

25 min. in a closed chamber where 400 kg fresh and graded (extra class) fruits were kept in 16 plastic crates of 25 kg each capacity. Uniform air circulation inside the chamber was duly ensured.

As a matter of fact, the effect of SO_2 on the surface growth of fruit, SO_2 injury to rind, and SO_2 residue level in the treated fruits depends on the fruit concentration of SO_2 applied and varies with cultivar, particular crop (locality wise), and the duration of fumigation. That is why it is necessary to work out the specific dose of sulfur for fumigation of litchi. After the sulfur fumigation worked out at Uttarakhand state of India, the maximum residue level (MRL) of 250 ppm SO_2 in the fruit rind and 10 ppm in aril was found at the time of fruit consumption which is within nonhazardous and in permissible limits.

3.9 Quality Parameters of Litchi Fruit

(a) Litchi is true to type, preferably with lower waste index.
(b) Litchi fruit should be well developed (wt. 15–30 g) depending upon cultivar, may be 59–65 days after fruit set, and clean and fully colored with tubercle tip not fully flat.
(c) It must be mature, sweet (TSS 16–21%) depending upon cultivar, healthy, sound, and firm with good aroma.
(d) It must be completely free from insect, fungal, as well as bacterial infection and foreign materials.
(e) Stem portion (fruit stalk) should be intact and not broken.
(f) It must be completely free from cracks and splits.
(g) It must be completely free from any physiological disorder such as dull or brown color, off-flavor, oozing of liquid and chilling injury symptoms.
(h) The fruit should have firm flesh (white to creamy color) depending on the cultivar without any discoloration.

References

Bradley JD (1953) Some important species of the genus *Cryptophlebia* Walsingham, 1899, with descriptions of three new species (Lepidoptera: Olethreutidae). Bull Entomol Res 43:679–689

Bradley JD (1986) Identity of the South-East Asian cocoa moth, *Conopomorpha cramerella* (Snellen) (*Lepidoptera: Gracillariidae*), with descriptions of three allied new species. Bull Entomol Res 76:41–51

Butani DK (1977) Pests of litchi in India and their control. Fruits 32:269–273

Choudhary BN, Rahman MF, Bora A (2004) Diversity of nematodes in four districts of Assam. Indian J Nematol 34:222–224

Ghosh B, Mitra SK (1990) Effect of varying levels of nitrogen, phosphorus and potassium on yield and quality of litchi (*Litchi chinensis* Sonn.) cv. Bombay. Haryana J Hortic Sci 19:7–12

Huang PY, Scott KJ (1985) Control of rotting and browning of litchi fruit after harvest at ambient temperatures in China. Trop Agric 62:2–4

Huang XM, Wang HC, Yuan WQ (2003) Effects of twig girdling at different stages on new shoot growth and carbon nutrient reservation. Acta Hortic Sin 30:192–194

Jiang Y, Duan X, Joyce D, Zhang Z, Li J (2004) Advances in understanding of enzymatic browning in harvested litchi fruit. Food Chem 88:443–446

Kiers ET, Duhamel M, Beesetty Y, Mensah JA, Franken O, Verbruggen E et al (2011) Reciprocal rewards stabilize cooperation in the mycorrhyzal symbiosis. Science 333:880–882

Kudachikar VB, Semeerbabu MT, Revathy B, Ushadevi A, Ramana KVR, Matches RS (2007) Effect of post harvest treatments on shelf life and quality of litchi fruit stored under modified atmosphere at low temperature. J Food Sci Technol 44:106–109

Kumar D, Mishra DS, Chakraborty B, Kumar P (2011) Pericarp browning and quality management of litchi fruit by antioxidants and salicylic acid during ambient storage. J Food Sci Technol, http://dx.doi.org/10.1007/s13197-011-0384-2

Kumar S, Mishra BB, Saxena S, Bandyopadhyay N, More V, Wadhawan S, Hajare SN, Gautam S, Sharma A (2012) Inhibition of pericarp browning and shelf life extension of litchi by combination dip treatment and radiation processing. Food Chem 131:1223–1232

Menzel CM, Simpson DR (1986) Description and performance of major lychee cultivars in subtropical Queensland. Queensland Agric J 112:125–136

Nath RC, Mukherjee B, Dasgupta MK (1996) Plant parasitic nematodes associated with major agricultural and horticultural crops in Tripura. Curr Nematol 7:89–104

Nijjar GS (1981) Litchi cultivation. Punjab Agricultural University, Ludhiana, India, p 11

Saxena M et al (2014) Facts and figure. In: Indian horticulture database. National Horticulture Board, Ministry of Agriculture, Government of India, pp 3–302

Shu B, Li W, Liu L, Wei Y, Shi S (2016) Transcriptomes of arbuscular mycorrhizal fungi and litchi host interaction after tree girdling. Front Microbiol 7:408

Tian SP, Li BQ, Xu Y (2005) Effect of O_2 and CO_2 concentrations on physiology and quality of litchi fruit in storage. Food Chem 91:659–663

Underhill SJR, Simons DH (1993) Lychee (*Litchi chinensis* Sonn.) pericarp desiccation and the importance of postharvest micro-cracking. Sci Hortic 54:287–294

Underhill SJR, Bagshaw J, Prasad A, Zauberman G, Ronen R, Fuchs Y (1992) The control of litchi postharvest skin browning using Sulphur dioxide and low pH. Acta Hortic 321:732–741

Waite GK, Gerson U (1994) The predator guild associated with *Aceria litchi* (Acari:Eriophyidae) in Australia and China. Entomophaga 39:275–280

Waite GK, Huwer RK (1998) Host plants and their role in the ecology of the fruitspotting bugs *Amblypelta nitida* Stal and *Amblypelta lutescens lutescens* (Distant) (Hemiptera: Coreidae). Aust J Entomol 37:340–349

Waite GK, McAlpine JD (1992) Honey bees as carriers of litchi erinose mite *Eriophyes litchi* (Acari: Eriophyidae). Exp Appl Acarol 15:299–302

Yuan RC, Huang HB (1993) Regulation of roots and shoots growth and fruit-drop of young litchi trees by trunk girdling in view of source-sink relationships. J Fruit Sci 10:195–198

Yueming J, Jianrong L, John S (2008) Enhanced effect of l-cysteine and citric acid combination on browning inhibition and quality maintenance in harvested litchi fruit. J Food Sci Technol 45:75–77

Zauberman G, Ronen R, Akerman M, Weksler A, Rot I, Fuchs Y (1991) Postharvest retention of the red colour of litchi fruit pericarp. Sci Hortic 47:89–97

Determination of Substrate Medium for Litchi Marcot Establishment in Nursery

4

Prity Sagar, Anfal Arshi, and Awadh Kishore Roy

Abstract

Vermicompost contains a large amount of micro- and macronutrients and also promotes the soil growth potential of beneficial rhizosphere microbes, including plant growth-promoting rhizobacteria (PGPR) and plant growth-promoting fungi (PGPF). Vermicompost is known to be a suitable media for gardening and farming in addition to supplying nutrients regularly. Litchi was selected as a referral host for the mass inoculum production of vesicular arbuscular mycorrhizae (VAM) because of its high affinity with symbionts. Of seven varieties of litchi, the variety Ref. II showed higher colonization and spore population, that is, 98% and 25 spores/g dry soil, respectively. The test plants treated with AM fungi in the mixture of soil and vermicompost (1:3 ratio) showed the best performance in terms of root dry matter (2.5 g) and root volume (22.5 ml) after 3 months of plant growth. Arbuscular mycorrhizal fungi (AMF) colonization and spore populations were also found to be influenced by the treatments.

Keywords

Vermicompost • *Litchi chinensis* • Vesicular arbuscular mycorrhizae

P. Sagar
Centre for Ecological Sciences, Indian Institute of Science, Bangalore, India

A. Arshi (✉)
Defence Research & Development Organisation, Defence Institute of Bio-Energy Research (DIBER), Haldwani, Uttarakhand 263 139, India
e-mail: anfalarshi@gmail.com; anfal_arshi@rediffmail.com

A.K. Roy
University Department of Botany, TM Bhagalpur University, Bhagalpur 812002, India

M. Kumar et al. (eds.), *Lychee Disease Management*,
DOI 10.1007/978-981-10-4247-8_4

4.1 Introduction

The beneficial effect of arbuscular mycorrhizal (AM) fungi in relationship to improvements in plant health and productivity are well documented (Wani and Lee 1992; Asif et al. 1995). AM fungi are an obligate symbiont requiring association with a living host for multiplication (Al-Raddad 1995). Although several culture techniques are in practice for commercial-scale production of the VAM inoculum, because of certain limitations no method is universally accepted for mass production of inocula in the laboratory. AM fungal culture production on a large scale can be achieved by a soil bed containing sand and soil with a suitable host (Ojala and Jarrell 1980; Schenck and Smight 1982; Mosse and Thompson 1984; Thompson 1986). In addition there are other techniques, such as hydroponic (Mosse and Thompson 1984), aeroponics (Sylvia and Jarstfer 1992, 1994), and root organ culture (Miller-Wideman and Watrude 1984), but all these processes of AM multiplication require several months for inoculum production as well as being expensive (Wood 1991). As such, development of rapid and more efficient culture systems still remains a challenge for commercialization. However, among all available culture methods, sterilized sand and soil have been used invariably for the production of AM inoculum.

However, for obtaining AM inocula in large quantities, it is necessary to select a host plant. It has been found that AMF grows rapidly with fibrous roots such as those of litchi, Sudan grass, onion, clover, cheena, Bahia grass, Rhodes grass (Howler et al. 1987; Srinivas and Bagyaraj 1988), and *Conchrus ciliaris* (Ranga Swami 1990).

Of the 8312 varieties of litchi (decoded for unpublished data), Ref. II (decoded for unpublished data) and Ref. III–VI are cultivated commonly in Bihar. So, there is an essential requirement for the selection of an efficient variety of litchi wherein AMF colonize quickly.

After selection of the varieties, for maximum inocula determination of a suitable soil testing mixture is needed. Several greenhouse and field studies have already proved the beneficial effect of vermicompost in the establishment of plantlets and seed (Gaddie and Douglas 1975; Sevugaperumal et al. 1998; Atiyeh et al. 1999; Buckerfield et al. 1999) because vermicompost supplies micronutrients and macronutrients in simpler form consistently and uniformly as well, as its effect does not burn the juvenile roots of plantlets and germinating seeds as do chemical fertilizers. However, there are only a few reports (Chan and Griffiths 1988; Edwards and Burrows 1988; Wilson and Carlile 1989; Mba 1996; Subler et al. 1998; Buckerfield and Webster 1998; Atiyeh et al. 2000) regarding the potential use of vermicompost as a soil testing media for mass inoculum production of VAM fungi. Therefore, standardization of the optimal dose of vermicompost for obtaining large amounts of inocula is required.

4.2 Selection of Host Plants

To obtain mass inocula of AM fungi, cultivars were put under study for the selection of host plants. These plants were grown in experimental plots to determine the intensity of VAM colonization under natural habitats at different growth stages. Root-based inoculum (RC, 100%) of AMF was layered just below the upper layer of soil in plots. Water was sprinkled as per requirement. Percentage root colonization and spore population were determined at each stage.

4.3 Selection of Variety of Litchi

The seven varieties of litchi, that is, referral litchi testing plants (decoded for unpublished data), were planted in a plot of the University Department of Botany, Bhagalpur in natural habitats to determine VAM affinity with individual varieties. Root infectivity and spore density were determined after a 3-month interval by the standard method of Phillips and Haymen (1970) and Gerdamann and Nicolson (1963).

4.4 Determination of Doses of Vermicompost with Mixed AMF

To evaluate the effect of different doses of vermicompost in the presence of mixed VAM fungi on litchi saplings as a test plant, the soil experiment was conducted without any addition of chemical fertilizer. The earthen soil beds (diameter 1 m) were filled with a 4-kg mixture of vermicompost and soil in the ratios 1:1, 1:2, and 1:3 with 40 g root-based inoculum of AMF. Ten seeds were sown in each soil bed; after the emergence of seedlings, only three seeds were allowed to grow per soil bed. Each soil bed was kept under similar conditions and watered equally as per requirement. The test plants were allowed to develop up to 3 months. Thereafter, root dry weight, root volume, and AMF infectivity were determined to establish the optimum dose of vermicompost in the presence of AM fungi.

The following sets of experiment were conducted: (1) soil (control); (2) soil + AMF; (3) vermicompost + AMF; (4) soil + vermicompost (1:1) + AMF; (5) soil + vermicompost (1:2) + AMF; (6) soil + vermicompost (1:3) + AMF.

4.4.1 Determination of Root Volume

The roots were carefully separated from the uprooted plants and washed under a stream of tap water. The roots were then pressed between clean dry filter papers to remove the excess water. On the other side, the water was taken in a measuring cylinder, and the roots immersed: the volume of water was noted before immersing (Initial) and after immersing (Final). The root volume was calculated by this formula:

Root volume (V) = Final reading (F) – Initial reading (I).

4.4.2 Estimation of Root Biomass

The roots were separated and washed thoroughly in slow-running tap water to remove the adhering soil particles. The roots were then pressed between pieces of clean dry filter paper to remove the excess water. The roots were carefully packed in separate papers and kept in an oven at 60 °C for 3 days and weighed until a constant weight was obtained.

4.4.3 Selection of Hosts

The results depicted in Table 4.1 show that maximum percentage of root colonization (R.C., 98%) and number of spores (S.P., 27 spore/g dry soil) were found in litchi cultivars (Ref. I, II at flowering stage followed by Ref. III, i.e., 95% R.C. and 23 S.P.). However, a lower degree of VAM infectivity, that is, root colonization (71%, 82%, 67%) and spore population (11, 14, 10), was recorded in host plants. On the basis of the present findings, the litchi was selected as the most efficient and susceptible host plants for mass inoculum production of AM fungi.

4.4.4 Selection of Litchi Variety

The results indicated that the root infective percentage and spore numbers were higher, that is, 98% and 25 spores/g dry soil, respectively, in variety Ref. II followed by Ref. III Parthasarthi and Ranganathan (1999). The variety 8312 showed minimum affinity with VAM colonization; however, spore density was minimal in the tested variety of litchi (Ref. I, unpublished). The results clearly indicate that BM-II was the suitable variety for mass production of AMF inoculum (Table 4.2).

4.4.5 Determination of Doses of Vermicompost and Root Inocula

The findings clearly indicate that the litchi treated with AM fungi in the mixture of soil and vermicompost at the 1:3 ratio showed the best performance in terms of root dry matter (2.5 g) and root volume (22.5 ml^3), followed by the treatments of vermicompost + VAM mixture; 1.2 g root dry matter and 10 ml^3 root volume, respectively, were recorded in the 1:1 ratio of soil + vermicompost with AMF whereas 1.6 g and 14.8 ml, respectively, were found in the 1:2 ratio (Edwards and Bohlen 1996). Edwards and Bates (1992) have also found that root biomass is greater in seed grown in earthworm castings.

Table 4.1 Showing variation in spore population (S.P.) and root colonization (R.C.) at different growth stages of host plants

Growth stage	Pre-flowering stage		Flowering stage		Post flowering stage	
Host plant	S.P./g (dry soil)	R.C. (%)	S.P./g (dry soil)	R.C. (%)	S.P./g (dry soil)	R.C. (%)
Tested plant-I (Litchi)	11	71	23	95	20	89
Tested plant-II (Litchi)	14	82	27	98	23	94
Tested plant-III (Litchi)	10	67	21	92	19	84

Table 4.2 Showing variation in spore population (S.P.)and root colonization (R.C.) in different varieties of litchi plants

Serial no.	Varieties of litchi[a]	R.C. (%)	S.P./g dry soil
1.	8312	75	17
2.	Ref-I	98	25
3.	Ref-II	79	16
4.	Ref-III	82	14
5.	Ref-IV	94	19
6.	Ref-V	91	15
7.	Ref-VI	96	21

[a]Data unpublished

Mycorrhizal colonization and spore population were also influenced by the treatments: 8%, 21%, and 28% increase in root infectivity and 7, 9, and 12 in spore numbers/g dry soil were recorded in plants grown in 1:1, 1:2, and 1:3 ratios of soil and vermicompost, respectively, with AMF compared to controls. Similar results have also been observed by Nicole et al. (2003). Kale et al. (1987) reported an increase in AMF colonization of clover and cucumber roots when these plants were planted in a vermicompost mixture. Sylvia et al. (1998) have noticed that AMF become more efficient in nutrient uptake when mixed with vermicompost.

A comparative observation of results revealed that the spore population, root colonization, root dry matter, and root volume gradually increased with the dose of vermicompost; the best performance was recorded in the 1:3 ratio of soil and vermicompost with the AMF admixture (Table 4.3).

4.5 Statistical Analysis

The maximum infectivity and spore population were recorded at the flowering stage, showing that active multiplication of AMF fungi was favored at this stage of the plant life cycle. Singh and Verma (1987) have also reported maximum sporulation in *Cicer arietinum* at the flowering stage. The minimal activity of AMF recorded at the pre-flowering stage might be correlated with the need of the host plant at this stage of the growing phase for synthesized carbohydrates for itself;

Table 4.3 Showing % R.C., S.P., root dry matter, and root volume of litchi plants, under various doses of vermicompost (VC) with arbuscular mycorrhizal fungi (AMF)

Sr. no.	Different combination	R.C. (%)	S.P. (/g dry soil)	Roots dry matter (g)	Root volume (ml^3)
1.	Soil	60	04	0.2	2.1
2.	Soil + AMF	65	08	0.5	3.3
3.	VC + AMF	90	14	2.2	20.2
4.	Soil + VC (1:1) + AMF	68	11	1.2	10.0
5.	Soil + VC (1:2) + AMF	81	13	1.6	14.8
6.	Soil + VC (1:3) + AMF	88	16	2.5	22.5

therefore, its lesser availability may slow down the rate of multiplication. At the flowering stage the growth becomes slow and assimilated carbohydrates stored in plant tissues are utilized by AMF for their multiplication.

The association of AM fungi with the roots of all varieties was noted as a universal phenomenon, but spore population and percent root colonization were noticed to vary markedly, which might be correlated with soil profile, root exudates, and rhizosphere microbes other than symbiotic fungi. Some varieties showed more affinity with AM fungi, which might also be related to the physiology and morphology of the roots also.

Of the different treatments for determination of the vermicompost dose for VAM mass multiplication, the highest spore population, root dry matter, and root volume of the host plant (litchi) were recorded in the 1:3 ratio of soil and vermicompost with mixed AMF mixture. Similar findings have also been recorded by Ruz-Jevez and Rogjer (1992) and by Spain et al. (1992). The explanation might be that the vermicompost initiates the AMF colonization with feeder roots, making them more efficient for mineral absorption from soil. In addition, the vermicompost improves the physical properties of the soil (Edward and Burrow 1988), stimulates development of microbial flora (Parthasarathi and Ranganathan 2000), and directly or indirectly enhances phosphorous uptake along with that of other minerals (Buwalda and Goh 1982; Hall et al. 1984; Son and Smith 1988; Peng et al. 1993; Graham and Eissenstat 1998); it also improves growth-promoting substances, viz. indole acetic acid and gibberalic acid (Tomati et al. 1987, 1990). Promoting a high level of mycorrhizal development on the host plant and rapid multiplication of the AMF inoculum is an important challenge for commercial use in agriculture, horticulture, and silviculture fields. Thus, from these findings it can be concluded that soil + vermicompost (1:3) with VAM might be a suitable soil bed mixture for the mass production of AM fungal cultures under greenhouse conditions, and the best variety of litchi would be a reference test for mass multiplication.

References

Al-Raddad A (1995) Mass production of *Glomus mosseae* spores. Mycorrhiza 5:229–231

Asif M, Khan AG, Khan MA Kerck C (1995) In: Adholeya A, Singh S (eds) Mycorrhizae: biofertilizer for the future. Proceedings of the third national conference on mycorrhiza. Tata Energy Res Inst Publ, pp 431–437

Atiyeh RM, Subler S, Edwards CA, Metzger JD (1999) Growth of tomato plants in horticultural soil bed media amended with vermicompost. Pedobiologia 43:724–728

Atiyeh RM, Arancon N, Edwards CA, Metzger JD (2000) Influence of earthworm processed pig manure on the growth and yield of green house tomato. Bioresour Technol 75:175–180

Bucker Field JC, Webster KA (1998) Worm worked waste boosts grape yields: prospects for vermicompost use in wineyards. Wine Ind J B 13:73–76

Buckerfield JC, Tamara C, Kenneth F, Lee E, Webster AK (1999) Vermicompost in solid and liquid as a plant growth promoter. Pedobiologia 43:753–759

Buwalda JG, Goh KM (1982) Host fungus competition for carbon as a cause of growth depressions in vesicular arbuscular mycorrhizal rye grass. Soil Biol Biochem 14:103–106

Chan PLS, Griffiths DA (1988) The vermicomposting of pre-treated pig manure. Biol Wastes 24:57–69

Edwards CA, Bates JE (1992) The use of earthworms in environmental management. Soil Biol Biochem 14(12):1683–1689

Edwards CA, Bohlen PJ (1996) Biology and ecology of earthworms, IIIrd edn. Chapman & Hall, London/New York

Edwards CA, Burrows I (1988) The soil potential of earthworms composts as plant growth media. In: Enwards CA, Neuhauser E (eds) Earthworms in waste and environmental management. SPB Academic Press, The Hague, pp 21–32

Gaddie RE, Douglas DE (1975) Earthworms for ecology and profit, vol I. Scientific earthworm farming, p 175

Gerdamann JW, Nicolson TH (1963) Spores of mycorrhizal *Endogone* species extracted from soil by wet sieving and decanting. Trans Br Mycol Soc 46:235–244

Graham JH, Eissenstat DM (1998) Field evidence for the carbon cost of citrus mycorrhizas. New Phytol 140:103–110

Hall JR, John Stone PD, Dolby R (1984) Interactions between endomycorrhizas and soil nitrogen and phosphorus on the growth of ryegrass. New Phytol 97:447–452

Howler RH, Siverding E, Safir S (1987) Practical aspects of mycorrhizal technology in some tropical crops and pastures. Plant Soil 100:249–283

Kale RD, Bano K, Sreenivasa MN, Bagyaraj DJ (1987) Influence of wormcast (see Comp. E. VAS'83) on the growth and mycorrhizal colonization of two ornamental plants. South Indian Hortic 35(5):433–437

Mba CC (1996) Treated-cassava peel vermicomposts enhances earthworm activities and cowpea growth in field plots. Resour Conserv Recycl 17:219–226

Miller-Wideman MA, Watrude LS (1984) Sporulation of *Glomus mosseae* in root cultures of tomato. Can J Microbiol 30:642–646

Mosse B, Thompson JP (1984) Vesicular arbuscular and mycorrhizal inoculum production. I. Exploratory experiments with beans (*Phaseolus vuglaris*) in nutrient flow culture. Can J Bot 62:1523–1530

Nicole Cavender D, Rola Atiyeh M, Knee M (2003) Vermicompost stimulates mycorrhizal colonization of roots of *Sorghum bicolor* at the expense of plant growth. Pedobiologia 47:85–89

Ojala JC, Jarrell WM (1980) Hydroponic sand culture system for mycorrhizal research. Plant Soil 57:297–303

Parthasarathi K, Ranganathan LS (1999) Longevity of microbial and enzyme activities and their influence on NPK content in pressmud vermicasts. Eur J Soil Biol 35:107–113

Parthasarathi K, Ranganathan LS (2000) Aging effect on enzyme activities in pressmud vermicasts of *Lampita mauriti* (Kinberg) and *Eudrilus eugeniae* (Kinberg). Biol Fertil Soil 30:347–350

Peng S, Eissenstat DM, Graham JH, Williams K (1993) Growth depression in mycorrhizal citrus and high phosphorus supply: analysis of carbon costs. Plant Physiol 101:1063–1071

Phillips JM, Haymen DS (1970) Improved procedure for clear staining parasitic and vasicular arbuscular mycorrhizal fungi for rapid assessment of infection. Trans Br Mycol Soc 55:158–161

Ranga Swami G (1990) In: Bagyaraj DJ, Manjunath A (eds) Mycorrhizal symbiosis and plant growth. Mycorrhiza Network Asia and University of Agricultural Science Bangalore Publications

Ruz-Jevez BP, Rogjer T (1992) Laboratory assessment of nutrient release from a pasture soil receiving grass or clover residues, in the presence or absence of *Lumbricus rubellus* or *Eisenia foetida*. Soil Biol Biochem 24(12):1529–1534

Schenck NC, Smight GS (1982) Mycologia 70:126-77-92

Sevugaperumal R, Jaishankar K, Jayaraj K (1998) Comparative analysis of the effect of vermicopmost and *Azospirillum* on sorghum, *Sorghum vulgare* Linn. ANJAC J 15:18–21

Singh K, Verma AK (1987) Nat Mycor JNU New Delhi

Son CL, Smith SE (1988) Mycorrhizal growth responses: interactions between photon irradiance and phosphorus nutrition. New Phytol 108:35–314

Spain A, Lavelle P, Mariotti A (1992) Stimulation of plant growth by tropical earthworm. Soil Biol Biochem 16(2):185–189

Srinivasa MN, Bagyaraj DJ (1988) Plant Soil 106:289–290

Subler S, Edwards CA, Metzger J (1998) Comparing vermicomposts and composts. Biocycle 39:63–66

Sylvia DM, Jarstfer AG (1992) Sheared-root inocula of vesicular-arbuscular mycorrhizal fungi. Appl Environ Microbiol 58:229–232

Sylvia DM, Jarstfer AG (1994) In: Robson AD, Abbott LK, Malajezuk N (eds) Management of mycorrhizae in agriculture horticulture and forestry. Kluwer, Dordrecht, pp 231–238

Sylvia DM, Fuhrmann JJ, Hartel PG, Zuberer DA (1998) Principles and applications of soil microbiology. Prentice-Hall, Upper Saddle River, pp 408–426

Thompson JB (1986) Soil less cultures of vesicular arbuscular mycorrhizae of cereals. Effect of nutrient concentration and nitrogen source. Can J Bot 64:2282–2294

Tomati U, Grappelli A, Galli E (1987) The presence of growth regulators in earthworms worked wastes. In: Bonvicini Paglioi AM, Omodeo P (eds) On earthworms. Modena, Italy, pp 423–436

Tomati U, Galli E, Grappelli A, Dihena (1990) Effect of earthworm casts on protein synthesis in radish (*Rhaphanus sativum*) and lettuce seedlings. Biol Fertil Soil 9:228289

Wani SP, Lee KK (1992). In: Tandon HLS (ed) Fertilizers organic manure, recyclable wastes and biofertilizers. Fertilizer Development and Consultation Organization, New Delhi, pp 91–112

Wilson DP, Carlile WR (1989) Plant growth in soil testing media containing worm-worked duck waste. Acta Hortic 238:205–238

Wood T (1991) VA mycorrhizal fungi challenges for commercialization. In: Arora DK, Elander RP, Mukerji KG (eds) Fungal biotechnology: handbook of applied mycology, 4th edn. Dekker, New York, pp 823–847

Impact Assessment of Bio-inoculants on Growth of Litchi [*Litchi chinensis* (Gaertn.) Sonn.] Plants

5

Prity Sagar, Anfal Arshi, and Awadh Kishore Roy

Abstract

Mycorrhizal fungi are known to establish a strong mutualistic and symbiotic relationship with the roots of litchi plants to the extent of mycorrhizal dependency. Arbuscular mycorrhizal fungi (AMF) have the potential to provide better sustenance to the plants even in adverse conditions by enhancing the mobility of ions/nutrients from the depletion zone of the soil; moreover, dual inoculation with *Azospirillum brasilense* exerts a synergistic effect on plant growth. The present investigation has been carried out with an aim to develop a bio-inoculant package to obtain better growth of four varieties of litchi plants, Desi, China, Shahi, and Purbi, commonly grown in this region. The effect of bio-inoculants with vermicompost was evaluated on the growth of litchi plants. In a comparative evaluation of the results, the marcots treated with indigenous AMF and *A. brasilense* with equal doses of vermicompost showed best growth in all the varieties except Shahi. To correlate the treatment impact on growth and mycorrhizal colonization, the rootlets of all the litchi varieties were screened. Percentage of root colonization ranged between 28 and 90%.

Keywords

VAM • Vermicompost • *Azospirillum* • Bio-inoculant • Litchi • Marcots

P. Sagar
Centre for Ecological Sciences, Indian Institute of Science, Bangalore, India

A. Arshi (✉)
Defence Research & Development Organisation, Defence Institute of Bio-Energy Research (DIBER), Haldwani, Uttarakhand 263 139, India
e-mail: anfalarshi@gmail.com; anfal_arshi@rediffmail.com

A.K. Roy
Microbiology & Plant Pathology, University Department of Botany, TM Bhagalpur University, Bhagalpur 812002, India

M. Kumar et al. (eds.), *Lychee Disease Management*,
DOI 10.1007/978-981-10-4247-8_5

5.1 Introduction

Litchi (*Litchi chinensis* Sonn.), which is one of the most environmentally sensitive tropical horticultural crops, is an economically, ecologically, and ornamentally viable fruit tree of India. In India 4,28,900 metric tonnes of litchi are produced, grown in the states of Bihar, Tripura, West Bengal, Uttar Pradesh, Punjab, and Haryana. Bihar is the major litchi-producing state in India, contributing 74% of total production. The major production of litchi in Bihar provides livelihood security for millions of marginal farmers as they get an additional amount of Rs. 13,000/- to 15,000/- in selling litchi produce per year per hectare. There is still scope for the expansion of litchi orchards in Bihar because of the adaptability of this fruit towards existing edaphic-climatic conditions, which might be proved as a step ahead in employment generation. However, shrinkage of cultural land and generation of degraded land, lack of modern agro-plantation techniques, and dependency on chemical fertilizer decrease the enthusiasm of the growers. Litchi production may be increased if the lands of these areas are exploited with the help of bio-inoculants, that is, vesicular arbuscular mycorrhizae and *Azospirillum brasilense* combined with organic manure.

The application of bio-inoculants is an environment-friendly, energy-efficient, and economically viable approach for reclaiming degraded soil, increasing the growth and yield of plants. The vesicular arbuscular mycorrhizal fungi establish a nonpathogenic symbiotic mutual association with roots of the litchi plant (Coville 1912; Kadman and Slori 1974; Marloth 1947), and this fungal population has a major function in the establishment and survival of plants (Miller and Jastrow 1992; Kranabetter and Wylie 1998). Mycorrhizal fungi are the keystone of the evolutionary biological potential of the sustainability of the agricultural and horticultural ecosystem and as such are essential components for ecosystem functioning (Hart et al. 2001). The filaments of mycorrhizal fungi also act in the soil as binding agents and are thus implicated in soil aggregation process (Hamel et al. 1977; Miller and Jastrow 2000). Moreover, mycorrhizae are capable of reducing the incidence of disease (Miller and Jastrow 2000) and improving the stress resistance of the plants (Charest and Brown 1993; Subramaian and Charest 1997).

Azospirillum brasilense, an asymbiotic aerophilic diazotroph, is used as a nitrogen fixer on a commercial level. These bacteria can fix atmospheric nitrogen even in nonleguminous plants. *A. brasilense* produces phytohormones, such as gibberellin, indole acetic acid, auxin, and cytokinin (Cheryl and Glick 1996), which are also capable of enhancing the nutrient elements from nonusable to usable forms through a biological process. *A. brasilense* becomes more efficient when it is inoculated with vesicular arbuscular mycorrhizae (VAM) (Rai and Gaur 1982; Pacovsky et al. 1985; Subba Rao et al. 1979, 1985).

Further, the potential of these bio-inoculants may be increased by addition of vermicompost; it is helpful in supplying macro- and micronutrients to plants (Tomati and Galli 1995; Edwards and Bohlen 1996) and contains plant growth-promoting hormones (Parthasarthi and Ranganathan 1999, 2000). Organic amendment with microbial input reduces the reliance on chemical fertilizer and improves the fertility of poor agricultural and horticultural soils.

The objective of the present investigation is to assess the impact of bio-inoculants on the growth of four varieties of litchi fruit trees in this region of Bihar: Desi, China, Shahi, and Purbi.

5.2 Ecological Conditions

5.2.1 Soil and Organic Amendments

Soil was obtained from degraded land from a depth of 0–15 cm with pH = 6.7, nitrogen = 0.003, phosphorus = 0.002, potassium = 0.271%, calcium = 0.041%, sodium = 0.003%, carbon = 0.13%, and organic matter = 0.224%, which was a poor nutrient profile. The soil was amended with VAM, *A. brasilense*, and vermicompost.

Rhizospheric soil (5 kg) with rootlets having AMF colonization (RC, 70–75%) was collected from a litchi orchard, and carrier-based (250 g) inoculum [sterilized soil and farmyard manure (FYM), 1:1] of *A. brasilense* (3×10^8 cell/g) including 5 kg vermicompost (pH = 7.20, OC = 11.1%, OM = 19.14%, P = 0.69%, K = 0.739%, Ca = 0.654%, N = 1.3%, Na = 0.443%) prepared by *Eisenia fetida* was used in treatments.

Experimental setup: The 0.5-m^3 pits were constructed in the nursery. Each pit was filled with sand and small gravel to 10 cm3 followed with a second layer of homogeneously sieved soil to 30 cm3; the remainder was left for treatments in the year 2002–2003.

Marcots of four varieties of litchi, Desi, China, Shahi, and Purbi, were procured from Bihar Agricultural College, Sabour, and planted in pits in the nursery with different treatments. Before planting, root infectivity was examined by the standard method of Phillips and Hayman (1970).

Experiments were carried out by providing four different treatment combinations with three replications. The control was fertilized with the recommended dose: Castorcake, 600 g; SSP, 750 g; potassium sulfate, 80 g; thymate 3 g; lime, 1 kg; and FYM, 12 kg. The treatments were T_1, control; T_2, control + AMF; T_3, vermicompost (VC); T_4, vermicompost + AMF; and T_5, vermicompost +AMF+ *A. brasilense*.

Assessment of growth parameters: The impact of bio-inoculants on growth of litchi plants was evaluated at 2-month intervals for a year in terms of plant height, shoot length, and shoot diameter; root colonization was again examined after a year.

The data were analyzed statistically by the standard methods of agricultural statistics (Rangaswamy 1995) for analysis of variance (ANOVA) and *F* ratio at 5% and 1% of significance level. For all data processing, Ms. Excel and Kheitan were used.

Litchi plants of the different varieties (Desi, China, Shahi, and Purbi) were screened for AMF colonization, which ranged from 14% to 34%, before planting in the pits.

5.3 Growth Performance

5.3.1 Plant Height

The varieties Desi and China showed the maximum percentage increase in height, 77% and 158%, respectively, in tripartite combination (VC + VAM+ *A. brasilense)* compared to the control, whereas the varieties Purbi and Shahi achieved the highest level, that is, 138.25% and 46.641% increase, respectively, in C + VAM treatment compared to control. Further, it was noticed that plant height followed a slow pace, as much as 6 months, in each variety and treatment (Fig. 5.1).

5.3.2 Shoot Length

Results presented in Fig. 5.2 show that the varieties Desi and Purbi achieved the highest percentage increase in shoot length, 66.52% and 195.07%, respectively, in combination with vermicompost with the microbial consortia (*A. brasilense* + AMF) compared to control; however, China showed the maximum in C + AMF and Shahi in vermicompost.

It was also observed that, when AMF was mixed with the control of each variety, the subsequent increase in shoot length was 28%, 90%, 21%, and 92.2% in varieties Desi, China, Shahi, and Purbi, respectively, compared to the control, whereas when AMF was added with vermicompost in each variety, only two varieties, Desi and Purbi, showed 12% and 141% increment in shoot length, respectively.

5.3.3 Diameter

The findings presented in Fig. 5.3 show that the experimental varieties of litchi, that is, China, Shahi, and Purbi, achieved the maximum percentage increase in stem width, 26.92%, 45.48%, and 33.09%, respectively; in a mixture containing VC + VAM + *A. brasilense* compared to control; however, Desi showed the maximum increase, 42.08%, in C + VAM.

Native fungi when mixed with either control or vermicompost mixture in case of all varieties of litchi always showed the positive response towards thickness of stem. The variety Desi exhibited the maximum percentage increase in diameter, 42.08%, followed by China at 22.15%, compared to control when treated with AMF only, whereas when AM fungi were added with vermicompost, the variety China revealed the highest percentage increase in thickness of stem, 6.92%, followed by

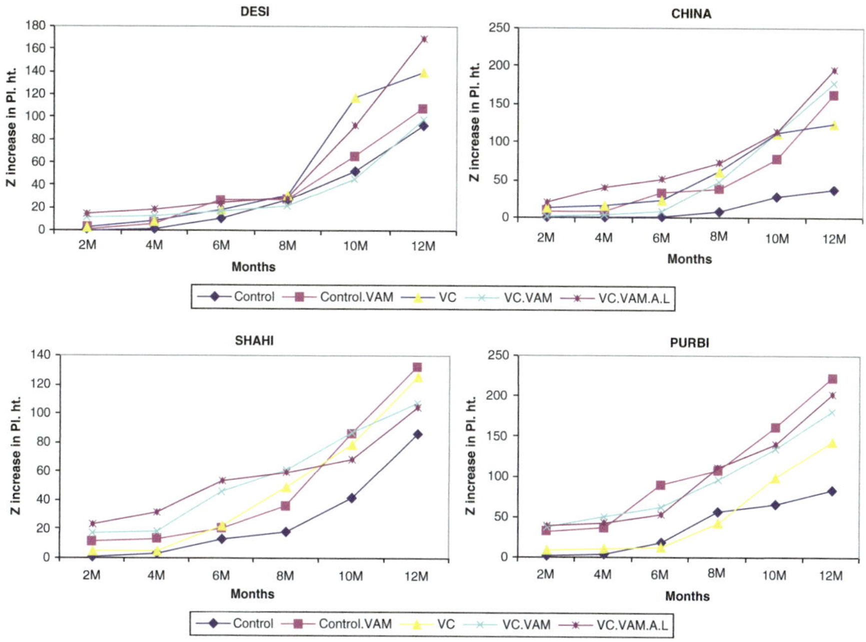

Fig. 5.1 Percentage increase in plant height (cm) in litchi varieties under different treatments

Shahi (1.19%) compared to control, and the minimum was found in the variety Purbi (0.87%), followed by Desi.

5.4 Statistical Analysis

The analysis of variance (ANOVA) (Table 5.1) showed that the main effect of varieties (V), treatments (T), and duration (D) and the interaction effect on plant height, shoot length, and diameter were found to be significant at the 1% level of significance. Following are the main and interaction effects of two-way ANOVA of variety (V), treatment (T), and duration (D).

5.4.1 Main Effect

1. Variety (V): of tested varieties, China attained the maximum height (60.25 cm), shoot length (64.72 cm), and diameter (1.40 cm) (Fig 5.4).
2. Treatment (T): among all treatments, the tripartite combination or T_5 treatment was more efficient for plant height (64.72 cm), shoot length (67.19 cm), and diameter (1.41 cm) (Fig. 5.5).
3. Day (D): significant effects of days on plant height (90.37 cm), shoot length (90.34 cm), and diameter (1.58 cm) were observed throughout the duration. The

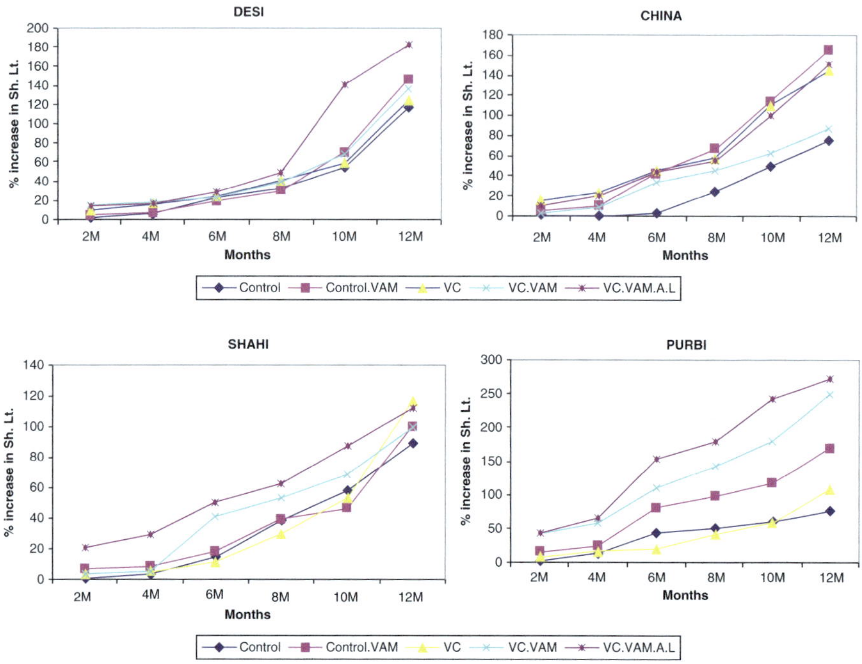

Fig. 5.2 Percentage increase in shoot length (cm) in litchi varieties under different treatments

growth of plants increased gradually with time and maximum was achieved after 1 year (Fig. 5.6).

Interaction effect:

4. Variety × treatment (VXT): treatment T_5 showed the maximum significant effect on plant height, shoot length, and diameter in all four varieties of litchi with few exceptions.
5. Variety × day (VXD): plant height, shoot length, and diameter of each variety were found to increase nonsignificantly until 4 months after planting; thereafter, the increase was noticed to be significant.
6. Day × Treatment (DXT): all the tested varieties of litchi showed significant growth in the tripartite combination VC + VAM+ *A. brasilense* (T_5) at 12 months.

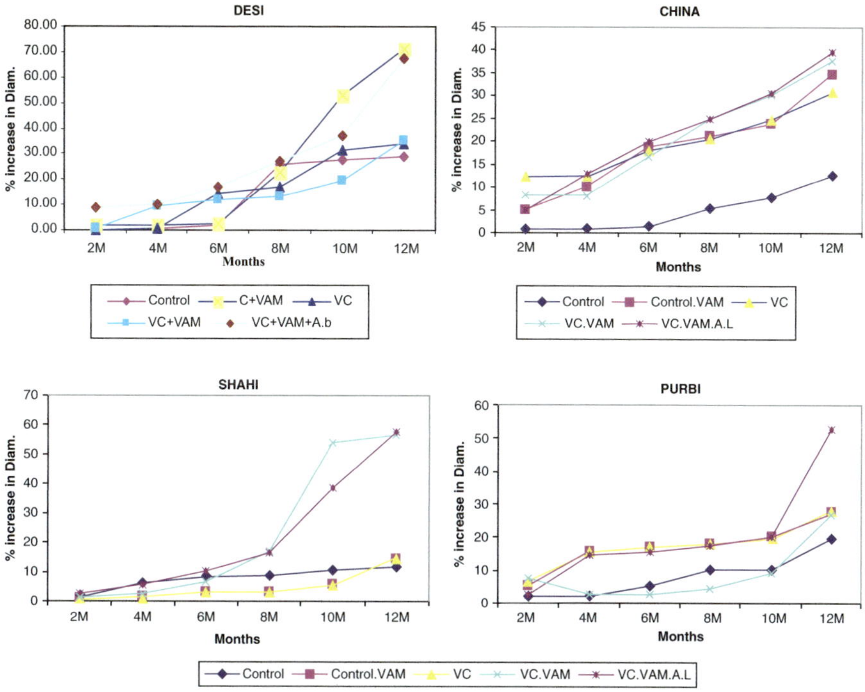

Fig. 5.3 Percentageincrease in diameter (cm) in litchi varieties under different treatments

Table 5.1 Analysis of variance (ANOVA) for plant height, shoot length, and diameter

Source of variations	Plant height		Shoot length		Diameter	
	Mean squares	*F* ratio	Mean squares	*F* ratio	Mean squares	*F* ratio
Replicates	205.3	21.8**	21.8	3.89*	0.05	4.21*
Variety (V)	2957.1	313.97**	4777.3	853.21**	0.72	63.63**
Treatment (T)	3804.6	403.95**	3529.2	630.30**	0.32	28.88**
Duration (D)	20,208.3	2145.62**	19,681.8	3515.08**	1.36	121.19**
V X T	1231.8	130.78**	787.6	140.67**	0.25	22.00**
V X D	113.5	12.05**	188.6	33.68**	0.04	3.91**
T X D	301.7	32.02**	497.4	88.84**	0.03	2.97**
V X T X D	129.7	13.77**	124.5	22.23**	0.02	2.29**
Mean	C.V.	5.4%		4.0%		8.04%

*Significant at 5% level; **significant at 1% level; *C.V.* coefficient of variance

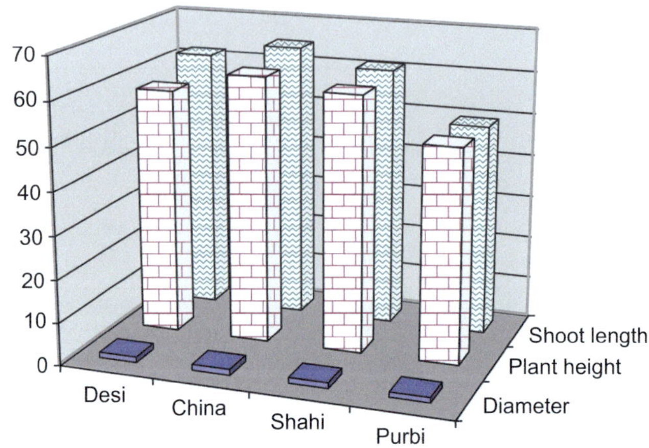

Fig. 5.4 Comparative performance of varieties in different growth parameters

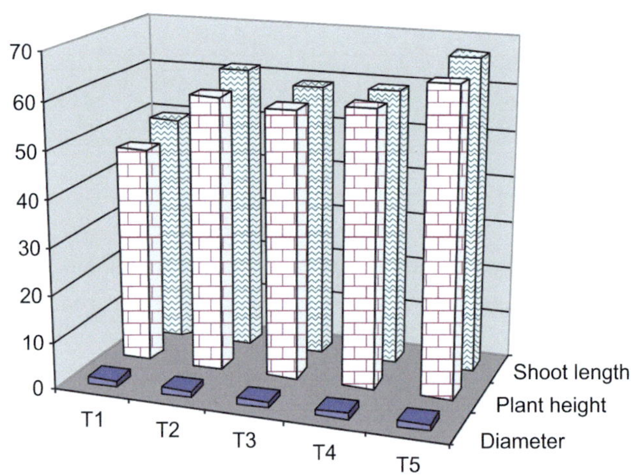

Fig. 5.5 Comparative performance of treatments in different growth parameters

5.5 Five Impacts of Treatments on Root Colonization

The root colonization of AMF was again examined after 1 year. Considerable variation was noticed in AMF infectivity under the five different treatments among all varieties. The results are depicted in Table 5.2 and Fig. 5.7.

The varieties Desi, China, and Shahi had 70%, 72%, and 45% root colonization, respectively, in tripartite combination (VAM + vermicompost + *A. brasilense*) after 1 year, followed by vermicompost + VAM treatment, whereas **Purbi** had the

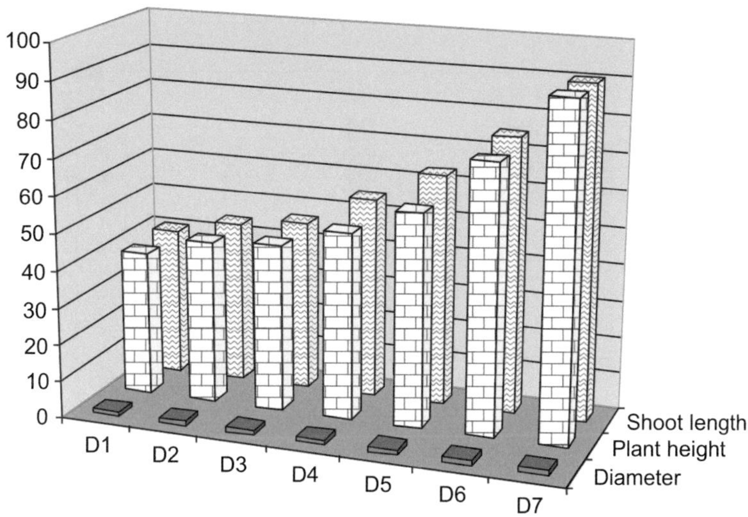

Fig. 5.6 Comparative performance of days in different growth parameters

Table 5.2 Percentage of root colonization in different varieties of litchi under different treatments after 1 year

Root colonization (RC), %				
Treatments	*Desi* (RC-34%)	*China* (RC-28%)	*Shahi* (RC-30%)	*Purbi* (RC-14%)
Control	52	45	28	48
C + VAM	66	52	30	70
VC	46	58	40	**90**
VC + VAM	64	65	42	80
VC + VAM + *Azospirillum brasilense*	70	72	45	54

RC – indicates percentage root colonization at the time of planting in pits

highest root colonization, that is, 90%, when treated with vermicompost only. The effect of different treatments on growth performance showed that Desi, China, and Purbi attained better plant height, shoot length, and diameter in the T_5 treatment; however, Shahi had no definite pattern of treatment effect.

The AMF symbiont is one of the important components of this combination, with a significant role in the maintenance of rhizosphere diversity and the development of soil, which has long been hidden by the great impact that these organisms have on individual plant growth and crop productivity (van der Heijden et al. 1998); as well, the symbiont solubilizes the complex phosphate molecules to make these available to the plants, causes the root system to proliferate in a wider zone of soil to absorb the maximum nutrients and available water, and thus to establish a nutrient pool between host and fungi. The *A. brasilense* bacteria was shown to enhance their

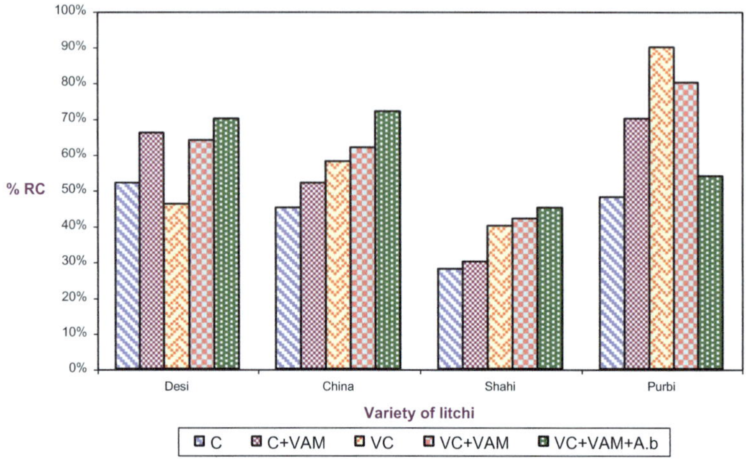

Fig. 5.7 Percentage root colonization in different varieties of litchi under different treatments after 1 year

production of auxin, gibberellin, and cytokinin-like substances (Tien et al. 1979; Wani 1990; Andreeva et al. 1993). It is known that growth substances produced by these bacteria are continuously released from the root surfaces into the inner rhizosphere where *Azospirillum* grows and is directly supplied to the photosynthate by the autotropic host (Patriquin and Dobereiner 1978; Diem and Dommergues 1980; O'Hara et al. 1981), and it also enhances the total assimilation of N, P, K, Rb^+ , and Fe^+ (Okon 1982; Lin et al. 1983; Wani and Lee 1991) and helps in nitrogen fixation (Nery et al. 1977; Sano et al. 1981; Boddey et al. 1983). The synergistic effect of dual inoculation might be the reason for the enhancement of growth of litchi plants (Tilak 1995; Prabhakaran et al. 1995). Additionally, organic fertilizer vermicompost helps in plant growth; it is known as a complete nutritional source for plant growth (Edwards 1995). As well, it initiates the AMF colonization, improves the physical properties of the soil (Edwards and Burrows 1988), stimulates development of microbial flora (Son and Smith 1988; Peng et al. 1993; Graham and Eissenstat 1998), and increases the water-holding capacity. The best results, as found in the combination of VAM + *A. brasilense* and vermicompost, could be derived from their combined positive effects exerted towards the growth of plants.

The results clearly showed that the tripartite combination accelerated mycorrhizal colonization in all the varieties of litchi plant (except Shahi), which leads to enhancement of plant growth (Nicole et al. 2003).

On the basis of the foregoing discussion, it is apparent that synergisms of arbuscular mycorrhizae with microbial populations such as bacteria and P-solubilizing fungi provide soils and plants with additional tools for improving plant growth (Kucey 1987; Kucey et al. 1989; Germida and Xavier 2001). Therefore, a bio-inoculant package containing the VAM *Azospirillum brasilense* with vermicompost might be the best suitable combination for a litchi plantation in this region.

References

Andreeva IN, Red' Kina TV, Izmailov SF (1993) The involvement of IAA in stimulation of *Rhizobium* legume symbiosis by *Azospirillum brasilense*. Russ J Plant Physiol 406:780–784

Boddey RM, Chalk M, Victoria R, Matsui E (1983) The ^{15}N isotope dilution technique applied to the estimation of biological nitrogen fixation associated with *Paspalum notatum* in the field. Soil Biol Biochem 15:25–32

Charest C, Brown A (1993) The effect of vesicular arbuscular mycorrhizae and chilling on two hybrids on *Zea mays* L. Mycorrhiza 4:89–92

Cheryl P, Glick B (1996) Bacterial biosynthesis of indole-3-acetic acid. Can J Microbiol 42:207–220

Coville FV (1912) The lychee *Litchi chinensis*, a mycorrhizal plant. In: Groff GM (ed) Appendix to the lychee and lungan. Canton Christian College and Grange Jud Company, New York

Diem HG, Dommergues YP (1980) Significance and improvement of rhizospheric nitrogen fixation. In: Subba Rao NS (ed) Recent advances in biological nitrogen fixation. Arnold, London, pp 190–226

Edwards CA (1995) Historical overview of vermicomposting. Biocycle 366:56–58

Edwards CA, Bohlen PJ (1996) Biology and ecology of earthworms, 3rd edn. Chapman & Hall, London/New York, p 426

Edwards CA, Burrows I (1988) The potential of earthworms composts as plant growth media. In: Enwards CA, Neuhauser E (eds) Earthworms in waste and environmental management. SPB Academic Press, The Hague, pp 21–32

Germida JJ, Xavier LJC (2001) Bacteria associated with *Glomus clarum* spores influence mycorrhizal activity. In: Third international conference on Mycorrhiza, Adelaide, Australia, 8–13 July 2001, pp 1–179

Graham JH, Eissenstat DM (1998) Field evidence for the carbon cost of citrus mycorrhizas. New Phytol 140:103–110

Hamel C, Dalpe Y, Furlan V, Parent S (1977) Indigenous populations of arbuscular mycorrhiza fungi and soil aggregate stability are major determinants of leek *Allium porrum* L. response to inoculation with *Glomus intraradices* Schenck & Smith or *Glomus versiforme* Karsten Berch. Mycorrhiza 7:187–196

Hart MM, Reader RJ, Klironomos JN (2001) Life history of arbuscular mycorrhizal fungi in relation to their successional dynamics. Mycologia 93:1186–1194

Kadman A, Slor E (1974) Experiments with propagation of the litchi *Litchi chinensis* in Israel. Ind J Horticult 31:28–33

Kranabetter JM, Wylie T (1998) Ectomycorrhizal community structure across forest openings on naturally regenerated western hemlock seedlings. Can J Bot 76:189–196

Kucey RMN (1987) Increased phosphorus uptake by wheat and field beans inoculated with a phosphorus-solubilizing *Penicillium bilajo* strain and with vesicular arbuscular mycorrhizal fungi. Appl Environ Microbiol 53:2699–2703

Kucey RMN, Janzen HH, Leggett ME (1989) Microbially mediated increases in plant-available phosphorus. Adv Agron 42:199–228

Lin W, Okon Y, Hardy RWF (1983) Enhanced mineral uptake by *Zea mays* and *Sorghum bicolor* roots inoculated with *Azospirillum brasilense*. Appl Environ Microbiol 20:387–389

Marloth RH (1947) The litchi in South Africa. I. Varieties and propagation. II. Cultural practices and marketing. Farming South Africa 22(824–30):863–870

Miller RM, Jastrow JD (1992) The application of VA mycorrhizae to ecosystem restoration and reclamation. In: Allen MF (ed) Mycorrhiza functioning. Chapman & Hall, London, pp 438–467

Miller RM, Jastrow JD (2000) Mycorrhizal fungi influence soil structure. In: Kapulanik Y, Douds D (eds) Arbuscular mycorrhizas: physiology and function. Kluwer, Dordrecht, pp 3–18

Nery M, Abrantes SGTV, Dos D, Dobereiner J (1977) Fixacaoda nitrogenio em Trigo. Rev Bras Clene Solo 1:15–20

Nicole CD, Rola AM, Michael K (2003) Vermicompost stimulates mycorrhizal colonization of roots of *Sorghum bicolor* at the expense of plant growth. Pedobiologia 47:85–89

O'Hara GW, Davey MR, Luca JA (1981) Effect of inoculation of *Zea mays* with *Azospirillum brasilense* strains under temperate condition. Can J Microbiol 27:871–877

Okon Y (1982) *Azospirillum*: physiology properties, mode and association with roots and its application for the benefit of cereal and forage grass crops. Isr J Bot 31:214–220

Pacovsky RS, Pawar EA, Bethlenfalvay GJ (1985) Nutrition of *Sorghum* plants fertilized with nitrogen or inoculation with *A. brasilense*. Plant Soil 85:145–148

Parthasarthi K, Ranganathan LS (1999) Longevity of microbial and enzyme activities and their influence on NPK content in pressmud vermicasts. Euro J Soil Biol 35:107–113

Parthasarthi K, Ranganathan LS (2000) Aging effect on enzyme activities in pressmud vermicasts of *Lampito mauritii* Kinberg and *Eudrilus eugeniae* Kinberg. Biol Fertil Soil 30:347–350

Patriquin DG, Dobereiner J (1978) Light microscopy observations of tetrazolium reducing bacteria in the endorhizosphere of maize and other grasses in Brazil. Can J Microbiol 24:734–742

Peng S, Eissenstat DM, Graham JH, Williams K (1993) Growth depression in mycorrhizal citrus and high phosphorus supply: analysis of carbon costs. Plant Physiol 101:1063–1071

Phillips JM, Hayman DS (1970) Improved procedure for clear staining parasitic and vasicular arbuscular mycorrhizal fungi for rapid assessment of infection. Trans Br Mycol Soc 55:58–161

Prabhakaran J, Arjunan G, Ravi KB, Ramamoorthi N (1995) Mycorrhizae: biofertilizer for the future. In: Adholeya A, Singh S (eds) Proceedings of third national conference on Mycorrhiza. Tata Energy Research Institute, pp 348–357

Rai SN, Gaur AC (1982) Nitrogen fixation by *Azospirillum* spp. and effect of *Azospirillum lipoferum* on yield and N-uptake of wheat crop. Plant Soil 69:233–283

Rangaswamy R (1995) A text book of agricultural statistic. Wiley, New Delhi

Sano T, Fujii T, Iyama S, Hirota Y, Kamagata K (1981) Nitrogen fixation in the rhizosphere of cultivated and wild rice strains. Crop Sci 21:758–761

Son CL, Smith SE (1988) Mycorrhizal growth responses: interactions between photon irradiance and phosphorus nutrition. New Phytol 108:305–314

Subba Rao NS, Tilak KVBR, Singh CS, Kumari KM (1979) Response of the few gramineous plants to inoculation with *Azospirillum brasilense*. Curr Sci 48:133–134

Subba Rao NS, Tilak KBVR, Singh CS (1985) Effect of combined inoculation of *Azospirillum brasilense* and vesicular arbuscular mycorrhiza on pearl millet *Pennisetum americanum*. Plant Soil 822:285–286

Subramaian KS, Charest C (1997) Nutritional, growth, and reproductive responses of maize *Zea mays* to arbuscular mycorrhizal inoculation during and after drought stress at tasselling. Mycorrhiza 7:25–32

Tien TM, Gaskin MH, Hubbell DH (1979) Plant growth substances produced by *A. brasilense* and their effect on growth of pearl millet. Appl Environ Microbiol 37:1016–1024

Tilak KVBR (1995) Vesicular arbuscular mycorrhiza and *Azospirillum brasilense* rhizocoenosis in pearl millet *Pennisetum americanum* L. Lecke in semi-arid tropics. In: Adholeya A, Singh S (eds) Mycorrhizae: biofertilizer for the future. Proceedings of third national conference on Mycorrhiza. Tata Energy Research Institute, pp 177–179

Tomati U, Galli E (1995) Earthworms: soil fertility and plant productivity. Acta Zool Fenn 196:11–14

Van der Heijden MGA, Boller T, Wiemken A, Sanders IR (1998) Different mycorrhizal fungal species are potencies determinants of plant community structure. Ecology 79:2082–2091

Wani SP (1990) Inoculation with associative nitrogen-fixing bacteria. Role in cereal grain production improvement. Ind J Microbiol 30:363–393

Wani SP, Lee KK (1991) Role of biofertilizers in upland crop production. In: Tandon HLS (ed) Fertilizers, organic manures, recyclable wastes and biofertilizers. Fertilizer Development and Consultation Organization, New Delhi, pp 91–92

Important Diseases of Litchi and Their Management

6

Ajit Kumar Dubedi Anal, Vinod Kumar, Mritunjay Tripathi, and Vishal Nath

Abstract

Litchi (*Litchi chinensis* Sonn.) is a juicy fruit belonging to the Sapindaceae family and is one of the most important evergreen fruit trees. Diseases are one of the constraints on the production of litchi fruits. They indirectly reduce yield by debilitating the tree and directly reduce the yield or quality of fruit before and after harvest. Diseases are more important after harvest, although undoubtedly many of the fruits are infected before picking. Some of the pathogens infect leaves, flowers, and fruit, and a few others are associated with tree decline and tree deaths. Some pathogens also affect multiple phenophases of this fruit crop. So far, only diseases caused by fungal pathogens have been reported. This chapter provides information on the major diseases of litchi in terms of their importance, symptoms, and management strategies.

Keywords

Litchi • *Litchi chinensis* • Diseases • Fungal pathogens • Disease management • IDM

6.1 Introduction

Litchi or lychee (*Litchi chinensis* Sonn.) is a juicy fruit belonging to the Sapindaceae family and is one of the most important evergreen fruit trees. Litchi is a cherished tropical fruit with a luscious red peel, delicate whitish pulp, and a unique perfumed flavour. Litchi contains many minerals and vitamins, along with its delicious taste and flavour, which makes it a perfect anytime treat for the senses. Litchi originated in China, where it has been cultivated for 300 years, and was

A.K.D. Anal (✉) • V. Kumar • M. Tripathi • V. Nath
ICAR-National Research Centre on Litchi, Mushahari, Muzaffarpur 842002, Bihar, India
e-mail: ajitnrcl@gmail.com

M. Kumar et al. (eds.), *Lychee Disease Management*,
DOI 10.1007/978-981-10-4247-8_6

introduced to Burma and India by the end of the seventeenth century. It is a highly priced, popular, and major table fruit in India, known as the 'queen of fruits,' which is cultivated over an area of about 84,000 ha with production around 5.85 lakh metric tonnes and productivity about 7.0 metric tonnes/ha (NHB 2015). Litchi contributes significantly to the growers' economy in Bihar, West Bengal, Assam, and Jharkhand states of India that accounts for 78% of the total production in the country. Bihar produces 45% of total litchi and occupies nearly 40% of the area in India. In Bihar, Muzaffarpur, Vaishali, Samastipur, East Champaran, West Champaran, and Sitamarhi are major litchi-producing regions in India.

Diseases are one of the constraints to production of litchi fruits. They indirectly reduce yield by debilitating the tree and directly reduce the yield or quality of fruit before and after they are harvested. The occurrence and severity of disease vary according to location, age of tree, cultural practices, and the orchard ecosystem. Integrated disease management (IDM) systems are being developed with their main goals to eliminate or reduce the initial inoculums, reduce the effectiveness of initial inoculums, increase the resistance of the host, delay the onset of disease, and slow the secondary cycles of the disease. Litchi plants as compared to many fruit-bearing species are least affected by diseases. Diseases are more important after harvest, although undoubtedly many of the fruits are infected before picking. There are a few organisms that infect the leaves, flowers, and fruit, and a few others are associated with tree decline and tree deaths. Some leaf spot diseases are becoming important that are caused by fungal pathogens. No bacterial or viral infections have been reported so far. A few reports of algal infections are also available. So, it is necessary to know about diagnosis and approaches for the management of these diseases.

6.2 Major Diseases of Litchi

6.2.1 Twig Blight

Twig blight is one of the emerging disease problems in litchi trees (Kumar et al. 2014a, b, 2011). The symptoms appear as the death of leaves on new shoots and a foliar blight and tip dieback that is difficult to separate. The afflicted leaves give an impression as if they were scorched by the sun. Twig blight is caused by *Colletotrichum gloeosporioides* Penz. and *Gloeosporium* sp. (Table 6.1).

6.2.1.1 Management of Twig Blight

- Fungicidal spray of copper oxychloride (0.25%) or carbendazim (0.1%) can be done if disease severity increases.

Table 6.1 Major diseases of litchi and their management in a nutshell

Number	Name of disease	Causal organism	Symptoms	Management
1.	Leaf spot	*Botryodiplodia theobromae*, *Pestalotia pauciseta*, *Colletotrichum gloeosporioides*, *Microdiplodia litchii*	*Botryodiplodia* and *Colletotrichum* start from either tip or margin of lamina. These spots are dark, chocolate in colour in earlier stage and in later stage, brown, irregular in outline.	Foliar spray of mancozeb (0.25%) or copper oxychloride (0.25%) or thiophanate methyl (0.15%)
			Pestalotia spots are light coloured and present on both sides of leaf.	
			Microdiplodia leaf spot is yellowish brown to brick red in colour, mostly around the margin.	
2.	Anthracnose	*Colletotrichum gloeosporioides*	On the fruits, brown pinhead lesions appear that later turn to circular dark brown to black sunken lesions on mature fruits.	Spray either with copper oxychloride (0.25%) or carbendazim (0.1%) or difenoconazole (0.05%) or azoxystrobin (0.023%). Pre-harvest spray of fungicides helps in extending postharvest life
3.	Wilt	*Fusarium* sp.	Young trees of litchi, often less than 5 years in age, wilt in less than 1 week. The first symptoms appear as yellowing of foliage, drooping leaves followed by gradual wilting and drying, and leading to complete death of the plant within 4–5 days.	Application of castor cake or neem cake as manure with biocontrol agents such as *Trichoderma harzianum*, *T. viride*, or *Pseudomonas fluorescens* has been proven effective in managing the disease. In the absence of a biocontrol agent, drenching rhizosphere soil with hexaconazole or carbendazim (0.1%) may be done.

(continued)

Table 6.1 (continued)

Number	Name of disease	Causal organism	Symptoms	Management
				Growers are also advised not to plant litchi trees on waterlogged soils or in low-lying fields often receiving flood water
4.	Red rust	*Cephaleuros virescens*	Red, circular to semicircular spots appear mostly on the leaves and sometimes on tender stems. Orange-yellow to pink velvety coating is formed on the spots on formation of sporangia of the algae. In older leaves, the lesions turn light brown to brick red	Foliar spray of copper oxychloride (0.3%) may be done in July and October if severity increases
5.	Fruit rots	*Alternaria alternata*, *Colletotrichum gloeosporioides*, *Aspergillus niger*, *Aspergillus flavus*	Initially perceptible on injured portion of the fruits. Decayed areas become depressed, and rot gradually penetrates deep into the pulp. Fruits emit an odour of fermentation.	One spray of carbendazim (0.1%) 15–20 days before harvest. Prevent from physical injuries during harvesting. Prompt precooling and maintenance of optimal temperature and relative humidity needed during storage and transport of fruits. Sulphur fumigation if allowed by importing countries. For this, fruits are placed in a closed chamber where 50–100 g sulphur/m ^{3} of air is burnt for 20–30 min. For transportation, use corrugated-fibre boxes of 2-kg capacity properly unitized for stacking.

6.2.2 Anthracnose

Anthracnose caused by *Colletotrichum gloeosporoides* has been a major disease of litchi fruits in India; however, in recent years, its incidence has decreased. It is mainly a fruit disease but also affects the leaves, twigs, and flowers of litchi. Infected fruit becomes unmarketable. The disease first appears as brown pinhead lesions, usually on the top of semi-mature fruit. Circular dark brown to black sunken lesions later become easily visible on mature fruits. More spots appear on the top and sides of the fruit and slowly cover the fruit surface (Kumar 2016). High temperature and high relative humidity are highly conducive for the development and spread of anthracnose on litchi fruits. Outbreaks are common after warm wet weather. The highest latent infection rate of anthracnose of fruit causes more serious postharvest decay and browning of the fruits. The rate of latent infection and the postharvest decay and browning of the fruits could be evidently controlled by integrated management of the disease in the growing season. Usually superficial skin blemishes do not affect production and fruit quality, but marketability is affected.

6.2.2.1 Management of Anthracnose

- Avoidance of overcrowding of trees and branches in orchard.
- Pruning of affected plants and burning have been suggested to minimize the chance of fresh infections.
- Spraying of copper oxychloride (0.25%) or carbendazim (0.1%) or chlorothalonil (0.15%) or difenoconazole (0.05%) or azoxystrobin (0.023%). Pre-harvest spray of fungicides helps in extending post-harvest life, but, while doing so, care must be taken for residual toxicity of the chemical applied.

6.2.3 Leaf Spots

Usually, a mixed infection of different pathogens appears. Older leaves show higher infection. According to Prasad (1962), three pathogenic fungi causing leaf spots of litchi were reported from Muzaffarpur (Bihar) India, namely, *Pestalotia pauciseta*, *Botryodiplodia theobromae*, and *Colletotrichum gloeosporioides*. In nature, occurrence of the infection may be either all these three organisms or, more commonly, by *Botryodiplodia theobromae* and *Colletotrichum gloeosporioides*. Details of the leaf spots are given next.

6.2.3.1 *Botryodiplodia* Leaf Spots

The disease is caused by *Botryodiplodia theobromae*. The symptoms of the disease start from the tip or margin of the lamina. These spots are deep chocolate in colour. The limiting margins of the spots become irregular. Pycnidia appear on both surfaces of the leaf but mainly occur in the upper side of the leaf surface. The incubation period ranges from 3 to 7 days.

6.2.3.2 *Pestalotia* Leaf Spots

The spots caused by *Pestalotia pauciseta* are light coloured and appear on both sides of the leaf. The size of the spots varies from 0.5 to 2 cm × 0.2 to 2 cm. Later, the spots coalesce to form larger lesions, and the colour of the spots changes from brown to russet or marsh brown. The incubation period is from 5 to 10 days (Mishra and Pandey 2001).

6.2.3.3 *Colletotrichum* Leaf Spots

The disease is caused by *Colletotrichum gloeosporioides*. The symptoms start from the tip of the lamina and extend towards the base. The spots are irregular and are brown in colour with prominent marsh brown colour encircling them. The incubation period of *Colletotrichum gloeosporioides* is 4–8 days.

6.2.3.4 *Microdiplodia* Leaf Spots

The disease is caused by *Microdiplodia litchii*. According to Pathak and Desai (1971), *Microdiplodia* leaf spot was first reported from Udaipur (Rajasthan) in India. The diseased leaves show yellow-brown to brick-red areas mostly towards the margin. The coloured areas gradually become light brown and show black dot-like pycnidia. The pycnidia are ostiolate and up to 108 μm in diameter. Conidia are olivaceous and uniseptate and measure 8.2–10.9 × 2.4–4.6 μm.

6.2.3.5 Management of the Leaf Spots

- Deterioration of infected leaves persists throughout the year and develops vigorously after onset of the monsoon; hence, destruction of infected leaves as far as possible may be done.
- Apply foliar spray of mancozeb (0.25%) or copper oxychloride (0.25%) or thiophanate methyl (0.15%) if severity of the disease increases.

6.2.4 Algal Leaf Spot

Sharma et al. (1972), Mishra et al. (1973), and Gupta (1992) reported algal leaf spot in litchi from Punjab, Bihar (Pusa), and Uttar Pradesh (Saharanpur), respectively. The infection reduces the vitality of the plant by hampering photosynthetic activity. The growth of the plant is retarded by defoliation which indirectly affects yield. The causal organism of the disease is *Cephaleuros virescens*. This disease appears mainly in the rainy season and continues until early winter. The disease is characterized by the red, circular, to semi-circular spots (3.0–6.5 mm) which appear mostly on the leaves and sometimes also on the tender stems. An orange-yellow to pink velvety coating is formed on the spots on formation of sporangia of the algae. In older leaves the lesions turn light brown to brick red. *Cephaleuros virescens* has long, unbranched, reddish brown, stout, erect filaments measuring 52.05–235.96 × 10.41–15.61 μm, emerging through the thalloid disc, and 5- to 12 stalked. Sporangia are purple brown, subspherical to oval and smooth, measuring 19.08–27.76 × 15.61–20.82 μm (Mishra et al. 1973).

6.2.4.1 Management of Algal Leaf Spot

- Foliar spray of copper oxychloride (0.3%) may be used in July and October if severity increases.

6.2.5 Wilt

Young trees of litchi, often below the age of 5 years, wilt in less than 1 week time (Kumar et al. 2011). The first symptoms appear as yellowing of foliage and drooping leaves, followed by gradual wilting and drying, leading to complete death of the plant within 4 to 5 days. Some brown spots appear on the root crown and lateral root phloem, spreading later to the xylem. The causal organism of the wilt is mainly *Fusarium solani*. The mode of survival is by chlamydospores in the soil. The pathogen spreads by movement of contaminated soil from diseased plant to healthy plant rhizosphere.

6.2.5.1 Management of Wilt

- Apply castor cake or neem as manure.
- Apply *Trichoderma harzianum*, *Trichoderma viride*, *Pseudomonas fluorescens*, etc. as biocontrol agents that also act as plant growth promoters.
- In the absence of a biocontrol agent, drench rhizosphere soil with hexaconazole or carbendazim (0.1%).

6.2.6 Fruit Rots

Fruit rot of litchi has been a serious problem. Litchi is host to range of post-harvest pathogens, often with quite different modes of infection. Several fruit rot diseases were reported by Prasad and Bilgrami (1973) in India. A wide range of fungi, such as *Alternaria*, *Colletotrichum*, *Botryodiplodia*, *Aspergillus*, *Fusarium*, and *Penicillium* sp., reported from India by Prasad and Bilgrami (1973) and Kumar et al. (2016), causes post-harvest fruit rots if the fruits are not handled properly. A mean pathological infestation to 23.2% during 2012 and to 17.9% during 2013 in the supply chain of litchi in India has been reported (Kumar et al. 2016), the highest being at the retail level. Initially, the disease symptoms are perceptible on the injured portion of the fruits. With the advance of the disease, the decayed areas become depressed and the rot gradually penetrates deep into the pulp. Ultimately, the rind of infected fruits cracks off, exposing the pulp which subsequently is covered with thick cottony mycelium. Such affected fruits emit an odour of fermentation.

6.2.6.1 *Colletotrichum gloeosporioides* Rot

Initially, black dots appear small, circular, slightly depressed, and water soaked. Subsequently, they coalesce to form larger spots and become irregular in shape.

After 5–6 days, acervuli appear as dark-coloured dot-like bodies in the infection course. The acervuli are provided with deep multicellular setae.

6.2.6.2 *Aspergillus niger* Rot

The disease appears first at the stalk end of the fruit as a light brown lesion encircling the stalks, which later change to dark brown. Black conidial heads are distinct 4–5 days after infections. Some whitish deposits are often observed in the infected regions.

6.2.6.3 *Aspergillus flavus* Rot

At the earlier stages of infection, the diseased fruits show brownish discoloration at the border of infected regions. Gradually, the diseased area turns velvety and becomes depressed. After 6–8 days, grass-green conidial heads appear that mask the major portions of the fruits. In advanced stages, the infected portion looks dirty, yellowish green and powdery in appearance.

6.2.6.4 Management of Fruit Rots

The incidence of fruit rot caused by *Colletotrichum* sp. is reduced by dosage of 75 and 300 Gy irradiation of litchi fruits with no adverse effect on quality when stored at 5 °C for 3 weeks (McLauchlan et al. 1992). According to Brown et al. (1984), rotting of litchi is also much reduced by dipping the fruits in a solution of 0.125–0.25 g/l prochloraz at room temperature (25 °C) for 5 s to 5 min. Low-temperature storage is the most successful means of slowing rot development. For example, fruit stored at 22 °C rotted three times more quickly than fruit stored at 5 °C. The recommended management strategies for fruit rots are as follow.

- One spray of carbendazim 15–20 days before harvest. Prevent from physical injuries during harvesting.
- Prompt precooling and maintenance of the optimal temperature and relative humidity during storage and transport of fruits.
- Sulphur fumigation if allowed by importing countries. For this, fruits are placed in a closed chamber in which 50–100 g sulphur/m^3 of air is burnt for 20–30 min.
- For transportation, use corrugated-fibre boxes of 2-kg capacity properly unitized for stacking.

6.3 Conclusions

Considerable scope exists for research into the cause, epidemiology, and management of diseases in litchi crops. The lack of basic understanding of the causal agents of the various diebacks and root rots suggests that improved understanding of pathogen biology and the formulation of more effective disease management strategies should be possible.

Options for disease management in the future include a range of strategies. We will probably see a shift towards the use of environmentally benign fungicides, coupled with more targeted chemical applications that use disease-forecasting models and improved application technology. Studies of natural defence systems in this crop may lead to opportunities for enhancement or induction of antifungal compounds using defence elicitors. Nutrition has been shown to be important in disease resistance in other crops such as avocado (Willingham et al. 2001), particularly the concentrations of nitrogen and calcium in fruit tissue. The role of these nutrients in the development of fruit disease in litchi should be investigated. Although there has been very little success in biological control of plant diseases, this strategy, when combined with other control methods, should not be discounted. The development of more disease-resistant cultivars through conventional breeding or biotechnology is another option.

References

Brown BI, Scott KJ, Mayer DG (1984) Control of ripe fruit rots of guava, lychee and custard apple by postharvest prochloraz dips. Singap J Prim Ind 12:40–49

Gupta JH (1992) Chemical control of algal rust of litchi caused by *Cephaleuros virescens*. Progress Hortic 24:109–110

Kumar V (2016) Diseases in litchi management. In: Nath V, Feza Ahmad M, Manoj Kundu HM, Sahay S, Pandey SD, Srivastava K, Pongener A (eds) Litchi global perspectives. Excel India, New Delhi, pp 101–105

Kumar V, Kumar A, Nath V (2011) Emerging pests and diseases of litchi (*Litchi chinensis* Sonn.) Pest Manag Hortic Ecosyst 17:11–13

Kumar V, Anal AKD, Nath V (2014a) Prevalence of some threatening pests and disease of litchi (*Litchi chinensis* Sonn.) in Bihar state of India. J Appl Hortic 16:235–240

Kumar V, Kumar A, Nath V, Kumar R (2014b) New threats of insect pests and disease in litchi (*Litchi chinensis* Sonn.) in India. Acta Hortic 1029:417–424. doi:10.17660/ActaHortic.2014.1029.53

Kumar V, Purbey SK, Anal AKD (2016) Losses in litchi at various stages of supply chain and changes in fruit quality parameters. Crop Prot 79:97–104. doi:10.1016/j.cropro.2015.10.014

McLauchlan RL, Mitchell GE, Johanson GI, Nottingham SM, Hammerton KM (1992) Effect of disinfection dose of irradiation on the physiology of Taiso lychee. Postharvest Biol Technol 1:273–281

Mishra AK, Pandey BK (2001) Disease of litchi and their management. In: Thind TS (ed) Diseases of fruits and vegetables and their management, 1st edn. Kalyani, Ludhiana, pp 150–155

Mishra B, Prakash O, Mishra AP (1973) A serious leaf spot disease of litchi caused by *Cephaleuros virescens*. Indian J Mycol Plant Pathol 3:219–220

NHB (2015) Indian Horticulture Database 2014. National Horticulture Board, Ministry of Agriculture, Government of India, Gurgaon, India. http://nhb.gov.in/area-pro/NHB_Database_2015.pdf. Accessed 6 Oct 2016

Pathak VN, Desai BG (1971) A new species of *Microdiplodia* on *Nephelium litchi* Camb. Sydowia 24:73–74

Prasad SS (1962) Two new leaf spot diseases of *Nephelium litchi* Camb. Curr Sci 31:293

Prasad SS, Bilgrami RS (1973) Investigation on diseases of litchi. III. Fruit rots and their control by post harvest technology. Indian Phytopathol 25:523–527

Sharma VK, Srivastava AK, Chohan JS (1972) A new host record for parasitic alga *Cephaleuros virescens* Kunze. (litchi). Sci Cult 38:39–40

Willingham SL, Pegg KG, Cooke AW, Coates LM, Landon PWB, Dean JR (2001) Rootstock influences postharvest anthracnose development in 'Hass' avocado. Aust J Agric Res 52:1017–1022

Tissue Culture: Present Scenario and Future Prospects in Litchi Management

7

Vishal Nath, Manvendra Singh, Alok Kumar Gupta, Evening Stone Marboh, and Karmabeer Yadav

Abstract

The present context elaborates the potential of tissue culture in production of quality planting materials in litchi (*Litchi chinensis* Sonn.). The plant is native to China but also grows well in India, Vietnam, Thailand, Taiwan, parts of Africa, Israel, Australia and at higher elevations in Mexico and Central and South America. However, its commercial cultivation is limited to only a few subtropical countries in the world due to its specific climatic requirements. With increased commercial importance coupled with rapid expanding market at both the national and international level, it is imperative to assure sufficient production of quality planting materials in litchi. In this review several in vitro culture techniques in litchi will be highlighted.

Keywords

Litchi • Tissue culture • In vitro • Clonal propagation

7.1 Introduction

Litchi (*Litchi chinensis* Sonn.) is a subtropical evergreen sapindaceous fruit crop. Its commercial cultivation is restricted to only a few subtropical countries in the world owing to its specificity in climatic requirements (Leenhouts 1978; Menzel and Simpson 1994). The litchi-growing countries are China, Israel, Australia, Thailand, Taiwan, India, Vietnam, parts of Africa and at higher elevations in Mexico and Central and South America (Menzel and Waite 2005). Globally litchi

V. Nath • M. Singh (✉) • A.K. Gupta • E.S. Marboh
Crop Improvement Section, ICAR-NRC on Litchi, Muzaffarpur 42002, Bihar, India
e-mail: manumbbt@gmail.com

K. Yadav
ICAR-Indian Institute of Soil & Water Conservation, Dehradun-248195, Uttarakhand, India

M. Kumar et al. (eds.), *Lychee Disease Management*,
DOI 10.1007/978-981-10-4247-8_7

production is around 2.11 million tons, with more than 95% of the area and production share of Asia. China followed by India is the lead producer which accounts for 91% of the world litchi production, but it is mainly marketed locally. The other major litchi-producing countries are Taiwan, Thailand and Vietnam (Singh et al. 2012). A relatively small amount of litchi is produced in the United States, Mexico and Central and South America. India enjoys a prominent position in the litchi map of the world both in terms of production and productivity and accounts for about one-fifth of the global production. Over the years, India has recorded significant growth in production and productivity of litchi, coupled with rapid expanding market at both the national and international level.

Litchi has bright red fruit with an attractive pericarp surrounding a white and translucent fleshy, sweet and rosy odour aril (Bhoopat et al. 2011). The fruit can be eaten fresh or processed into juice, vinegar, jelly and wine (Chen et al. 2001; Alves et al. 2011; Saxena et al. 2011) and has established popularity with high demand in the international market. However, the cultivation and commercialization of litchi have not achieved the expected heights as the conventional propagation methods lack in providing rapid quality planting material to the farmers. Looking at the present burgeoning population, future demand will need to be supported by high-yielding varieties with improved fruit traits that can be assured through good-quality planting material (Singh et al. 2012; Menzel and Waite 2005). This emphasizes the need to standardize in vitro methods for mass production of litchi using tissue culture techniques. The purpose of this overview is to summarize the existing literature on in vitro culture of litchi that can serve as the baseline for further improvement in clonal production of litchi.

7.2 Conventional Propagation Methods in Litchi

Litchi can be propagated by seed as well as vegetative means. Propagation by seed results in seedling variability (Pandey and Sharma 1989; Thakur et al. 2001) in addition to poor germination rate, slow growth and lengthy juvenile period. Besides, owing to its recalcitrant nature, the seed loses viability in a very short period once separated from the fruit (Xia et al. 1992; Fu et al. 1990). Since embryo abortion is common in litchi, selection of well-formed mature seeds for propagation purposes (Bolt and Joubert 1968; Ray and Sharma 1987; Hartman et al. 1990) seems crucial. Hence, seed propagation is not a recommended practice in litchi as plants raised by this method fail to bear true-to-type and quality fruits (Kuhn 1962; Xia et al. 1992, 1993).

There are available literatures on successful propagation of litchi via cuttings (Paxton et al. 1978; Hartman and Kester 1986; Bose et al. 1985; Leonel et al. 1994; Menzel and Waite 2005; Singh et al. 2012). But planting of soft terminal cuttings in the field has limited success due to poor rooting. Moreover, a huge number of cuttings are required for mass propagation of plants which is an obstacle as a number of mother plants need to be vanishing. Air layering or 'gootee' is widely accepted for propagating litchi but, however, faces serious drawbacks due to poor survival of layers in the field (Sharma and Grewal 1989; Sharma et al. 1990a, b;

Kaundal et al. 1993; Cull and Lindsay 1995; Subhadrabandhu 1990; Wong 2000; Xu and Zheng 1999; Menzel and Waite 2005; Singh et al. 2012). Besides, it also damages the mother plant when a large number of layers are required to be made (Cull and Lindsay 1995; Kanwar and Kahlon 1986; Menzel and Waite 2005; Singh et al. 2012). Plants prepared by this method are delicate and difficult to transport (Menzel and Waite 2005; Singh et al. 2012). Efforts on grafting in litchi have been attempted by several workers (Pandey and Sharma 1989; Ou et al. 1993; Yee 1972; Menzel and Waite 2005; Singh et al. 2012). However, the success rate is variable and often low, because of stock-scion incompatibility, poor cambial contact, grafting at the wrong physiological stage and poor post-grafting management (Ruan 1988; Menzel and Waite 2005; Singh et al. 2012).

Although, till date, vegetative propagation methods have been the only option for the mass propagation of litchi, ample supply of planting material is still inadequate because vegetative propagation is limited by production of less number of individuals multiplied from a single plant.

7.3 In vitro Tissue Culture of Litchi

Litchi seeds have poor germination (Xia et al. 1992, 1993), and conventional methods produce insufficient amount of plants (Bolt and Joubert 1968; Ray and Sharma 1987; Hartman et al. 1990; Menzel and Waite 2005; Singh et al. 2012). Plant tissue culture technique offers a viable option in accelerating high frequency mass multiplication of litchi from single explant. Biotechnological interventions for in vitro regeneration and mass micropropagation methods in fruit tree species have been practised with success. There are available published reports on in vitro regeneration and growth of litchi through embryogenesis or organogenesis (Kantharajah et al. 1992; Das et al. 1999; Yu et al. 2000; Puchooa 2004; Kumar et al. 2004; Khan and Ahmad 2005; Kumar et al. 2006; Raharjo and Litz 2007; Ma et al. 2009). Nevertheless, explant source, their disinfection process, media formulations, culture conditions, accumulation of phenolics in media and media discoloration significantly influence shoot regeneration even from different genotypes of the same species (Kumar et al. 2004, 2006; Das et al. 1999).

7.3.1 Explant Source

The type of explants for in vitro regeneration influenced the mass multiplication and proliferation. Successful in vitro regeneration in litchi has been reported using various explants, for instance, immature embryo (Kantharajah et al. 1992; Yu et al. 2000), seed (Das et al. 1999; Khan and Ahmad 2005), cotyledonary node (Das et al. 1999; Khan and Ahmad 2005), nodal segments (Kumar et al. 2004, 2006) and leaf (Puchooa 2004; Raharjo and Litz 2007; Ma et al. 2009). Somatic embryogenesis in litchi was previously reported from leaves (Puchooa 2004; Raharjo and Litz 2007; Ma et al. 2009). Ma et al. (2009) made a report on callus induction from leaf explant and established suspension culture. Immature embryo was also used for embryogenesis and protoplast isolation.

7.3.2 Basal Media and Carbohydrate Sources

Vegetative and reproductive plant parts of various fruit crops are used as explant source in tissue culture. The primary aims of tissue culture in fruit crops are proliferation of axillary and adventitious buds, induction of rooting in microshoots and production of callus. The most commonly used carbon source in culture medium is sucrose (2–5%) and is indispensable for various metabolic activities. It is required for differentiation of xylem and phloem elements in culture cells, provides energy to the tissue and maintains an optimum level of osmotic potential (Bhojwani and Razdan 1996). Different types of media used in litchi for regeneration, callus induction and somatic embryogenesis are presented in Table 7.1.

Murashige and Skoog (MS) media were mostly used in plant tissue culture. Kantharajah et al. (1992) reported that MS solid medium with 2% sucrose supplemented with 150 ml/l coconut water was most effective in stimulating the germination of immature litchi embryos. Das et al. (1999) showed the effectiveness of culture supplemented with MS liquid medium when supported on a filter-paper bridge technique for direct seed germination. Yu et al. (2000) used immature zygotic embryos as explant and cultured on induction medium, which consisted of MS salts and B_5 vitamins and sucrose (50 g/l). In vitro plantlet regeneration of litchi was achieved from callus cultures derived from young, tender leaf explants on MS medium (Puchooa 2004). Multiple shoot formation was obtained on modified WPM medium (Kumar et al. 2004, 2006). Raharjo and Litz (2007) made a recipe for embryogenic cultures of litchi on B5 medium containing 400 mg/l glutamine, 200 mg/l casein hydrolysate, 30 g/l sucrose, 4.52 mM 2,4-D, 9.30 mM kinetin and 3 g/l gellan gum in darkness. Ma et al. (2009) reported formation of friable callus in MS media that subsequently gave rise to globular embryo.

7.3.3 Effect of Plant Growth Regulators on Organogenesis

The development of adventitious shoots directly from the explants is categorically a more suitable approach for clonal propagation of plant, when compared to the callus-mediated regeneration. Callus recurrently produces asynchronous plants, whereas adventitious shoots form homogeneous diploid individuals (Bhojwani and Razdan 1996). In view of the fact that adventitious shoot formation can be achieved directly from tissue lacking a conciliator callus phase, it is an opposite method to produce large-scale true-to-type plants. So far various media formulations fortified with variable sources and doses of plant growth regulators have been explored for shoot regeneration in litchi using different explants (Table 7.1). Multiple shoot induction was achieved in litchi through nodal explants (Kumar et al. 2004, 2006). Savlon (1.0% v/v), bavastin (1.0% w/v) and mercuric

Table 7.1 In vitro tissue culture medium of litchi

S. no.	Explant	Media	Response	References
1.	Immature embryos	MS + 2% sucrose + 150 ml/l + 100 mg/l BAP	Multiple shooting	Kantharajah et al. (1992)
		MS + 0.5 mg/l NAA + activated charcoal	Rooting	
2.	Seed	MS+ 20 mg/l BAP	Seed germination	Das et al. (1999)
	Cotyledonary node	BAP (100 μg on alternate days)	Shooting	
		0.25 mg/l IBA (pulse treatment)	Rooting	
3.	Immature embryo	MS + 1.0 mg/l 2, 4-D + 0.2 mg/l NAA+ 250 mg/l glutamine +0.5 mg/l zeatin	Somatic embryo, protoplast culture	Yu et al. (2000)
4.	Young, tender leaf	MS + 1.5 mg/l 2,4-D	Callus shooting	Puchooa (2004)
		MS + 2.0 mg/l BAP		
		MS + 3.0 mg/l IAA		
		MS + 2.0 mg/l IBA	Rooting	
5.	Nodal	WPM + 2.5 mg/l BAP + 0.2 mg/l KIN + 0.1 mg/l GA3	Shooting	Kumar et al. (2004)
		WPM (semi-solid) + 2.5 mg/l BAP + 0.1 mg/l GA3	Shoot elongation	
		WPM (semi-solid) + 2.5 mg/l BAP+ 400 mg/l casein hydrolysate		
		MS (liquid) + 1.5 mg/l BAP + 300 mg/l casein hydrolysate		
6.	Seed	MS (liquid) + 20 mg/l BAP	Seed germination	Khan and Ahmad (2005)
	Cotyledonary node	BAP (100 μg on alternate days)	Shooting	
7.	Nodal	WPM + BAP (11 μM) + KIN (0.5 mg/l) + GA3 (0.2 mg/l) + CW (15%)	Shoot bud formation	Kumar et al. (2006)
		MS + BAP (6.6 μM) + GA3 (0.15 μM) + SN (30 μM) + CH (300 mg/l)	Multiple shooting	
		½ MS + IBA (20.6 μM) + litchi seed powder (1 g/l)	Rooting	
8.	Leaf	B5 + glutamine (400 mg/l) + casein hydrolysate (200 mg/l) + sucrose (30 mg/l) + gellan gum (3 g/l) + 2,4-D (4.52 μM) + KIN (9.30 μM) 2,4-D (4.52 μM) + zeatin (0.91 μM)	Embryo induction	Raharjo and Litz (2007)
			Embryo maintenance	
		MS	Embryo maturation	
		½ MS + 200 mg/l	Rooting	
9.	Leaf	2,4-D	Callusing and suspension culture	Ma et al. (2009)

chloride (0.2% w/v) surfactants were used for sterilization of nodal explants. The most preferable time of cultures was March–May due to least contamination and the best shoot growth recorded during this season. Semi-solid medium used with the mixture of activated charcoal (0.2%) and PVP (0.2%) was also effective, but rapid subculturing was essential to solve the browning problem. Filter-paper bridge technique was also found to be very convenient and an effective method to eliminate the browning problem from the culture (Kumar et al. 2004, 2006). Multiple shoot formation and shoot elongation medium are shown in Table 7.1.

7.3.4 Effect of Plant Growth Regulators on Callogenesis and Somatic Embryogenesis

Plant cells which multiply in a disorganized fashion gave rise to coherent and amorphous mass of tissue termed as callus. Conversely, callus when cultured in the reverse direction can result in the differentiation of adventitious shoots, roots or even embryos. Culture systems can be utilized for studying the cytological or biochemical processes of growth which are associated with cell division, cell enlargement and differentiation. Additionally, the dissimilar performance of genotypes of one species during in vitro culture has often been explained and is presumed to be owing to either different endogenous phytohormone levels or their variable ability to react to exogenously applied PGRs (Bhojwani and Razdan 1996).

Culturing of immature embryos of different sizes and ages in a range of different media produces multiple embryonic shoots in different litchi varieties with differential response. Up to 15 adventitious buds were produced in Bengal litchi after pretreatment with 100 mg/l BAP for 3 h (Kantharajah et al. 1992). Yu and Chen (1997) cultured zygotic embryos of litchi and observed that the so-obtained embryogenic cultures maintained on semi-solid medium consisted of a mixture of PEMs and maturing somatic embryos. Silver thiosulphate was also incorporated in the medium to inhibit somatic embryo maturation and induce friable cultures (Yu and Chen 1997). Das et al. (1999) reported two pathways for shoot induction in litchi. The first one was direct germination of litchi seeds in 6-benzylaminopurine (20 mg/l)-supplemented MS liquid medium and supported on a filter-paper bridge, and the second method was in treatment with 6-benzylaminopurine (100 μg on alternate days) of the axillary bud regions of plants germinated. Yu et al. (2000) obtained white non-hyperhydric somatic embryos from protoplast-derived microcalli or proembryos which were alternatively maintained in liquid and on solid media containing silver thiosulphate.

Auxins especially 2,4-D (1.5 mg/l) promote callogenesis. Media supplemented with 2,4-D and BAP produce nodular compact callus which proliferated and differentiated into shoots (Puchooa 2004). Raharjo and Litz (2007) reported embryogenic cultures were induced from leaflets excised from new vegetative flushes of mature 'Brewster' litchi trees, and recovery of plants from somatic embryos was improved with GA_3 on half-strength MS medium with activated charcoal. Suspension culture was also established from friable calli on MS medium supplemented with BAP (2 mg/l) and IAA (3 mg/l) which subsequently forms globular embryos (Ma et al. 2009).

7.3.5 Effect of Plant Growth Regulators on Root Induction

In some woody species, rooting of microcuttings may require a high concentration of auxin. Prolonged exposure to high auxin levels, however, has some undesirable effects, such as callusing, leaf chlorosis and inhibition of root elongation and shoot tip dormancy which is difficult to overcome in the acclimatization stage (Bhojwani and Razdan 1996). Generally, auxin is essential for rooting either at high concentration for a short pulse or in optimum concentration for an optimum duration in the medium. Literatures revealed the wide use of IBA and NAA for root induction in litchi (Kantharajah et al. 1992; Das et al. 1999; Kumar et al. 2004, 2006; Puchooa, 2004; Khan and Ahmad 2005). Adventitious rooting was reported in 65% of shoot culture on MS medium supplemented with NAA (0.5 mg/l) and activated charcoal (Kantharajah et al. 1992). Das et al. (1999) reported shoot elongation and rooting through pulse treatment with IBA (25 mg/l) for 15 min. Kumar et al. (2006) observed 70% rooting of shoots on MS medium supplemented with IBA (20 μM) + litchi powder (1 g/l) after a pulse treatment of IBA for 15 min. Prominent shoots were produced when regenerated shoots were transferred to MS medium supplemented with 2.0 mg/l IBA (Puchooa 2004).

7.4 Acclimatization

Micropropagation is the most important phase of tissue culture procedure. Success rate of micropropagation on a commercial scale depends on the ability to establish the plants out of culture on a large scale, at low cost and with high survival rates (Kumar et al. 2004). In the case of litchi, few published reports of micropropagation are available (Kantharajah et al. 1992; Das et al. 1999; Kumar et al. 2004, 2006; Puchooa 2004; Khan and Ahmad 2005). Rooted plantlets of litchi were successfully transferred and grown in the glasshouse (Kantharajah et al. 1992; Yu et al. 2000). Das et al. (1999) reported combined method of rooting and hardening phase, thus reducing the time required for mass multiplication, and the plantlets were successfully transferred to the soil. Puchooa (2004) reported that IBA was better than NAA in improving the survival rates of plantlets during the hardening phase due to better and quality rooting. Kumar and co-workers successfully transferred plantlets in pots containing garden soil mixed with sand and vermiculite (1:1:2). Raharjo and Litz (2007) established plantlets with well-developed root systems in pots containing Promix BX and Perlite (1:1) and reported up to 38% survival rate in greenhouse.

7.5 Future Perspective

In vitro propagation is an important alternative approach to conventional propagation and breeding procedures in litchi. The above review explored the potential of tissue culture methods in producing somatic embryogenesis and organogenesis from litchi explants. The type of explants and type and levels of plant growth regulators used determine the success of regeneration of litchi through micropropagation. In most cases, auxins like 2,4-D and NAA were indispensable for callusing and somatic embryogenesis in litchi. BAP and GA_3 were also important for growth and multiple shoot formation. Till date literatures on in vitro intervention in litchi had been majorly focused on the development of somatic embryo and their physiological as well as morphological aspects. The efficiency and reproducibility of the basic protocol of micropropagation still need to be emphasized and developed which will serve as a platform for transmitting economically imperative traits through genetic engineering, inducing somaclonal variations, in vitro mutations, double haploids and development and utilization of somatic hybrids in litchi.

References

Alves JA, deOliveira LLC, Nunes CA, Dias DR, Schwan RF (2011) Chemical, physical-chemical, and sensory characteristics of lychee (*Litchi chinensis* Sonn) wines. J Food Sci 76:S330–S336

Bhojwani SS, Razdan MK (1996) Plant tissue culture: theory and practice – errata slip inserted. Elsevier, Amsterdam

Bhoopat L, Srichairatanakool S, Kanjanapothi D, Taesotikul T, Thananchai H, Bhoopat T (2011) Hepatoprotective effects of lychee (*Litchi chinensis* Sonn.): a combination of antioxidant and anti-apoptotic activities. J Ethnopharmacol 136:55–66

Bolt LC, Joubert AJ (1968) Litchi grafting trials. Farmg S Afr 44:11–13

Bose TK, Dhua RS, Mitra SK, Sen SK (1985) Synergism of phenolic substances with IBA in regeneration of root from litchi cuttings. Indian J Hortic 42:153–155

Chen GL, Chen S, Sun YM, Huang XY, Yang M, Xie ZH (2001) Studies on the application of composite enzyme in clarification of litchi juice. Acta Hortic 558:393–398

Cull BW, Lindsay P (1995) Fruit growing in warm climates for commercial growers and home gardens. Reed Books, Chestwood, 192 pp

Das DK, Prakash NS, Bhalla-sarin N (1999) Multiple shoot induction and plant regeneration in litchi (*Litchi chinensis* Sonn.) Plant Cell Rep 18:691–695

Fu JR, Zhang BZ, Wang XP, Qiao YZ, Huang XL (1990) Physiological studies on desiccation, wet storage and cryopreservation of recalcitrant seeds of three fruit species and their excised embryonic axes. Seed Sci Technol 18:743–754

Hartman HT, Kester DE (1986) Plant propagation: principles and practices. Prentice-Hall of India, New Delhi, 727 pp

Hartman HT, Kester DE, Davies FT (1990) Theoretical aspects of grafting and budding. In: Hartman HT, Kester DE (eds) Plant propagation: principles and practices. Prentice Hall, Englewood Cliffs, pp 305–348. http://www.ajol.info/index.php/ajb/article/view/15021

Kantharajah AS, Mcconchie CA, Dodd WA (1992) *In Vitro* embryo culture and induction of multiple shoots in lychee (*Litchi chinensis* Sonn.) Ann Bot 70:153–156

Kanwar JS, Kahlon GS (1986) Propagation studies in litchi (*Litchi chinensis* Sonn.) J Res Punjab Agric Univ 23:33–39

Kaundal GS, Grewal GPS, Deol IS (1993) Studies on propagation and post transplanting success of litchi (*Litchi chinensis* Sonn.) layers. Sci Hortic 3:31–36

Khan MA, Ahmad I (2005) Multiple shoot induction and plant regeneration in litchi (*Litchi chinensis* Sonn.) Int J Agric Biol 7(3):524–526

Kuhn GD (1962) Dehydration studies of litchi fruits. Proceedings of the Florida State Horticultural Society 75:273–277

Kumar M, Prakas NS, Muthusamy A, Prasad US, Bhalla-Sarin N (2004) Problems and perspectives of mass scale production Litchi (*Litchi chinensis* Sonn.) using *in vitro* cultures. In: Proceeding of the National Seminar on recent advances in production and post harvest technology of litchi for export. BCKV. West Bengal. June 24–26, pp 12–17

Kumar M, Praksh NS, Bhalla-Sarin N (2006) A novel approach of regeneration from nodal Explants of field-grown litchi (*Litchi chinensis* Sonn.) fruit trees. J Plant Sci 1(3):240–246

Leenhouts PW (1978) Systematic notes on the Sapindaceae-Nephelieae. Blumea 24:395–403

Leonel S, Rodrigues JD, Cereda E (1994) Effect of growth regulators and boric acid on lychee cuttings. Científica 22:105–110

Ma XY, Yi GJ, Huang XL, Zeng JW (2009) Leaf callus induction and suspension culture establishment in lychee (*Litchi chinensis* Sonn.) cv. Huaizhi. Acta Physiol Plant 31:401–405

Menzel CM, Simpson DR (1994) Lychee. In: Schaffer, B. and Andersen, P.C. (eds)

Menzel CM, Waite G (2005) Litchi and longan: botany, production, and uses. CABI Pub, Cambridge, MA. http://www.SLQ.eblib.com.au/patron/FullRecord.aspx?p=297524

Murashige T, Skoog F (1962) A revised medium for rapid growth and bioassays with tobacco tissue cultures. Physiol Plant 15:473–497

Ou LX, Zhong YW, Li JS (1993) Preliminary report of a study on the compatibility of litchi scion–rootstock combinations. China Fruits 2:30–31

Pandey RM, Sharma HC (1989) The litchi. ICAR, New Delhi, 80 pp

Paxton B, Saranah J, Chapman KR (1978) The propagation of litchis by cuttings. Biennial Report Maroochy Horticultural Research Station 1:34

Puchooa D (2004) In vitro regeneration of lychee (*Litchi chinensis* Sonn.) Afr J Biotechnol 3 (11):576–584

Raharjo SHT, Litz RE (2007) Somatic embryogenesis and plant regeneration of litchi (*Litchi chinensis* Sonn.) from leaves of mature phase trees. Plant Cell Tissue Organ Cult 89:113–119

Ray PK, Sharma SB (1987) Growth, maturing, germination and storage of litchi seeds. Sci Hortic 33:213–221

Ruan ST (1988) A discussion on litchi grafting. Litchi Tech Commun 4:4

Saxena S, Hajare SN, More V, Kumar S, Wadhawan S, Mishra BB, Parte MN, Gautam S, Sharma A (2011) Antioxidant and radioprotective properties of commercially grown litchi (*Litchi chinensis*) from India. Food Chem 126:39–45

Sharma RC, Grewal GPS (1989) A note on propagation studies in litchi (*Litchi chinensis* Sonn.) Haryana J Hortic Sci 18:74–76

Sharma SB, Ray PK, Singh BK (1990a) A note on the effect of defoliation on survival of litchi layers in a nursery. Haryana J Hortic Sci 19:129–130

Sharma SB, Ray PK, Singh BK (1990b) Note on rooting and survival of litchi layers. Indian J Hortic 47:174–176

Singh G, Nath V, Pandey SN, Ray PK, Singh HS (2012) Litchi. New Delhi, FAO

Subhadrabandhu S (1990) Lychee and longan cultivation in Thailand. Rumthai Publishers, Bangkok, 40 pp

Thakur S, Sharma RN, Sharma RK (2001) Breeding. In: Chauhan KS (ed) Litchi: botany, Production and Utilization. Kalyani Publishers, Ludhiana, pp 28–36

Wong KC (2000) Longan production in Asia. Food and agriculture Organization of the United Nations, Bangkok, Thailand, 44 pp

Xia QH, Chen RZ, Fu JR (1992) Effects of desiccation, temperature and other factors on the germination of litchi (*Litchi chinensis* Sonn.) and longan (*Euphoria longan* Steud.) seeds. Seed Sci Technol 20:119–127

Xia QH, Chen RZ, Fu JR (1993) Studies on physiology of lychee (*Litchi chinensis* Sonn.) seeds at different developmental stages. Acta Scientiarum Naturalium Universitatis Sunyatseni 32:80–86

Xu XD, Zheng SQ (1999) Photographic illustration of high yielding longan cultivation. China Agricultural Press, Beijing, pp 5–29

Yee W (1972) The lychee in Hawaii. College of Tropical Agriculture, University of Hawaii, Honolulu Circular 366: 1–4

Yu CH, Chen Z (1997) Induction of litchi embryogenic calli by immature embryos and anthers culture *in vitro*. J Fujian Agric Univ 26:168–172

Yu C, Chen Z, Lu L, Lin J (2000) Somatic embryogenesis and plant regeneration from litchi protoplasts isolated from embryogenic suspensions. Plant Cell Tissue Organ Cult 61:51–58

Systemic Information for Future Perspectives in Litchi Crop Improvement

8

Alok Kumar Gupta, Vishal Nath, Awtar Singh, Manvendra Singh, Evening Stone Marboh, Swapnil Pandey, and Ashish Pathak

Abstract

Litchi (*Litchi chinensis* Sonn.) is a subtropical evergreen fruit crop of the Sapindaceae family. Over the years, significant growth in production and productivity coupled with the fast-expanding market of litchi has been recorded at both the national and international level. Most commercial cultivars in litchi have been clonally selected under Chinese or Indian conditions and have been adapted to a limited climatic condition. Disease infestation in litchi has been a long-term debate in the horticulture and agriculture sector. Keeping the damage factor as a prime concern, clonal selection as the basis of cultivar selection must rely on limited characteristics such as systemic susceptibility in terms of fruit size, quality, and period of maturity, which narrows down the diversity, focusing on only a few commercial traits. Hence, creation of variability within the litchi gene pool is of paramount importance to yield desirable characters such as precocity, dwarfness, regularity of bearing, wider adaptability, and resistance and avoidance of pests and disorders. The heterozygosity of litchi produces a wide extent of variability, which serves as the baseline for new selections through harnessing precocious genes and exploiting natural hybrid vigour and other genetic manipulations. Different strategic efforts on a breeding programme need to be undertaken on a large scale with considerations of a comprehensive survey of various genotypes and trait inheritance patterns, raising a large population of open pollinated seeds with known parentage, mutation breeding because of obvious difficulties with traditional litchi breeding and the lack of pure lines. This context provides the basic information for further improvement and genetic enhancement of the breeding programme in litchi disease resistance.

A.K. Gupta (✉) • V. Nath • A. Singh • M. Singh • E.S. Marboh • S. Pandey • A. Pathak
Crop Improvement Section, ICAR-National Research Centre on Litchi, Muzaffarpur 842002, Bihar, India
e-mail: alokguptabhu@gmail.com

M. Kumar et al. (eds.), *Lychee Disease Management*,
DOI 10.1007/978-981-10-4247-8_8

Keywords
Litchi chinensis • Variability • Heterozygosity • Genetic enhancement

8.1 Introduction

Litchi (*Litchi chinensis* Sonn.) is a subtropical evergreen fruit crop of the Sapindaceae family (Leenhouts 1978). Its climatic requirements are highly specific, and thus its commercial cultivation is restricted to only a few subtropical countries in the world. The major litchi-producing countries are China, India, Taiwan, Thailand and Vietnam. A relatively small amount of litchi is produced in the United States, Mexico and Central and South America (Menzel and Waite 2005). World production of litchi is estimated to be around 2.11 million tons, with more than 95% of the area and production shared by Asia. India and China account for 91% of the world production, but it is mainly marketed locally. Over the years, significant growth in production and productivity of litchi have been recorded coupled with a fast-expanding market at both the national and international level (Singh et al. 2012).

There has been a substantial increase in the area and production of litchi in India during the past 50 years. The area has increased from 9400 ha (1949–1950) to 84,950 ha (2014–2015). The contribution of litchi to the total area under fruit cultivation has increased from 0.75% to 1.70%. Increase in area during 1991–1992 to 2014–2015 (22 years) has been more than 80%, whereas production increase during the same period is to the tune of more than 150% (NHB 2015). Productivity also recorded an increase of about 50% during the same period. Evidently, production and productivity of litchi are constantly increasing in the country. The fresh fruit market dominates the trade, followed by dried and canned fruit. The main importers are the European Union, the United States, Hong Kong, Singapore, Japan and Canada, and China, Taiwan, Thailand, Madagascar, South Africa, Australia and Mexico are the main exporting countries.

Globally, the litchi industry is beset with myriads of problems. Low success rate of establishment, lesser fruiting span, low and irregular yields because of poor flowering and fruit set, fruit cracking, browning and rotting of fruits, fruit borers and mites, poor shelf life, and lack of suitable varieties with early and late maturity and good-quality fruits are some of the factors encountered.

8.2 Origin and Distribution

The *Litchi* genus contains three subspecies: *Litchi chinensis* ssp. *chinensis*, *L. chinensis* ssp. *philippinensis*, and *L. chinensis* ssp. *javensis* (Leenhouts 1978). The subspecies *philippinensis* is found growing wild in the Philippines from sea level to 500 m altitude (Sotto 2002). The fruits are distinguished from cultivated litchi by their long and oval shape, with long thorny protuberances and inedible flesh that partially covers the seed (Menzel 1991). Fruits of the subspecies *javensis* from the Malay Peninsula and Indonesia produce an aril thinner than that of the cultivated litchi. Neither of these two subspecies is grown commercially. The subspecies *chinensis*, a litchi of commerce, originated in Southern China and Northern Vietnam from wild populations of these regions (Wu 1998b; Hai and Dung 2002). In China, enormous diversity has been observed in the moist forests of Hainan from low elevations up to 1000 m, below 500 m in hilly areas of western Guangdong and eastern Guangxi, and in valley or hilly regions of southern Yunnan below 1000 m (Wu 1998a). Litchi dominates most of these forest compositions and often grows in mixed stands with *Vatica astrotricha* (green plum), *Hopea hainanensis*, *Heritiera parvifolia* of the Chinese parasol family, *Coelodepas hainanensis*, *Polyathia laui*, and *Diospyros hainanensis*, which belongs to the Ebenaceae family. The small area of wild litchis have also been found growing in the natural rainforest of Vietnam at low elevation in the Ba Vì mountains and in the forests of Tam Dao (Vinh Phuc Province) and Tuyên Hóa of Quang Binh Province (Hai and Dung 2002). Generally, in Hainan the wild trees look similar to cultivated litchi and produce edible fruit, but the flesh is relatively thin and sour. Fruits are variable and can be characterized into three groups based on their shape and arrangements of skin segments and protuberances: group 1 possesses sharp and pointed protuberances, group 2 has protruded but obtuse skin segments, and group 3 has flat skin segments. The wild types evolved in two directions, with skin segments becoming protruded and long, as in 'Dazao' ('Tai So') and 'Guiwei' ('Kwai May'), or flattened, as in 'Sanyuehong' ('Sum Yee Hong'), 'Shuidong' ('Souey Tung'), 'Nuomici' ('No Mai Chee'), and 'Huaizhi' ('Wai Chee') (Wu 1998a). In Yunnan, a population of wild litchis was identified in which the trees required less cool weather to initiate flowering. These trees mature earlier and produce a crop in warmer climates than can the traditional subtropical ecotypes. Flowers from these wild specimens have sepals with brownish pubescence, which gave the species the name 'brown-hair litchi' or 'Hemaoli.'

Litchi, a luscious fruit crop, originated from the Kwantung and Fukien Provinces of Southern China (Tao 1955). According to de Candolle (1909), 'Chinese authors living at Peking knew about litchi only late in the third century of our era.' Although controversial, the first citation for this fruit in the literature probably can be traced back to as early as 1766 BC. However, a clear reference has been mentioned in the literature of the Han dynasty (140–86 BC). Possibly the first complete book on litchi in English (Groff 1921) was a monograph published by Ts'a' Hsiang (1059 AD).

Until the tenth century, litchi was propagated through seedlings. Later, vegetative propagation through air layering or marcotting became widely accepted among the growers, being first practised in the fourth century AD. By the sixteenth century, grafting was also recorded (Anonymous 1978). However, their detailed application was first documented in the Registers by Xu Bo in 1579 and by Deng Qingcai in 1628, respectively.

Until the late seventeenth century, litchi cultivation was restricted to Southern China and Northern Vietnam (Tindall 1994; Hai and Dung 2002). It then spread to other regions through the route suggested by Galán Saúco and Menini (1989). Apparently, it reached Burma and Eastern India in the late seventeenth century (Hayes 1957). Subsequently, by the end of the eighteenth century, it made its way to Bengal, whence it diversified to other parts of India on a commercial scale and later to Nepal (Budathoki 2002) and Bangladesh (Abu Baker Siddiqui 2002). The litchi industry has grown significantly in these countries. Litchi was first introduced into Thailand from China 300 years ago by merchants who carried the fruit with them. Some plants adapted successfully to the tropical conditions of the central region of the country. Promising seedlings were selected and accordingly named by local growers as lowland litchi or tropical litchi, as the trees do not require a long period of cold to initiate flowers (Subhadrabandhu 1990). In Chiang Mai, litchi was planted around 1890, through air layers brought by emigrants who migrated from Yunnan through Laos or Myanmar (Boonrat 1984). These are truly subtropical types, requiring a longer period of low temperature for flowering. Many of them still retain their Chinese names, such as 'O-Hia,' 'Hong Huay,' and 'Kim Cheng'; however, the Thai spellings and pronunciations are different (Subhadrabandhu and Yapwattanaphun 2001a). Litchi reached the Philippines from China before 1916 but failed to flower and set fruit at low altitudes until 1931 (Sotto 2002). Later, in 1931, trees were introduced from other sources that bear fruit and gave hope for litchi growing in the more elevated areas. The crop was introduced in Australia through Chinese migrants attracted by gold rushes around 1854 (Menzel et al. 1988) and reached Southern Africa about 50 years earlier.

In Madagascar the arrival of litchi dates back to 1802 and the 1940s, with many old and relatively few new plantations still in existence at a commercial level (Jahiel and Abraham 2001). Evidence revealed that the first litchi trees were imported into South Africa from Mauritius in 1876. However, trees had already been observed in Natal in 1875, suggesting there must have been earlier imports (Oosthuizen 1991). From Natal, trees traveled to the Transvaal Lowveld as well as to other suitable frost-free areas. The species arrived in Florida in the 1880s, but commercial cultivation started in the 1940s. 'Brewster' ('Chenzi') is the main cultivar of the industry in the region, but currently it is replaced by 'Mauritius' ('Dazao,' 'Tai So') (Knight 2001). The first trees, known as 'Afong,' which were transported to Hawaii in 1873 by Chinese merchants, later were identified as being similar to 'Dazao' (Nakasone and Paull 1998a). In Israel litchi was introduced in the 1930s from different parts of South Africa ('Mauritius'), California ('Floridian'), and India ('Bengal'); however, a commercial industry developed only after the 1980s (Goren et al. 2001).

At a later event, litchi was introduced to several other countries. In Hawaii, it was probably successfully introduced in 1873. Its cultivation in the West Indies and Natal, South Africa, dates back to 1775 and 1869 (Marloth 1947), respectively. Commercial cultivation of litchi in Queensland (Australia) was of late occurrence (Batten and Lahav 1994), although it was introduced as early as 1854. In the United States, it reached from Saharanpur (India) to Florida in 1883, to California in 1897, and subsequently again to Florida from Fukien Province of China in 1906 (Singh and Singh 1954), where it was named 'Brewster' litchi. Early in the nineteenth century, it also reached England and France but was unsuccessful (Pandey and Sharma 1989).

8.3 Classification of Litchi

The majority of species in the Sapindaceae are either trees or shrubs native to Asia, although there are a few species in South America, Africa, and Australia (Bailey 1949; Leenhouts 1971, 1978, 1986). The family derived its name from the soap-berry, *Sapindus saponaria*, whose fruit is used as a soap substitute in the tropics (Nakasone and Paull 1998b). Litchi relatives from Southeast Asia include rambutan, *Nephelium lappaceum*, and pulasan, *Nephelium mutabile*. These species differ from litchi in having long hairs or spinterns instead of protuberances. Litchi and longan are long-lived evergreen trees producing new leaves, flowers and fruit on terminal shoots. The inflorescences produce hundreds of functionally male and female flowers, which bear 5 to 80 fruits at harvest. Other tropical species of local significance of the subfamily Sapindoideae include tuan, dawa, or Fiji longan, *Pometia pinnata*, from Southeast Asia and the Pacific (tribe Nephelieae), mamoncillo, *Melicoccus bijugatus*, from the Caribbean (tribe Cupanieae); ackee, *Blighia sapida*, from West Africa (tribe Sapindaceae); and guarana, *Paullinia cupana*, from the Amazon basin (tribe Paullinieae) (Yeap 1987; Menzel et al. 1993). The sapindaceous trees were first described by Cambessedes in 1828; however, detailed systematic study was published only after the start of the twentieth century.

Classification of litchi by Radlkofer (1932) was based on a wide range of evidence including the presence or absence of a terminal leaflet, the number of ovules per carpel, the structure of the fruit, the presence or absence of an aril, and pollen morphology. Although several revisions on the classification of the Sapindaceae have been made, the scheme of Radlkofer is essentially accepted with only minor modification. According to plant characteristics, pollen morphology, and geography, the Sapindaceae are grouped into two subfamilies: Dodonaeoideae (Austral distribution) and Sapindoideae. The latter can be separated into three main groups centred around Sapindeae (pantropical) or Cupanieae (pantropical) and a third group separating into Thiouinieae and Paullinieae, both predominantly American (Leenhouts 1971, 1978, 1986). The majority of the cultivated species in the Sapindaceae belong to Litchi, Nephelium, Dimocarpus, and Blighia, with their horticultural classification based largely on fruit characteristics (Tindall 1994). Minor attention has been paid to leaf and flower structures. Subsequently, Leenhouts (1971, 1978, 1986) and Choo and Ketsa (1991) reviewed the taxonomy of litchi.

According to Leenhouts (1978), there are three subspecies of *Litchi chinensis* based on the thickness of the twigs, arrangement of the flowers, number of stamens, and fruit characteristics. *Litchi chinensis* subspecies (subsp.) *chinensis* is the commercial litchi that grows wild in Southern China, Northern Vietnam, and Cambodia (Groff 1921). The tree possesses slender twigs and bears flowers in lax cymules.

The flowers usually have six stamens. The fruit is smooth or with protuberances up to 2 mm high (Groff 1921; Singh and Singh 1954; Menzel and Simpson 1990; Tindall 1994). The two other subspecies are not commercialized (Menzel et al. 1993). *Litchi chinensis* subsp. *philippinensis* is rarely cultivated although it is common and known in the Philippines and Papua New Guinea as alupag, arupag, or mamata. The taxonomic classification of *Litchi* is described as follows:

Kingdom	Plantae
Subkingdom	Tracheobionta
Superdivision	Spermatophyta
Division	Magnoliophyta
Class	Magnoliopsida
Subclass	Rosidae
Order	Sapindales
Family	Sapindaceae – Soapberry family
Genus	*Litchi* Sonn.
Species	*Litchi chinensis* Sonn.

8.4 Flower Biology

Litchi requires a period of cool weather (15°–20°C) for successful flower initiation. The flowering period in litchi varies with the genotype and environmental condition (Zhang 1997; Chen and Huang 2001; Ghosh 2001; Singh et al. 2012). Generally, the flower bud differentiation in litchi starts in December and is completed by the end of January (Singh et al. 2012). The flower panicle emerges from January to the end of February. Before floral bud initiation, the twig produces vegetative growth that subsequently throws a reproductive bud in the form of a racemose panicle (Menzel et al. 2000; Subhadrabandhu and Yapwattanaphun 2001; Stern and Gazit 2003; Menzel and Waite 2005).

Litchi bears a determinate panicle either terminally or axillarily composed of several branches produced on the current season's wood. The panicles are normally produced terminally in clusters of 10 to 20. The terminal panicles are more robust and productive compared to lateral panicles (Shukla and Bajpai 1974; Scholefield 1982; Naphrom et al. 2001; Singh et al. 2012). Litchi produces mixed panicles with the lowest buds producing leaves only, the middle buds producing floral buds in the axils of the leaves, and the topmost buds producing mostly floral branches (Menzel and Waite 2005). The number and percentage of different types of flowers in litchi vary with cultivar, environmental conditions, tree and panicle within a tree (Menzel and Waite 2005; Singh et al. 2012). The proportion of functional female flowers varies between 10% and 60%, depending on tree age. Each panicle produces several small, white, greenish or yellowish flowers (Zheng et al. 2001a, b; Batten and McConchie 1995; Menzel and Waite 2005; Singh et al. 2012).

8.4.1 Flower

The flower is terminal or auxiliary, arising from the axil of the first or second uppermost leaf of the new shoot. It bears tiny, greenish white or yellowish flowers and is branched racemosely. The inflorescence varies from 20 to 60 cm in length and from 15 to 30 cm in width. Litchi flowers measure 3–6 mm wide when fully open and the rest, on pedicels, measure approximately 1.5 mm. Flowers are apetalous and possess a cup-shaped calyx with four or five short, serrated sepals (Liu 1954; Costés 1988; Goren et al. 1998). Three types of flowers open in succession on the same panicle, and they vary in terms of degree of sexual development pertaining to length and functionality of the stamens and development and functionality of the pistil (Menzel and Waite 2005; Singh et al. 2012).

8.4.1.1 Male or Staminate Flowers (Type I or M1)

These flowers possess a structure which contains a pink, pubescent protuberance and very rudimentary pistil lacking both stigma and style (Joubert 1985; Costés 1988; Robbertse et al. 1995). The pistil is surrounded by 4 to 12 stamens (usually 8) with hairy filaments. The filament varies from 5 to 6 mm in length (Fig. 8.1a–c). In some flowers, the filaments are two to three times as long than the anther, whereas in some the filaments are equal to the length of the anthers (Liu 1954; Mustard et al. 1953). Type I is defined as the nonfunctional male, which opens for 10 days. Stamens are long with thin creamy white filaments which are deflexed and inserted into an angular glabrous nectary disc. Anthers are small, two-celled, elliptical, emarginated at apex, fixed and longitudinally dehiscent; initially pale white, they become brownish at maturity (Das and Choudhury 1958; Joubert 1985; Stern and Gazit 1996). During anthesis, the pollen sacs mature and the anther dehisces longitudinally and successively. Pollen grains are small and triangular, having nipple-like angles with three germinating pores. The ovary is very small, reddish white and abortive. The flowers possess nectar discs at their base, although they are not well developed (Menzel and Waite 2005; Singh et al. 2012).

8.4.1.2 Hermaphrodite Female Flower (Type II or F)

These flowers resemble the hermaphrodite male flower with a short filament except that the lobes of the stigma open down the vertical cleft (Fig. 8.1d–f). Called type II or F flowers, they appear and remain functional only for 2 days. These flowers have a small but well-developed pistil attached to a short peduncle (Joubert 1985; Costés 1988; Robbertse et al. 1995). The ovary has two to four carpels, each containing an ovule. The surface of the ovary is pubescent with protuberances persisting throughout the fruit development period. The ovary is long and composed of a short style with a bilobed, white and sticky stigma. The nectar disc at the base of the ovary produces abundant secretions, attracting insects. Generally, only one of the ovary lobes develops into a fruit; the rest abort. Occasionally, however, two lobes may develop, producing two fruits each containing a seed, which remain embedded at their bases (Das and Choudhury 1958; Joubert 1985; Stern and Gazit 1996). The pistil is usually surrounded by five to eight stamens with very short filaments (less

Fig. 8.1 Pattern of flowering in litchi. **a–c** Male flower (M_1) in full bloom with ripe pollen and unopened females. **d–f** Fully opened hermaphrodite female flowers (F) with well-developed stigma, style and rudimentary stamens. **g–i** Fully developed hermaphrodite male flowers (M_2) with rudimentary stigma and degenerated ovary in the centre of the stamens

than 1.5 mm long). The anthers do not normally dehisce and contain little viable pollen, thereby ensuring cross-pollination (Menzel and Waite 2005; Singh et al. 2012).

8.4.1.3 Hermaphrodite Male Flower (Type III or M2)

Type III is a functional male flower with more hermaphrodite features than the type I flower. Both stamens and pistil are present, but the lobe of the stigma fails to open to permit pollination. The filaments vary in length. Stamens arise in two forms, of which those with long filaments are fertile. The pistil consists of a short undivided and nonfunctional style. Type III flowers have an ovary with certain resemblances to both the aforementioned types (Fig. 8.1g–i). Anthesis occurs for 7 to 10 days. The flower is surrounded by six to ten stamens which are similar to those of type I flowers, possessing a rudimentary style, stigma and a nectary disc (Liu 1954; Costés 1988; Goren et al. 1998). Anthesis of the three flower forms usually occurs in succession on the same tree, usually on different panicles or branches. However, the first flowers to appear on any tree may not necessarily be type I. There seems to be a tendency in young trees for type II flowers to open first (Costés 1988; Menzel and Waite 2005; Singh et al. 2012).

The duration of flowering (anthesis to pollination) in litchi is between 20 and 45 days, depending on the cultivar, season and location. Anthesis occurs during both the day and night, with peak opening in the early hours of the day (0600). Temperatures below 8 °C suppress flower opening, but this occurs in the rainy and dry seasons. Under very dry conditions, the young flowers may dehydrate and fail to develop. High night temperatures (approximately 21 °C) reduce the duration of female flower anthesis (Mustard et al. 1953; Menzel and Waite 2005; Singh et al. 2012).

Anthers dehisce 1 day after anthesis and continue for up to 3 days, occurring continuously and more frequently between 0800 and 1000, with no apparent environmental, cultural or genetic effect. The pollen of type III flowers is more viable than that of type I flowers, which are, in turn, more viable than that of type II flowers. The stigma becomes receptive as soon as the lobes separate and remains receptive up to 3 days after anthesis. Maximum receptivity (75%) occurs 1 day after anthesis. When the stigmas cease receptiveness, they become dry and brown and lose their glossy appearance. Abundant and maximum viable pollens are produced by M2 flowers. Estimated pollen viability ranges from 4% to 40% at the time of pollen release and decreases rapidly thereafter (Wang and Qiu 1997; Goren et al. 1998; Menzel and Waite 2005; Singh et al. 2012).

8.4.1.4 Sex Ratio

Sex ratio in litchi varies with cultivar and the environment. The proportion of F, M1, and M2 flowers ranges from 1.00:0.16:0.32 ('Mountain Litchi') to 1.00:2.30:3.90 ('Brewster') litchi in Florida (Mustard et al. 1953) and from 1.00:0.16:0.90 ('Calcuttia Late') to 1.00:1.60:2.40 ('Dehradun') in India (Chadha and Rajpoot 1969). In 'Mauritius' the ratio varies from 1.00:1.10:1.10 in Israel (Stern et al. 1993) to 1.00:1.20:1.50 in Reunion (Costes 1988). Chaturvedi (1965) observed 32% female flowers in 'Early Large Red' in India; the proportion of female flowers in different cultivars at several parts of Queensland, Australia, varied from 16% to 43%, with the extent of variation as much as observed within cultivars (Menzel and Simpson 1992). 'Fay Zee Siu' in China produced a high proportion of male flowers in long inflorescences, whereas a high proportion of female flowers was noted in short inflorescences (Wang and Qiu 1997). Pruning the inflorescence 6 weeks before anthesis increased the female:male ratio and initial fruit set (Wu et al. 2001). Sex ratio is also affected by temperature. When five cultivars of age 2 years were kept at 15 °/10 °, 20 °/15 °, 25 °/20 °, or 30 °/25 °C, the highest proportion of female flowers (72%) was noted from the lowest temperature regime. At high temperature (25 °/20 ° or 30 °/25 °C), 'Bengal,' 'Souey Tung' and 'Wai Chee' fail to set female flowers (Menzel and Simpson 1991; Menzel and Waite 2005; Singh et al. 2012).

8.5 Pollination

Litchi is a highly cross-pollinated crop. Although anthesis of functional male and female flowers coincides, the presence of sterile pollen grains necessitates cross-pollination (Dhaliwal et al. 1977; Du Toit 1994; King et al. 1989; Butcher 1957; Pandey and Yadava 1970). The nectar produced by flowers attracts pollinators such as honey bees, flies, ants and wasps, facilitating cross-pollination (Stern and Gazit 1996). A nectary disc occurs on every flower as a large fleshy crenulate gland within the calyx to which the stamens and pistils are inserted. Nectar, secreted only in the morning, is highly attractive to honey bees and flies (Das and Choudhury 1958; Chadha and Rajpoot 1969; Wang and Qiu Wang and Qiu 1997). At the time of flowering, installation of honey bee boxes in a litchi orchard increases the yield up to 30–40%. Pollinators forage primarily between 0600 and 1200, although foraging continues later in the day at much lower levels (Stern and Gazit 1996).

Honey bees are the major pollinators in litchi (Fig. 8.2). Other pollinators include screwworm, ants, wasps, coleopterans, hemipterans, homopterans, and lepidopterans. Pollinator activity is affected by extremes of temperature, cloud cover, heavy rain and strong winds, and in the presence of insecticides. The presence of 10 to 12 standard bee colonies is sufficient to ensure good pollination and fruit setting in a 1-ha area (Menzel and Waite 2005; Singh et al. 2012).

8.6 Fruit

Botanically, litchi fruit is pendant in a loose cluster (up to 0.6 m long) of several dozen fruits. Fruits are round, ovoid, heart shaped or even kidney shaped and vary in size depending on the cultivar but can measure up to 5 cm in length and up to 4 cm in diameter. The fruit possesses shoulder, suture, apex and skin segments as observed externally (Deng et al. 1999; Huang et al. 2005a). Different parts of the fruits are briefly described in following paragraphs (Menzel and Waite 2005; Singh et al. 2012).

8.6.1 Epicarp

The rind of the fruit (pericarp) is thin, tough, hard, brittle and frangible. At maturity, its colour changes from green to bright red. The intensity of redness varies among cultivars, and fruits may even be yellow or green depending upon cultivars. The skin possesses prominent protuberances but is less marked than in related species such as rambutan and pulusan. When overmature, the skin assumes a dirty brown colour and loses its extensibility (Huang et al. 2005b); this occurs rapidly, even when the fruit is still perfectly edible (Menzel and Waite 2005; Singh et al. 2012).

Fig. 8.2 **a–d** Pollinators collecting nectar from the calyx gland of litchi flowers. **e** Flower drop in bagged and unpollinated litchi panicles

8.6.2 Mesocarp

The mesocarp tissue has a loose consistency without clear demarcation between the epicarp and the mesocarp. In the initial stages, this tissue is composed of ordinary parenchymatous cells which become separated with time as large intercellular spaces develop. Botanically, the edible part of the fruit is known as the aril. In a developing fruit, the aril grows continuously as an outgrowth of the outer cells of the seed coat (outer integument); the rate of outgrowth varies greatly among cultivars (Huang 2005). The aril is translucent, white, slightly acid, juicy and sweet, with a faint and pleasant aroma. Many cultivars of litchi can be distinguished by their flavour and aroma (Menzel and Waite 2005; Singh et al. 2012).

8.6.3 Endocarp

Two distinct tissues have been noted in the endocarp. Lying next to the aril is a dual layer of rectangular cells which are considerably thickened and lignified; just above it is the tissue composed of thick, elongated parenchymatous cells with tapering walls (Huang 2005; Menzel and Waite 2005; Singh et al. 2012).

8.6.4 Seed

The fruit contains a single dark brown seed 6 to 12 mm wide and 10 to 23 mm long. Some cultivars have a high proportion of aborted seeds (chicken-tongued seeds) which are shrivelled and nonviable (Huang 2005). The seed is surrounded by a creamy and pulpy edible aril. It is cylindrical, compressed piano convex or concavo convex, exalbuminous and chocolate in colour (Menzel and Waite 2005; Singh et al. 2012).

8.7 Leaf

Litchi produces compound leaves with 7 to 12 lamellae, alternate, coppery red or red to pale green, creamy when young and bright and dark green when matured. Leaf shape can be lanceolate, oblong lanceolate or even elliptical, measuring 7.5–20 cm in length and 2.5–6 cm in width (petiole and rachis), smooth but coriaceous in texture with rounded base. Eight to ten pairs of leaflets may be arranged along the rachis directly or opposite to each other, with petiole length ranging between 3 and 25 mm. Leaflets are glossy with short internodal length, dark green adaxially and dull and waxy on the abaxial surface. Low-vigour varieties have small leaves with shorter internodal length. In some genotypes, young flushes tend to produce attractive reddish-bronze-coloured leaves (Menzel and Waite 2005; Singh et al. 2012).

8.8 Cytogenetics

There is a dearth of information on the aspects related to the cytogenetics of litchi (Liu 1954; Chapman 1984) and this has not drawn much attention among litchi researchers. The species is probably a natural hybrid involving more than one wild progenitor. Variable haploid chromosome numbers of 14, 15, 16, and rarely 17, have been reported in litchi, thus suggesting its polyphyletic origin (Chapman 1984; Sarin et al. 2009).

8.9 Germplasm Collection and Conservation

Historically, litchi cultivar breeding relied primarily on seedling selection. To date, more than 300 cultivars have been planted in the National Litchi Germplasm repository located in Guangzhou, China, established in 1998, which is the largest litchi germplasm gene bank in the world (Wu et al. 2007). Globally, standardized protocols are being followed for cryopreserved tissues and excised embryos, recalcitrant seeds, and vegetative propagated plants of litchi for long-term storage at −196 °C. Besides seed banks, in vitro gene banks have also been established for germplasm conservation.

Selection of improved cultivars in litchi has been practised for thousands of years in Asia. Fay Zee Siu, Bah Lup, Lanzhu, Baitang-ying, Haak Yip, Kwai May (Red), No May Chee and Wai Chee, the main cultivars evolved in China, form the basis of the litchi industry in many other countries; for example, Tai So and Wai Chee in Thailand; Tai So, Kwai May Pink and Wai Chee in Australia; and local seedling selections of Chinese origin in Vietnam, India, Nepal, Bangladesh and Southern Thailand. Breeding effort in the past 50 to 60 years led to the development of popular seedless or small-seeded Dongguan Seedless and Hexiachuan and late maturity Maguili in Guangdong (China), SahKeng in Taiwan (China), Kom and Chacapat in Thailand, UPLB Red in the Philippines and Salathiel in Australia. There is a great potential for improving productivity through breeding new selections, with emphasis on traditional breeding rather than on biotechnology Bose (2001). According to the Chinese, litchi has more cultivars than any other fruit Bose (2001). Bah Lup, Baitang-ying, Haak Yip, Fay Zee Siu, Kwai May, No Mai Chee and Wai Chee are the important cultivars in Guangdong with Wai Chee accounting for more than 80% of the area in Guangxi. However, in Fujian, Lanzhu is the predominant cultivar. No Mai Chee and Kwai May with high proportion of chicken-tongued or aborted seeds and Fay Zee Siu having good fruit size (24–32 g) with excellent eating quality Bose (2001). Haak Yip dominates in the Taiwan Province of China and accounts for more than 50% of the area. Other important cultivars include Sum Yee Hong, Chong Yun Hong, No Mai Chee and SahKeng Bose (2001).

In Vietnam, 80% of the area is occupied by Vaithieu Bose (2001). Tai So (Hong Huay) is the main cultivar in Northern Thailand, besides Wai Chee, O-Hia (Baidum) and Chacapat (Chakrapad). Kom, Luk Lai, SampaoKaow, KalakeBaiYaow and Red China are ecotypes developed for Bangkok regions but of low fruit quality. In India, most cultivars are of seedlings origin introduced from China. Selections of more than 30 cultivars are grown in this region, but only Shahi (Muzaffarpur), China, Calcuttia, Bedana, Late Bedana and Longia, possessing large fruits of excellent quality, are commercially important. In West Bengal, Bombai, Shahi and Rose Scented yield up to 40 kg/tree compared to 15–25 kg/tree in many other cultivars Bose (2001). In Nepal, Majfpuri, Raja Saheb, Dehraduni, China and Calcuttia are the established cultivars probably introduced from India, whereas commercial production in the hilly tract is primarily based on seedling populations. Bombai, Muzaffarpuri, Bedana and China Number Three are the popular cultivars of Bangladesh. In the hilly areas of the Philippines, Mauritius and a local selection from China, Sinco, dominates production, and an introduction from Thailand, UPLB Red, is planted in the lowlands. In Australia, Kwai May Pink accounts for more than 50% of the area, besides Tai So, Souey Tung, Fay Zee Siu, Salathiel and Wai Chee Bose (2001).

Litchi has a narrow genetic base. However, ecological variations do exist in litchi, bringing the total of cultivars to more than 60, although commercially only a few varieties such as Shahi, China, and Bedana are popular. The National Active Germplasm Site at NRCL maintains more than 50 such variants, 20 clonal variants, and a large number of seedling populations arising from various parental combinations. The crop improvement programme in litchi started back in the 1950s at Sabour and Ranchi, resulting in promising cultivars such as Swarna Roopa, Sabour Bedana, and Sabour Priya. The main target of crop improvement in litchi is the integration of such characters as larger fruit size, smaller seed, high pulp ratio and acceptable pulp quality. In clonally propagated species, genetic characterization of clones held in gene banks is important for proper documentation and assessment of variability in the collection, thereby enhancing the effectiveness of gene conservation efforts. The pulp ratio of the existing varieties is moderate with associated problems related to irregularity in bearing, incidence of fruit cracking, and susceptibility to pest and disorder, thus affecting the overall yield potential. Development of improved cultivars coupled with efficient cultural practices is vital for improving crop production in a region. Improved varieties of litchi have been identified and evaluated for their potential from ICAR-NRC on Litchi, Muzaffarpur. The striking features of these varieties have been described.

Shahi Shahi is the most popular cultivar, grown in North Bihar, Jharkhand, Uttarakhand and Uttar Pradesh. Shahi fruits are known for excellent aroma and quality with 65–70% pulp recovery. Fruits are globose, heart or obtuse in shape; pulp is greyish white, soft, moderately juicy and sweet [19–21° B total soluble solids (TSS)]; seeds are large but 6–8% small shrunken seeds are produced. Besides high-quality fruit, Shahi possesses a rosy flavour and heavy fruit weight (22–26 g), earliest in maturity (20–30 May); this is the first variety to reach the market. Tree size is large with high-yield potential (140–150 kg^{-1} tree) and fruit is moderately prone to cracking (Fig. 8.3a).

China This variety has a very high yield potential and is tolerant to heat waves and fluctuations in soil moisture. It is medium to late in maturity (30 May–10 June), with comparatively compact trees and high-yield potential (150–160 kg^{-1} tree), but prone to alternate bearing. Fruits of China are large (22–25 g), oblong in shape and tyrant rose in colour with dark tubercles at maturity. The aril is creamy white, soft, juicy and sweet (18–19° B); seeds are large and less prone to cracking with good flesh recovery (60–67%). The maturity period of this variety coincides with pre-monsoon showers (Fig. 8.3b).

Swarna Roopa A late-maturing, cracking-resistant cultivar selected at Ranchi has attractive red-coloured fruits with small seeds and high (65–70%) aril content; fruits are medium in size weighing 15–17 g and have a high pulp content. The pulp contains high TSS and low acidity. The cultivar is suitable for extended harvest as it matures after China and is prized for its attractive fruit colour.

Purbi Purbi is mostly grown for table purposes in the eastern part of Bihar. Fruits are medium large, oblong conical in shape, and ripen at the end of May or first week of June. At maturity red tubercles appear on the pinkish-brown background. The average yield is 100–120 kg/tree.

Fig. 8.3 Important litchi varieties of ICAR-NRC on litchi, Muzaffarpur: Shahi (**a**), China (**b**), GandakiLalima (**c**), Gandaki Yogita (**d**), and Gandaki Sampada (**e**). (**f**) Seedlessness in litchi resulting from embryo abortion

Gandaki Sampada A late-maturing strain ripens during mid-June. Fruits are large in size, conical in shape and vermilion to carmine in colour at maturity with dark blackish-brown tubercles (Fig. 8.3e). Pulp is creamy white, soft and juicy, with large fruit size (35–42 g), cracking resistance, pleasant aroma and good yield potential (120–140 kg^{-1} tree). It has a very high percentage of shrivelled and small-seeded fruits, resulting in 80–86% pulp recovery. This variety has potential for export purpose.

Gandaki Yoita This is a dwarf plant, comparatively compact, and tolerant of heat waves and fluctuations in soil moisture. The fruits are free from fruit borers and are of very late maturity (5–15 June). Fruits are round in shape, tyrant rose in colour with dark tubercles and flexible seeds at maturity. The aril is creamy white, soft, juicy and sweet; fruit size ranges from 15 to 19 g with small seeds (Fig. 8.3d). Pulp recovery is 70–75%, possessing melting texture with pleasant aroma and a good blend of sugar and acid. This variety has good yield potential (70–80 kg^{-1} tree) and can be recommended for high-density planting as a speciality variety.

Gandaki Lalima Trees are moderately vigorous, attaining an average height of 10–12 m and almost uniform spread. It is the most nutrient-efficient strain having dark green leaves and the capacity to withstand climatic aberrations. Fruits are conical, bright marigold-orange red in colour with sharp tubercles. A late-maturing cultivar ripens in the second week of June; a heavy yielder has an average yield of 130–140 kg per tree and the fruit weighs between 28 and 32 g. The fruit pulp is creamy white in colour, sweet (18–19° B TSS), soft and juicy with agreeable flavour. Seeds are moderate in size and pulp recovery is above 60% (Fig. 8.3c).

This variety can offer a better alternative for commercial exploitation for export of litchi from India.

8.10 Breeding for Litchi Improvement

8.10.1 Breeding Objectives

Litchi, being a cross-pollinated crop, has a high degree of heterozygosity which does not produce true-to-type plants through sexual means. Most commercial cultivars clonally selected under Chinese or Indian conditions have been adapted to limited climatic conditions (Pandey and Sharma 1989). Clonal selections as the basis of cultivar selection rely on limited characters such as fruit size, quality, and period of maturity. Creation of variability within the litchi gene pool is of paramount importance for desirable characters such as precocity, dwarfness, regularity of bearing, wider adaptability in tree characteristics and resistance to physiological disorders. Cross pollination results gametic variations which serve as the baseline for new selections through genetic manipulation; this stresses the need for initiating a planned breeding programme and raising plants from seeds (Sarin et al. 2009; Singh et al. 2012). The literature reveals that less attention has been paid to breeding work for raising new varieties; although programmes were initiated and conducted sporadically at Hawaii (Storey et al. 1953) and Florida, USA (Knight 2001), Queensland, Australia (Cull 1977), and Saharanpur, India (Lal and Nirwan 1980). As a result, important cultivars such as 'Groff' and 'Brewster' were developed. A strategic effort for a breeding programme on a large scale is needed with consideration of a comprehensive survey of various genotypes and trait inheritance patterns because of the obvious difficulties in litchi breeding (Sarin et al. 2009; Menzel and Waite 2005; Singh et al. 2012).

8.10.2 Inheritance Pattern or Linkage

Studies on the trait inheritance pattern in litchi have been very limited, complicating our understanding of the segregation pattern of genetic traits (Galen Sauco 1989). Differentiating cultivars through genetic markers revealed that most cultivars shared similar parentage. Variability among litchi cultivars is still unknown because breeding for new cultivars is done by growers and is based on a small number of parents (Kumar et al. 2006). However, high heritability and high genetic advances were recorded for fresh seed weight, fruit weight, fruit volume and fresh and dry pulp weight. Fruit and seed weight had strong positive correlations with total sugars, ascorbic acid, protein and tryptophan content but a significant negative correlation with acidity and phenol content (Singh et al. 1987). A negative partial correlation between embryo and aril and a direct repressive effect of the former on the latter were confirmed. These results may be taken into account when breeding varieties resistant to cracking. In nature, the extent of outcrossing in litchi varies from 65% to 87% depending on the proximity of the pollen source. Inbreeding depression occurs on selfing with respect to fruit and seed weight (Stern et al. 1993).

Understanding the genetic diversity within a collection is of prime importance in facilitating germplasm utilization (Strauss et al. 1998). To identify genetic materials containing desirable traits, a systematic evaluation of genetic diversity is required to understand the genetic structure and relationships among accessions and their corresponding collecting site environment (Steiner and Greene 1996). Comparison of parents using different DNA markers is one of the approaches by which breeders can increase the probability of selecting parents with different gene sets. Cultivars with high percentages of aborted seeds (chicken tongue) are highly sought after in litchi. Even within the same cultivar, some trees consistently produce higher percentages of chicken-tongued seeds, and this stresses the needs to determine whether this phenomenon is attributable to environmental conditions or to genetic variation (Madhou et al. 2010).

8.10.3 Breeding Strategies

Up to the present, the majority of litchi breeding work has been attempted through conventional methods with minor attention paid to modern biotechnological tools. Reducing the breeding period in evolving new improved cultivars is very desirable from a breeder's view. Breeding methods combined with novel technologies including genetic engineering, in vitro mutagenesis and molecular assisted breeding would assist litchi breeders to develop new cultivars in a cost-effective manner.

8.10.4 Conventional Breeding

Seeds are the main source of variability. However, the lengthy juvenile period associated with seedling populations hinders the breeding progress. Breeding programmes have been undertaken in several countries within the past two decades. One was initiated during 1992–1993 at the Institute for Tropical & Subtropical Plants (ITSP), South Africa, for the purpose of cultivar selection for South African conditions (Froneman and Oosthuizen 1995). In China, breeding programmes and selection of cultivars with longer duration of bearing period and good-quality fruits have been initiated (Huang et al. 2005a). A seedless litchi variety was produced by conventional breeding in Hainan, China (Sarin et al. 2009; Menzel and Waite 2005; Singh et al. 2012).

Reciprocal controlled crosses among several existing cultivars have been reported in Australia (Dixon et al. 2005). Most seedlings bear fruit within 3 years under subtropical and tropical conditions, and one promising genotype producing 32 kg of fruits with an individual fruit weight of 35–45 g was identified. According to Miao et al. (1998), four new litchi selections were recommended for cultivation in Hainan Province of China. Of these, Aili is a dwarf selection producing an average fruit weight of 24.8 g, and Ziangxi has large fruits (39.1–59.5 g) with acceptable eating quality. Apart from yield and quality characters, there is still a great scope for evolving cultivars that are less prone to cracking and browning as these are directly related to fruit quality. Extension of bearing period, irregular bearing, poor shelf life and disease resistance are the other significant areas to be addressed in future programmes on litchi breeding (Sarin et al. 2009; Menzel and Waite 2005; Singh et al. 2012).

8.10.4.1 Crossing Techniques

Litchi requires cross-pollination, carried out primarily by bees. Although self-sterility has been reported in litchi, bagging of individual panicles is necessary to avoid outcrossing. Male flower development of litchi was defined in three stages. In the first stage, the anther extended just beyond the receptacle, and the filaments were not visible. In the second stage, the filaments extended by half. In the third stage, the filaments extended further before the anthers dehisced. The maximum number of mature pollen grains was found in the third stage before anther cracking (Wang et al. 2015); however, exclusive collection was inefficient because anther development in the same male flower was asynchronous. Therefore, anthers between stage 2 and stage 3 can be collected to ensure sufficient pollen for a crossing programme. For effective crossing, flowers should be collected at the second stage (half-extended filaments light yellow in colour), and at the third stage, bags are removed and fresh male flowers are collected when the anthers turn light yellow in colour. The pollen can be collected either directly from freshly dehisced anthers, if used immediately, or from indehiscent mature anthers when it will be stored for a day or two. Anthers are first separated from flowers by passing them over a screen, which allow them dehisce overnight in petri plates at room temperature. Moist cotton is placed inside the petri plates for obtaining fresh, quality and abundant pollens used for crossing. Pollen obtained in this way can be used directly in crosses and remains viable for 3–5 days at room temperature. It can remain for several weeks if refrigerated under low relative humidity. For long-term storage, litchi pollen can be stored at −86 °C for 2 years. Litchi pollen stored at 4°C remained viable after 52 days, which may meet the demand for artificial supplementary pollination in production, but the demand for artificial cross-breeding is not completely met (Wang et al. 2015).

8.10.4.2 Emasculation and Pollination

To emasculate flowers, scissors are used to remove excess flowers; only 10–15 flowers per panicle are retained for effective crossing, and flowers can be pollinated on the same or the next day. By dipping a small brush into vials or petri plates, pollens are placed on the stigma. Holding the dehisced anthers by their filaments using forceps can also be adopted to transfer pollen. After pollination, flowers are bagged. Successful fertilization occurs after a period of several days.

8.10.4.3 Technique/Procedure of Artificial Hybridization/Artificial Crossing

M_2 pollen has a higher germination rate over a wide range of temperatures. Hand pollination with M_2 pollen produces a satisfactory fruit set compared to M_1 pollen. Hot (32/27 °C) and warm (27/22 °C) temperature regimes during flower development had a pronounced effect on pollen viability compared with cool regimes (22/17 °C) (Stern and Gazit 1998). The pollen germination percentage varied greatly among cultivars and was also significantly affected by weather conditions before the opening of staminate flowers, especially during the anther development stage (Xian et al. 1994).

Fruit and leaf characters are widely used to identify litchi hybrids, although it is known that phenotypic traits can be influenced by environmental conditions and are not reliable indicators (Anuntalabhochai et al. 2002). Hence, along with morphological characterization, molecular studies are essential to confirm the identity of litchi cultivars for optimum germplasm management and establishment of appropriate breeding programmes.

8.10.4.4 Germplasm Evaluation and Clonal Selection

Because many of the existing cultivars originated from a relatively limited ancestral stock, the introduction of new germplasm from wild forms and varieties into the genetic composition of existing cultivars is of paramount importance for achieving the breeding objectives. The important genetic resources of litchi identified for different characters are listed in Table 8.1.

At present, most cultivars grown are of Chinese origin, and the genetic base of the commercial cultivars is relatively small. the majority of the cultivars arose through clonal propagation of high-performing parents. Wild forms or types of the three known litchi subspecies have been widely collected but have been minimally used for integration in breeding programmes, and genetic enhancements are largely based on collections of commercial cultivars selected from open-pollinated Chinese seedlings.

8.10.4.5 Hybridization

Selection for high-yielding and good-quality litchi has been attempted over a long period, but the breeding of new hybrids has not yet attained a significant impact. Recent breeding efforts at Sabour (India) led to the development of two hybrids, namely, Sabour Madhu and Sabour Priya (Table 8.2). The growth of the hybrid seedlings is slow, and only 4% of the total population reached maiden flowering at the age of 14 years (Thakur 1992). Thus, in addition to a short period of seed viability, the late-bearing habit of the seedlings poses serious problems for hybridization work. Besides, these hybrids showed erratic flowering which made it difficult to obtain pollen for breeding purpose.

8.10.4.6 Distant Hybridization

Hybridization was also attempted between litchi and longan, and seedling progenies were variable with small fruit size, which appeared to be a dominant

Table 8.1 Special characteristics of some of the litchi cultivars

Special traits	Source of traits
Small seed	Bedana, Nuomici, Lingshan Xiangli, Hainan Xiaodingxiang, Guangxi Zhangluoli
Crisp and sweet flavour with low tannin	Lingshan Xiangxi
Early maturity	Shahi, Shanyuedong Red, Early Bedana, Dehra Rose
Large fruit	Edanli (60–70 g), Vnsdasu (40–45 g)
High yields	Heiye (Black Leaf), Baitang-ying (White Super Poppy)
High temperature tolerance	Shanyehong: better flower bud differentiation at 20 °C as compared to 12 °C to common varieties
Good for canning purpose	Heiye, Xuangxi
Drought tolerance	Tianyan (Sweet Stone)
Good on-tree storage	Huaizhi: fully ripe fruits can be left on trees for fresh picking
Late maturity	Xuehuazi: ripening during early/mid-July, with high-yielding characteristics; Fijian Xiafanli (Fujian Xiafan litchi): maturing during late July or early August, enabling extended supply period in combination with early and intermediate maturity types; Longia, Kaselia, Yogda

Table 8.2 Litchi cultivars developed through hybridization

Number	Hybrids	Description
1	Sabour Madhu	This hybrid resulted from Purbi × Bedana. It has a higher number of fruits (24) per panicle and ripens 8 days later than another late-maturing cultivar, Kasba. It has higher TSS and aril percentage than Purbi. Fruit shape resembles Purbi.
2	Sabour Priya	This is a product of Purbi × Bedana. It has better fruit quality than Purbi in terms of higher aril percentage and TSS content. The fruit shape is intermediate between both parents. The fruit weight is higher than the better parent (Purbi).

character. Attempts on diallele crosses between *Nephelium lappaceum*, *N. rambutanakee*, *Dimocarpus longan*, and *Litchi chinensis* were made, and only intergeneric crosses between longan and litchi were successful. The pollen of both litchi and longan germinated on stigma of rambutan but was arrested in the embryo sac. However, there appears to be no breeding barrier between cultivars or species within a genus except when seedless fruits are commonly produced (Fig. 8.3f).

McConchie et al. (1994) attempted reciprocal crosses between commercial cultivars of litchi (Bengal and Kwai May Pink) and longan (McLeans Ridges and Duan Yu) and found that hybrid progeny developed only when litchi was used as the female parent. Morphologically the hybrid plants were similar to litchi, but the leaves were smaller. Three types of seeds developed in litchi following pollination with longan pollen: (1) normal seeds with well-developed testa and embryo, (2) seeds with aborted embryo but normal testa development, and (3) seedless, wherein the ovule development is arrested.

8.10.5 Characterization of Litchi Based on Morphological Traits

Litchi cultivars were classified into four categories according to fruit shape: long oval, heart shaped, short heart shaped and round (Li and Fang 1956). Guangdong litchi cultivars were grouped into seven types in 'The Flora of Guangdong Litchi' (Guangdong Academy of Agricultural Sciences 1978). These types include Guiwei, Xiaozhi, Jinfeng, Sanyuehong, Heiye, Nuomici and Huaizhi, based on the shape of tortoise shell-like cracking segments and pericarp with or without sharp protuberances as well as the shape of leaf, inflorescence and fruit, maturity time and quality of fruit. Later, in 1986, litchi cultivars were separated into three types and seven groups according to a four-rank classification criterion considering pericarp, fruit shape and other morphological characteristics. These classification principles were adopted in the subsequent 'The Flora of Chinese Fruit Tree, the Volume of Litchi' (Wu 1998b); Wu et al. (2016) classified litchi cultivars into three types based on the smoothness or roughness of the pericarp. Morphological characteristics such as tree height, canopy spread, tree shape, foliage texture and colour, leaf length, width, shape and orientation, intermodal distance, number of leaflets per leaf, number of leaves per flush, flush colour, panicle length, number of anthers and carpels per flower, filament and style size and fruit colour and size have been used to quantify genetic diversity in litchi (Khurshid et al. 2004).

8.10.6 Characterization of Litchi Based on Isozyme Analysis

Isozyme markers were effective in differentiating litchi accessions. Ambiguities resulting from synonyms and homonyms and sexual propagation of accessions have been resolved to some extent by comparing isozyme fingerprints of different accessions. However, an authentic collection and evaluation of native Chinese cultivars would help in establishing the identity of the present litchi collection. Liu et al. (1989) and Zhou et al. (2001) analysed the PER isozyme of 24 Guangxi litchi cultivars and 35 Guangdong litchi cultivars and produced almost identical results (Wu et al. 2007). Liu et al. (1989) also reported the stability of the litchi PER isozyme pattern as unaffected by the age of leaves. Aradhya et al. (1995) reported genetic diversity among 49 litchi (*Litchi chinensis* Sonn.) accessions using eight enzyme systems encoding 12 loci and revealed moderate to high levels of genetic variability. Comparison of isozyme fingerprints revealed different isozyme patterns in 'Nomai tsz', 'Kwai mi' and 'Hak ip,' believed to be synonyms, whereas some others with different names displayed identical patterns. Wu et al. (2016) reported the usefulness of morphological traits of leaf and branching to discriminate 146 Chinese litchi germplasms.

8.10.7 Characterization of Litchi Based on Molecular Markers

Molecular studies on genetic diversity of litchi are limited. The presence of variable litchi cultivars in China and India provides a good basis for development of new cultivars (Groff 1921). Recent development in molecular tools further enhances our efficiency in understanding the genomic variability and the diversity between and within different species of litchi.

Initially, studies on genetic diversity in litchi have been largely based on morphological characters. Development of molecular markers in the past few decades yielded accurate and precise information on the extent of genetic diversity in litchi. The most commonly used DNA marker is random amplification of polymorphic DNA (RAPD) (Ding et al. 2000; Chen et al. 2004, 2005; Liu and Mei 2005; Wang et al. 2006; Bajpai et al. 2016), followed by amplified fragment length polymorphism (AFLP) (Yi et al. 2003; Peng et al. 2006) and simple sequence repeat (SSR) (Li and Zheng 2004). However, in most of the tested litchi cultivars, contradictory results were obtained from different markers, and even those obtained from the same DNA marker are not totally identical. Results of RAPD and AFLP analysis indicated a low level of genetic diversity within litchi collections; several workers characterized litchi cultivars using molecular markers and reported the occurrence of synonyms (Ding et al. 2000; Yi et al. 2003; Chen et al. 2004) such as 'Nongmei No. 9' and 'Qiongshan No. 27' (Chen et al. 2004), 'Dazao' and 'Zaohong,' 'Baiye' and 'Guahong,' 'Feizixiao' and 'Zhimali,' 'Ziniangxi' and 'Zengchengdaguoli' (Liu and Mei 2005) and 'Fengshuang' and 'Tunchangfengshuang' (Wang et al. 2006). However, Ding et al. (2000) reported that 'Dazao' and 'Zaohong' are two entirely different cultivars. Viruel and Hormaza (2004) obtained 12 microsatellites enriched in CT repeats from a genomic library of the litchi cultivar 'Mauritius.' Assessment of genetic diversity of litchi at Kalyani, West Bengal (India), revealed the segregation of cultivars into two clusters which can give heterotic hybrids when intercluster cultivars are crossed (Dwivedi and Mitra 1995, 1996). Recently, Liu and Mei (2005) reported an average of 15.8% polymorphic and 0.10% monomorphic RAPD markers from a population comprising 60 litchi cultivars, one longan cultivar, and one tentative intergeneric hybrid of litchi and longan. Two accessions (LH80 and LH109) were genetically distinct as revealed by RAPD and AFLP markers. In Thai litchi cultivars, the percentages of polymorphic markers for RAPD and AFLP were 34.6% and 36.3%, respectively (Tongpamnak et al. 2002), and each marker system was able to differentiate all accessions. RAPD markers have been widely used to study the genetic relatedness among litchi cultivars (Kumar 2006); Ding et al. (2000) investigated the segregation patterns of RAPD markers in an F_1 population of litchi from a cross between 'Wuye' and 'Luhebao.' Bajpai et al. (2016) reported genetic diversity of 20 litchi cultivars from the Indian peninsula, and phylogenetic analysis based on RAPD and microsatellites revealed clustering of the cultivars into four major groups, although within a very narrow range (0.63–0.90) of similarity, viz. Seedless (i.e., Bedana), Mandarji, Shahi and China groups. Simple sequence repeats (SSRs) markers previously developed for litchi have also been evaluated for polymorphism in different populations (Ekuae et al. 2009). The Power Core programme based on 30 EST-SSRs and a combined dataset of the EST-SSRs and 16 phenotypic traits has been used to construct two core collections from 96 accessions (Sun et al. 2012). Madhou et al. (2013) conducted molecular characterization study of litchi accessions from Mauritius and Reunion and compared them with Spanish litchi cultivars. Liu et al. (2015) obtained the potential of single nucleotide polymorphisms (SNP) for the identification of 96 representative litchi accessions, and their genetic relationships in China were evaluated using 155 SNPs that were evenly spaced across the litchi genome. Ninety SNPs with minor allele frequencies above 0.05 and a good genotyping success rate were used for further analysis. A relatively high level of genetic variation was observed among litchi accessions, as quantified by the expected heterozygosity (He =0.305).

8.10.8 Mutation Breeding

So far, induced mutations have not been in vogue in litchi improvement. It was reported that seedlessness in the litchi cultivars Lanzhu and Luhebao might have resulted from mutation, gene interaction and long-term artificial selection and is not the result of variation in chromosome number or structure (Lu et al. 1987). Qinzhou red litchi, a new variety of litchi, was derived from spontaneous mutation of the cultivar Black Leaf (Peng et al. 2001).

8.10.9 Transgenic Approach

Green fluorescent protein (GFP) gene expression in leaf tissues of litchi after transformation is using *Agrobacterium* (Puchooa 2004a). In vitro, grown leaf tissues were used for transformation. After 4 weeks in culture, expression of GFP was apparent when the regenerated callus and the leaves were observed under a fluorescence microscope fitted with a blue exciter filter, a blue dichroic mirror and a barrier filter. Although no transformed litchi plantlets were regenerated, screening for GFP gene expression may prove useful to improve transformation efficiency and to facilitate the detection of transformed litchi plants. Ouyang et al. (1985) reported T-DNA transfer and tumour formation induced by *Agrobacterium tumefaciens* on litchi. Therefore, genetic transformation of litchi using *Agrobacterium* could be exploited in the future.

8.11 Constraints in Litchi Breeding

The development of improved cultivars in litchi is a lengthy and cumbersome process, and trees when fruited only produce less than 1% of the seedlings worthy of selection. Breeding for a new hybrid cultivar through conventional methods would require about 40 years in fruit trees such as litchi (Zheng et al. 2001). In addition to poor seed viability and wide variability in seedling progenies, erratic flowering poses a hindrance for pollen collection during the breeding process. Although crosses between litchi and longan are successful, the use of longan as parents in a hybridization programme is limited by its biennial bearing tendency. In crosses involving plants that have a tendency to produce chicken-tongued seeded fruits as the female parent, many of the most valuable progeny are lost before the harvest (Puchooa 2004b). Only a few cultivars have been bred through conventional methods because of the long juvenile period, the apparent lack of genetic variability in the existing germplasm, and the great expenditure that is required in terms of land, time and money (Litz et al. 2005). Thus, the improvement of litchi appears to be confined mainly to selections of improved chance seedlings or genotypes. Future efforts in plant breeding need to emphasize identification of potential parents followed by their reciprocal crossing.

8.12 Future Prospectives

Despite constraints and the long time frame needed for conventional breeding, opportunities do exist to develop new cultivars with improved traits that can improve productivity and fruit quality against the endogenous and exogenous attack of pathogens, mites and insects. New ideotypes through appropriate blending of traditional breeding and biotechnological tools can be used in the future for disease management in the litchi system. Recent and future development in high-throughput MAS can substantially improve the efficiency of the conventional litchi breeding to attain the resistance in susceptible litchi crops. Application of molecular markers in the creation of a genetic map and other pre-selection techniques has the potential to substantially reduce selection time frame in litchi breeding programmes. Identification and cloning of genes of horticultural interest from litchi and its wild relatives will enhance the rate of conservation of plant survival from germinated transgenic somatic embryos and would certainly revolutionize the genetic improvement of litchi cultivars. The advantage of mutation induction in obtaining the hidden genetic variation and improvement of vegetatively propagated plants when one or few characters of an outstanding cultivar are to be modified can also be exploited to a great extent. Biotechnological developments and statistical analysis of current breeding populations would immensely improve our understanding of the genetic traits and their inheritance in litchi which, in turn, make it easier for the breeders to select parents and design a systematic breeding programme with more specific breeding goals than has been pursued in the past. However, there is still significant constraint in controlled hybridization that needs to be addressed.

References

Abu Baker Siddiqui SBM (2002) Lychee production in Bangladesh. In: Papademetriou MK, Dent FJ (eds) Lychee production in the Asia-Pacific region. Food and Agricultural Organization of the United Nations, Bangkok, pp 28–40

Anonymous (1978) Guangdong litchi register. Guangdong Science Press, Guangzhou

Anuntalabhochai S, Chundet R, Chiangda J, Apavatjrut P (2002) Genetic diversity within lychee (*Litchi chinensis* Sonn.) based on RAPD analysis. Acta Hortic 1:253–260

Aradhya MK, Zee FT, Manshardt R (1995) Isozyme variation in lychee (*Litchi chinensis* Sonn.) Sci Hortic 63:21–35

Bailey LH (1949) Manual of cultivated plants. Macmillan, New York

Bajpai A, Muthukumar M, Singh A, Nath V, Ravishankar H (2016) Narrow genetic base of Indian litchi (*Litchi chinensis*) cultivars based on molecular markers. Indian J Agric Sci 86:448–455

Batten DJ, Lahav E (1994) Base temperatures for growth processes of lychee, a recurrently flushing tree, are similar but optima differ. Aust J Plant Physiol 21(5):589–602

Batten DJ, McConchie CA (1995) Floral induction in growing buds of lychee (*Litchi chinensis*) and mango (*Mangifera indica*). Aust J Plant Physiol 22:783–791

Boonrat S (1984) Lychee cultivation. Horticultural Research Institute, Department of Agriculture, Bangkok

Bose TK (2001) Fruit: tropical and subtropical, 1. Naya Yug, p 721

Budathoki K (2002) Lychee production in Nepal. In: Papademetriou MK, Dent FJ (eds) Lychee production in the Asia-Pacific region. Food and Agricultural Organization of the United Nations, Bangkok, pp 81–93

Butcher FG (1957) Pollinating insects on lychee blossoms. Proc Fla State Hortic Soc 70:326–328

Chadha KL, Rajpoot MS (1969) Studies on floral biology, fruit set, and its retention and quality of some litchi varieties. Indian J Hortic 26:124–129

Chapman KR (1984) Sapindaceae. In: Page PE (ed) Tropical tree fruits for Australia. Queensland Department of Primary Industry, Queensland, pp 179–191

Chaturvedi RB (1965) Preliminary studies in the sex distribution, pollination and fruit development in litchi (*Litchi chinensis* Sonn.) Allahabad Farmer 39:49–51

Chen H, Huang H (2001) China litchi industry: development, achievements and problems. Acta Hortic 558:31–39

Chen YY, Deng SS, Zhang X, Wei SX, Gao AP, Wang JB, Zhou B (2004) RAPD analysis of genetic relationship among partial litchi germplasms in Hainan island. Acta Hortic Sin 31 (2):224–226

Chen YT, Lai ZX, Chen JY, Guo YQ, Cai YQ, Wu JS (2005) RAPD analysis of genetic relationship among some ancient litchi genetic resources. J Fujian Agric For Univ 34 (4):458–463

Choo WC, Ketsa S (1991) *Dimocarpus longan* Lour. In: Verheij EWM, Coronel RE (eds) Plant resources of South-East Asia, No. 2: edible fruits and nuts. Pudoc, Wageningen, pp 146–151

Costés E (1988) Analyze architectural et modelisation du litchi (*Litchi chinensis* Sonn.).

Cull BW (1977) Report on overseas study tour Queensland Department of Primary Industries, pp 1–92

Das CS, Choudhury KR (1958) Floral biology of litchi (*Litchi chinensis* Sonn.) S Indian Hortic 6:17–22

De Candolle (1909) Origin of cultivated plants. International scientific series. Kegan Paul, Trench, Tribner & Co. Ltd., London Dept. Primary Ind., p 123

Deng XM, Han ZH, Li SH (1999) Fruit tree biology. Higher Education Press, Beijing

Dhaliwal HS, Srivastava S, Adlakha RL (1977) Insect pollination of lychee, *Litchi chinensis* Sonn. in the valley areas of the Indian Himalayas. In: Proceedings of the twenty-sixth international apiculture congress. Apimondia, Bucharest

Ding XD, Xin LL, Jing CX, Xiong G, Ding L, Lu XD, Chen X, Guan X (2000) Identifying litchi cultivars and evaluating their genetic relationships by RAPD markers. J Trop Subtrop Bot 8:49–54

Dixon T, Batten D, Mcconchie C (2005) Genetic improvement of lychee in Australia. ISHS II International Symposium on Lychee, Longan, Rambutan and Other Sapindaceae plants. Acta Hortic 665:107–110

Du Toit AP (1994) Pollination of avocados, mangoes and litchis. Plant Prot Res Inst S Afr Plant Prot News Bull 35:4–5

Dwivedi AK, Mitra SK (1995) Genetic diversity of fruit quality traits in litchi (*Litchi chinensis* Sonn.) Hortic J 8(2):113–118

Dwivedi AK, Mitra SK (1996) Divergence analysis of litchi (*Litchi chinensis* Sonn.) cultivars grown in West Bengal. Indian J Genet Plant Breed 56(4):486–489

Ekuâe MRM, Gailing O, Finkeldey R (2009) Transferability of simple sequence repeat (SSR) markers developed in *Litchi chinensis* to *Blighia sapida* (Sapindaceae). Plant Mol Biol Rep 27:570–574

Froneman JJ, Oosthuizen JH (1995) ETSG litchi breeding programme. Yearbook of South African litchi growers. Assessment 7:35–37

Galan Sauco V (1989) Litchi cultivation. Food and Agriculture Organization of the United Nations, Rome

Galán Saúco V, Menini UG (1989) Lychee cultivation. FAO Plant Production and Protection Paper No. 83. Food and Agricultural Organization of the United Nations, Rome

Ghosh SP (2001) World trade in litchi: past, present and future. Acta Hortic 558:23–30

Goren M, Degani O, Gazit S (1998) Flowering phenology of 10 lychee cultivars in the Volcani center, Israel. Alon Hanotea 52:59–63

Goren M, Tomer E, Gazit S (2001) Litchi cultivars in Israel. Acta Hortic 558:129–133

Groff GW (1921) The lychee and longan. Orange Judd Company Canton Christ College, China. Sci Cult 38:39–40

Guangdong Academy of Agricultural Sciences (1978) The flora of Guangdong litchi. Guangdong Scientific & Technical Press, Guangzhou, pp 25–36

Hai VM, Dung NV (2002) Lychee production in Viet Nam. In: Papademetriou MK, Dent FJ (eds) Lychee production in the Asia-Pacific region. Food and Agricultural Organization of the United Nations, Bangkok, pp 114–119

Hayes WB (1957) Fruit growing in India. Kitabistan, Allahabad, p 180

Huang HB (2005) Fruit set, development and maturation. A. Litchi. In: Menzel CM, Waite GK (eds) Litchi and longan: botany, production and uses. CABI Publishing, Wallingford, pp 115–137

Huang X, Subhadrabandhu S, Mitra SK, Ben-Arie R, Stern RA (2005a) Origin, history, production and processing. In: Menzel CM, Waite GK (eds) Litchi and longan: botany, production and uses. CAB International, Wallingford

Huang X, Zeng L, Huang HB (2005b) Lychee and longan production in China. ISHS II International Symposium on Litchi, Longan, Rambutan and other Sapindaceae Plants. Acta Hortic 665

Jahiel M, Abraham D (2001) The growth of the Malagasy litchi field face to the hardening of European control in the matters of sulphured residue. Acta Hortic 558:63–71

Joubert AJ (1985) *Litchi chinensis*. In: Halevy AH (ed) Handbook of flowering, 5. CRC Press, Boca Raton, pp 204–210

Khurshid S, Ahmad I, Anjum MA (2004) Genetic diversity in different morphological characteristics of litchi (*Litchi chinensis* Sonn.) Int J Agric Biol 6(6):1062–1065

King J, Exley EM, Vithanage V (1989) Insect pollination for yield increases in lychee. In: Batten DJ (ed) Proceedings of the fourth Australian conference on tree and nut crops. Exotic Fruit Growers' Association, Lismore, pp 142–145

Knight RJ Jr (2001) The lychee's history and current status in Florida. Acta Hortic 558:41–44

Kumar M (2006) Mass scale propagation of *Litchi chinensis* Sonn. through in vitro techniques. Ph.D. Thesis, Bhagalpur University, Bhagalpur

Kumar M, Gupta M, Shrivastava D, Prasad M, Prasad US (2006) Genetic relatedness among Indian litchi accessions (*Litchi chinensis* Sonn.) by RAPD markers. Int J Agric Res 1:390–400

Lal BS, Nirwan RS (1980) Saharanpur selection: a new litchi cultivar. Prog Hortic 12:75–77

Leenhouts PW (1971) A revision of *Dimocarpus* (Sapindaceae). Blumea 19:113–131

Leenhouts PW (1978) Systematic notes on the Sapindaceae-Nephelieae. Blumea 24:395–403

Leenhouts PW (1986) A taxonomic revision of *Nephelium* (Sapindaceae). Blumea 31:373–436

Li LR, Fang Y (1956) The research of litchi and longan. Scientific Press, Beijing, pp 1–10

Li MF, Zheng XQ (2004) Development of SSR markers in litchi (*Litchi chinensis*). Hereditas 26 (6):911–916

Litz RE, Raherjo S, Matsumoto K, Witjaksono (2005) Biotechnology. In: Menzel CM, Waite (eds) Litchi and longan: botany, production and uses CAB International, Cambridge, MA pp 49–58

Liu SY (1954) Studies of *Litchi chinensis* Sonn. PhD thesis, University of Michigan, Ann Arbor, Michigan

Liu C, Mei M (2005) Classification of lychee cultivars with RAPD analysis. Acta Hortic 665:149–159

Liu HB, Liu YQ, Liu RG (1989) The peroxidase isozymes analysis of litchi cultivar. J Guangxi Agric Sci 5:27–29

Liu W, Xiao Z, Bao X, Yang X, Fang J, Xiang X (2015) Identifying litchi (*Litchi chinensis* Sonn.) cultivars and their genetic relationships using single nucleotide polymorphism (SNP) markers. PLoS One 10(8):e0135390

Lu LX, Chen JL, Chen XJ (1987) A study on chromosome number and meiosis in pollen mother cells in *Litchi chinensis* Sonn. J Fujian Agric Coll 16(3):224–228

Madhou M, Bahorun J, Hormaza JI (2010) Phenotypic and molecular diversity of litchi cultivars in Mauritius. Fruits 65:141–152

Madhou M, Normand F, Bahorun T, Hormaza JI (2013) Fingerprinting and analysis of genetic diversity of litchi (*Litchi chinensis* Sonn.) accessions from different germplasm collections using microsatellite markers. Tree Genet Genomes 9:387–396

Marloth RH (1947) The litchi in South Africa. I. Varieties and propagation. II. Cultural practices and marketing. Farming S Afr 22:824–830. 863–870

McConchie CA, Vithanage V, Batten DJ (1994) Intergeneric hybridization between litchi (*Litchi chinensis* Sonn.) and longan (*Dimocarpus longan* Lour.) Ann Bot 74(2):111–118

Menzel CM (1991) *Litchi chinensis* Sonn. In: Verheij EW, Coronel, RE (eds) Plant resources of South East Asia, No. 2: Edible

Menzel CM, Olesen T, McConchie C, Wiltshire N, Diczbalis Y, Wicks C (2000) Lychee, Longan and Rambutan: optimizing canopy management. Rural Industries Research and Development Corporation, No. 00/29, Canberra, Australia

Menzel CM, Simpson DR (1990) Performance and improvement of litchi cultivars: a review. Fruit Varieties J 44:197–215

Menzel CM, Simpson DR (1991) A description of litchi cultivars. Fruit Varieties J 45:45–56

Menzel CM, Simpson DR (1992) Flowering and fruit set in lychee (*Litchi chinensis* Sonn.) in subtropical Queensland. Aust J Exp Agric 32:105–111

Menzel CM, Simpson DR, Watson BJ (1993) Fruits of tropical climates: fruits of the Sapindaceae. In: Caballero B, Trugo L Finglas P (eds) Encyclopedia of food science food technology and nutrition. Academic Press, London, pp 2114–2120

Menzel CM, Waite G (2005) Litchi and longan: botany, production, and uses. CABI, Cambridge

Menzel CM, Watson BJ, Simpson DR (1988) The lychee in Australia. Qld Agric J 114:19–27

Miao PS, Han J, Huang YQ, Lin D, Wang FN (1998) Introduction of four new litchi selections. China Fruits 2:53

Mustard MJ, Liu S, Nelson RO (1953) Observations of floral biology and fruit setting in lychee varieties. Proc Fla State Hortic Soc 66:212–220

Nakasone HY, Paull RE (1998a) Litchi, longan and rambutan. In: Tropical fruits. CAB International, Wallingford, pp 173–207
Nakasone HY, Paull RE (1998b) Tropical fruits. CAB International, Wallingford
Naphrom D, Potjanapimon C, Subhadrabandhu S (2001) Changes in cytokinin and gibberellin-like substances in stem apex of lychee cv. Hong Huay prior to flowering. Acta Hortic 558:199–203
NHB (2015) National Horticulture Database. Haryana Government of India, Gurgaon
Oosthuizen JH (1991) Lychee cultivation in South Africa Yearbook of the Australian Lychee Growers. Assessment 1:51–55
Ouyang S, Zheng X et al (1985) T-DNA transfer and tumor formation induced by *Agrobacterium tumefaciens* on *Litchi chinensis*. Acta Genet Sin 12:42–45
Pandey RM, Sharma HC (1989) The litchi. ICAR, New Delhi
Pandey RS, Yadava RPS (1970) Pollination of litchi (*Litchi chinensis*) by insects, with special reference to honeybees. J Apic Res 9:103–105
Peng HX, Cao HQ, Zhu JH, Liao HH, Huang FZ, Li JZ, Liang W (2006) Studies on the application of AFLP molecular markers on genetic diversity and classification of good and rare litchi resources in Guangxi. Southwest China J Agric Sci 19(1):108–111
Peng HX, Li YC, Huang DJ, Su WQ, Liu RG (2001) 'Qinzhou red litchi': a new big fruit from variety of litchi. Acta Hortic Sin 28(3):276
Puchooa D (2004a) Expression of green fluorescent protein gene in litchi (*Litchi chinensis* Sonn.) tissues. J Appl Hortic 6(1):11–15
Puchooa D (2004b) In vitro regeneration of litchi (*Litchi chinensis* Sonn.) Afr J Biotechnol 3 (11):576–584
Radlkofer L (1932) Sapindaceae. Pflanzenreich, Weinheim, 1539 pp
Robbertse H, Fivaz J, Menzel CM (1995) A reevaluation of tree model, inflorescence morphology, and sex ratio in lychee (*Litchi chinensis* Sonn.) J Am Soc Hortic Sci 120:914–920. Sapindaceae plants held at Chiang Mai, Thailand
Sarin N, Prasad U, Kumar M, Jain SM (2009) Litchi breeding for genetic improvement.
Scholefield PB (1982) A scanning electron microscope study of flowers of avocado, litchi, macadamia and mango. Sci Hortic 16:263–272
Shukla RK, Bajpai PN (1974) Blossom bud differentiation and ontogeny in litchi (*Litchi chinensis* Sonn.) Indian J Hortic 31:224–228
Singh A, Abidi AB, Srivastava S, Singh IS (1987) Correlations among physical and biochemical traits in litchi (*Litchi chinensis* Sonn.) Narendra Deva J Agric Res 2(2):160–163
Singh G, Nath V, Pandey SN, Ray PK, Singh HS (2012) Litchi, FAO, New Delhi
Singh LB, Singh UP (1954) The litchi. Lucknow Superintendent of Printing and Stationery, Uttar Pradesh
Sotto RC (2002) Lychee production in the Philippines. In: Papademetriou MK, Dent FJ (eds) Lychee production in the Asia-Pacific region. Food and Agricultural Organization of the United Nations, Bangkok, pp 94–105
Steiner JJ, Greene SL (1996) Proposed use of ecological descriptors and their utility for plant germplasm collections. Crop Sci 36:439–451
Stern RA, Gazit S (1996) Lychee pollination by the honeybee. J Am Soc Hortic Sci 121:152–157
Stern RA, Gazit S (1998) Pollen viability in lychee. J Am Soc Hortic Sci 123:41–46
Stern RA, Gazit S (2003) The reproductive biology of the lychee. Hortic Rev 28:393–453
Stern RA, Gazit S, El-Batsri R, Degani C (1993) Pollen parent effect on outcrossing rate, yield, and fruit characteristics of 'Floridian' and 'Mauritius' lychee. J Am Soc Hortic Sci 118(1):109–114
Storey WB, Hamilton RA, Nakasone HY (1953) Groff–a new variety of lychee. Circular Hawaii Agric Exp Sta 39:1–8
Strauss ML, Pino JA, Cohen JI (1998) Quantification of diversity in ex situ plant collections. Diversity 16:30–32
Subhadrabandhu S (1990) Lychee and longan cultivation in Thailand. Rumthai, Bangkok, p 40
Subhadrabandhu S, Yapwattanaphun C (2001) Lychee and longan production in Thailand. Acta Hortic 558:49–57

Sun Q, Bai L, Ke L, Xiang X, Zhao J, Ou L (2012) Developing a core collection of litchi (*Litchi chinensis* Sonn.) based on EST-SSR genotype data and agronomic traits. Sci Hortic 146:29–38

Tao R (1955) The superior lychee. Proceedings Florida Lychee Growers Association 2:73–74. Techniques. PhD Thesis, Bhagalpur University, Bhagalpur

Thakur S (1992) Hybridization in litchi. Souvenir. National Seminar on Research Developments in Litchi Production, RAU, Pusa, p 13

Tindall HD (1994) Sapindaceous fruits: botany and horticulture. Hortic Rev 16:143–196

Tongpamnak PA, Patanatra PS, Babprasert C (2002) Determination of genetic diversity and relationship among Thai litchi accessions by RAPD and AFLP markers. Kasetsart J (Nat Sci) 36:370–380

Viruel MA, Hormaza JI (2004) Development, characterization and variability analysis of microsatellites in lychee (*Litchi chinensis* Sonn., Sapindaceae). Theor Appl Genet 108 (5):896–902

Wang B, Qiu Y (1997) Growth and fruiting. In: Zhang ZW, Yuan PY, Wang BQ, Qui YP (eds) Litchi pictorial narration of cultivation. Pomology Research Institute, Guangdong Academy of Agricultural Sciences, Guangzhou, pp 66–88

Wang JB, Deng SS, Liu ZY, Liu LZ, Du ZJ, Xu BY, Chen YY (2006) RAPD analysis on main cultivars of litchi (*Litchi chinensis* Sonn.) in Hainan. J Agric Biotechnol 14(3):391–396

Wang L, Wu J, Chen J, Fua D, Zhanga C, Cai C, Ou L (2015) A simple pollen collection, dehydration, and long-term storage method for litchi (*Litchi chinensis* Sonn.) Sci Hortic 188:78–83

Wu D, Lin X, Ye Q, Wang W (2001) Improvement of fruit set in secondary panicles of 'Feizixiao' litchi by removal of the primary panicles. In: Huang HB, Huang XM (eds) First international symposium on litchi and longan programme and abstracts. South China Agricultural University, Guangzhou, p 42

Wu J, Zhanga C, Chena J, Caia C, Wanga L, Fua D, Oua L (2016) Morphological diversity within litchi (*Litchi chinensis* Sonn.) based on leaf and branch traits. Sci Hortic 207:21–27

Wu SX (1998a) Encyclopedia of China fruits: litchi. China Forestry Press, Beijing, 221 pp

Wu SX (1998b) The flora of China fruit tree, the volume of litchi. China Forestry Press, Beijing, pp 94–198

Wu Y, Yi G, Zhou B, Zeng J, Huang Y (2007) The advancement of research on litchi and longan germplasm resources in China. Sci Hortic 114:143–150

Xian X, Zhang ZW, Wang BQ, Yuan PY, Qiu YP (1994) Fertility and storability of lychee pollens. Guangdong Agric Sci 4:25–27

Yeap CK (1987) Sapindaceous fruits and nuts. Yearbook West Australia Nut and Tree Crops Assoc 12:16–33

Yi GJ, Huo HQ, Chen DC, Huang ZR, Cai CH, Qiu YP (2003) Studies on genetic relationship among litchi cultivars by using AFLP. Acta Hortic Sin 30(4):399–403

Zhang Z (1997) Local cultivars and selections with potential. In: Zhang ZW, Yuan PY, Wang BQ, Qui YP (eds) Litchi pictorial narration of cultivation. Pomology Research Institute, Guangdong Academy of Agricultural Sciences, Guangzhou, pp 48–57

Zheng Q, Davenport TL, Li YC (2001a) Stem age, winter temperature and flowering of lychee in South Florida. Acta Hortic 558:237–240

Zheng Q, Hu G, Chen D (2001b) Biotechnology in longan and litchi breeding: applications and prospects. Acta Hortic 558:143–148

Zhou JH, Luo HN, Huang JB, Chen CM (2001) The peroxidase isozymes analysis of 35 cultivars of lychee in Guangdong. J Zhongkai Agric Coll 14(2):20–25

Problems and Prospects of Lychee Cultivation in India

9

Amar Singh Kashyap and Nitika Thakur

Abstract

Lychee—an evergreen, climate-specific fruit crop, belonging to the family Sapindaceae—stands out as a most promising income source because it supports the livelihood of many small and marginal farmers, especially those residing in Bihar. It has adapted well to the climate in eastern India, that is, Bihar, Jharkhand, West Bengal, Tripura, Uttar Pradesh, Uttaranchal, Chhattisgarh, Punjab and Himachal Pradesh. Because of its excellent quality, pleasant favour, juicy pulp with attractive red colour and additional nutritional qualities, there is an increasing demand for its cultivation and production. The increasing demand for litchi cultivation has raised the standards for its production to be enhanced on a large area, but some major constraints pose great hindrance for its production. Concerns such as productivity losses and genetic parameters are still a threat to lychee cultivation. In addition, there is still a need for developing lines or hybrids with larger fruits, small/chicken-tongued seeds, and tolerance to pericarp splitting and have various maturity groupings. The need of the hour demands designing agro-techniques particularly for source and sink management, micronutrients, post-harvest technology and effective marketing which results in beneficial exchange of information among various countries by which progressive moves can be made for stepping ahead from the limited constraints.

Keywords

Lychee • Agro-techniques • Cultivation • Productivity

A.S. Kashyap • N. Thakur (✉)
ICCOA, Bangalore 560041, India

Department of Biotechnology, Shoolini University of Biotechnology and Management Sciences, Bajhol-campus, Solan 173212, HP, India
e-mail: nitikathakur45@gmail.com

M. Kumar et al. (eds.), *Lychee Disease Management*,
DOI 10.1007/978-981-10-4247-8_9

9.1 Introduction

The lychee (*Litchi chinensis* Sonn., family Sapindaceae) is an important subtropical evergreen fruit crop which is believed to have originated in China, where it has been grown in Southern Guangdong state for thousands of years. It is highly specific to climatic requirements, and probably for this reason, its cultivation remains restricted to a few countries in the world. In India litchi was introduced in Burma in the eighteenth century from where it spread to many countries. India and China contribute about 91% of the world lychee production. About 428,900 metric tons of lychee are produced annually from 56,200 ha in India. Lychee, being exacting in its climatic requirements, is confined to a few states with 74% of production recorded in Bihar (Fig. 9.1).

Most of the lychee production area is in North Bihar, comprising Muzaffarpur, Vaishali, Samastipur, Begusarai, East and West Champaran, and Bhagalpur districts; lychee is also commercially cultivated in Thane (Maharashtra), Saharanpur, Dehradun, Hoogli, Tripura, Assam and Punjab (Chadha 1968).

Lychee is widely distributed throughout the world's subtropical and tropical regions. Major centres of production are China, India, Vietnam, Taiwan, Thailand, Nepal, Australia, Africa, Israel, Mexico, Brazil and the USA. Small production areas include Australia, South Africa, the USA and Mexico, and Hawaii, New Caledonia and Tahiti are well known for their minor production. Lychee is a vigorous perennial tree attaining a height of about 5 m with a broad, round stalked crown having six to nine elliptical oblong and lanceolate, abruptly pointed leaves. Colour of the leaves varies from light to dark green, marked by greenish white or yellowish flowers born in clusters. Fruits are round or heart shaped with thin and leathery skin. The colour of fruits varies with cultivars, being distinctive. The edible portion of the fruit is the aril. Seeds are bold, but in some cultivars, seeds may be only partially developed because of the failure of pollination, yielding what are referred to as "chicken-tongue" seeds. The trees with small-seeded fruits are prized because of the greater portion of pulp. accounting for its excellent quality, pleasant

Fig. 9.1 Pictorial representation of lychee fruit

flavour, juicy pulp and attractive red colour. A highly flavoured squash is also prepared from its fruits. Because of the firmness of the shell of the dried fruits, they came to be nicknamed "lychee, or litchi, nuts" by the uninitiated, and this erroneous name has led to much misunderstanding of the nature of this highly desirable fruit. It is definitely not a "nut," and the seed is inedible. The fruits consist of 60% juice, 8% rag, 19% seed, 13% skin, and 40–90 mg/100 g vitamin C. On the other hand, it contains insignificant amounts of protein (0.8–0.9%), fat (0.3%), pectin (0.43%) and minerals, especially calcium, phosphorus, and iron (0.7%). Litchi or lychee, a fruit with a rough brown pericarp surrounding a white flesh (aril), is popular for its delicious taste and possible health benefits (Jiang et al. 2013). The litchi fruit pericarp (LFP), which accounts for approximately 15% by weight of the whole fresh fruit, is usually discarded as waste during processing. Recent studies have reported that the litchi pericarp contains significant amounts of polyphenols, flavonoids, anthocyanins and polysaccharides (Duan et al. 2007a; Duan et al. 2007b; Zhao et al. 2006). The major phenolics in LFP tissues were identified as epicatechin, procyanidin B4 and procyanidin B2. These phenolics from LFP have good antioxidant activity and antiinflammatory, anticarcinogenic and immunomodulatory properties (Li et al. 2012; Prasad et al. 2009). Considering the importance of this fruit, efforts are made to provide technological support through research, promoting production, post-harvest management and marketing. A large market base and climate sensitivity make litchi production attractive. As the litchi crop in India matures early in comparison to other litchi-growing countries, this offers better domestic and export markets. Accordingly, there is a potential for an additional 100,000 hectares to be brought under litchi cultivation with strategic planning and improved production systems.

9.2 Nutritional Profile of Lychee

Lychee fruit contains 66 calories/100 g comprising good amounts of dietary fibre, vitamins and antioxidants with no saturated fats or cholesterol.

Recent research indicates a low molecular weight polyphenol, oligonol, this is present in lychee fruit. It is thought to have excellent antioxidants and antiinfluenza virus action. It also improves the blood flow to organs, reduces weight and protects the skin from harmful UV (Sakurai et al. 2008). Studies suggest that lychee is a source of vitamin C; 100 g fresh fruits provide 71.5 mg or 119% the daily recommended value. This vitamin in turn helps the human body develop resistance against infectious agents and scavenge harmful, pro-inflammatory free radicals.

Further, it is a very good source of B-complex vitamins such as thiamin, niacin and folates and minerals such as potassium and copper which are helpful in controlling blood pressure and are required in the production of red blood cells.

9.3 Climate and Soil

A moist atmosphere free from frost and with abundant rainfall is the first choice for the litchi fruit to flourish. The plant generally grows luxuriantly at 30 °C. The maximum temperature during flowering and fruit development varies from 21 °C in February to 38 °C in June in Bihar. Another important factor determining the fate of litchi is humidity as it limits the expansion of litchi cultivation. Therefore, wet springs, dry summers and light winters are desirable conditions for litchi fruiting.

The soil type preferred for litchi cultivation includes fairly deep, well-drained loam soil enriched in organic matter, and light sandy loam is ideal. Lychee cultivation is highly successful in areas having a minimum temperature of 10 °C from December through February and 38 °C from April to June. A dry climate free from rain for about 2 months before flowering induces flower bud differentiation, blossoming, and consequently provides high production. The subtropical to mild temperate climates in the foothills and the valleys of the Himalayas are also suitable for lychee cultivation. In Chotanagpur, the fog-free dry winter, mild subtropical summer and intermittent pre-monsoon showers during April–May have been observed to be highly favourable for blossoming and improvement in fruit quality. The time of flowering and maturity is determined depending on the temperature rise after winter. No fruiting has been recorded when litchi has been grown in tropical conditions.

9.4 Propagation

Lychee can be propagated by seeds, but this method is not reliable because trees raised from seeds require a long juvenile period, and thus fail to produce true-to-type plants, which in turn often produce fruits of inferior quality.

Another propagation method is by vegetative cutting mode, but as the plants produced are less vigorous this method is generally not recommended. The most common method adopted by vegetative propagation is air layering. The stooling method of propagation is becoming popular because of a higher success rate when compared to the air-layering method (Pandey and Sharma 1989).

Air layering or *gootee* is a widely accepted method of propagation in India. In this method a healthy and vigorous upright twig about 1-year old and 2.5–4 cm in diameter is selected. A circular strip of bark about 2 cm wide just below a bud is completely removed from the selected twig. This portion is packed with moist sphagnum moss and is tied with a polythene sheath to prevent loss of moisture. In about 6 weeks, when the roots are visible through the polythene bag, the rooted branch is detached from the parent plant and potted in the nursery.

Stooling involves cutting a 2.5-year-old lychee plant, in the month of February, at 25 cm from ground level. After a month, six to eight side shoots appear, and all but one shoot are selected for this procedure in the month of June. A ring of a bar 3 cm wide, 20 cm from the tip of the shoot, is removed. A paste of IBA (25 mg) and lanolin (10 g) is applied to the ring area, and 10 days later soil is mounded around the base of the newly developed shoot to cover 10–15 cm of the stem above the ring to encourage adventitious roots. The rooted shoots are separated from the mother plant in the month of September and are immediately planted in nursery beds or pots. This method of propagation is adventitious over the air-layering method because of the higher survival rate (81–82%) as compared with (40–50%) air layers.

9.5 Present Scenario of Lychee Cultivation in the Country

In India, lychee ranks seventh in the area and ninth in production among the fruit crops, but in value terms it ranks sixth. In Bihar State lychee is considered to be the most important fruit as it contributes significantly to total fruit production. During the past 50 years there has been substantial increase in the area and production of lychee, accounting for an increased area from 9400 hectares in 1949–1950 to 56,000 hectares in 1998–1999 (Singh and Yadav 1992). An increase in productivity of about 52.91% was recorded during the same tenure. On the whole it is evident that the production and productivity of lychee is constantly increasing throughout the country. Of the total production of lychee in India, 74% is contributed by Bihar, with second place occupied by West Bengal followed by Tripura and Assam. An interesting feature of distribution of lychee in India is that maturity commences first in Tripura, followed by West Bengal then Bihar.

There is a sizeable increase in acreage and production of litchi in India. Cultivation of litchi has increased, from nearly 72,000 ha in 2008–2009 to 84,170 ha in 2013–2014. In terms of production, however, it has increased from 4,23,000 to 5,85,300 tons during the same period. The total production of litchi is concentrated mainly in Bihar, West Bengal, Assam and Jharkhand and to a smaller extent in Tripura, Punjab, Uttarakhand and Odisha. Bihar is the leading state in litchi production (234,200 tons), followed by West Bengal (93,900 tons) and Jharkhand (58,240 tons). Export of litchi has increased from 161.5 tons in 2007–2008 to 794.86 tons in 2012–2013, but it is evident from Table 9.2 that the quantum has gone up to 1546 tons in 2008–2009, which shows very good potential for export. In recent years, the domestic market for litchi has increased tremendously and good price has been realized in the distant south Indian market. Export of litchi is mainly confined to SAARC countries, notably Bangladesh, Nepal, Maldives, and Bhutan, and the UAE. There is excellent potential for export of litchi to Gulf Cooperation Council countries at competitive prices compared to Thailand, as India produces excellent quality litchi fruits and it is nearer to Gulf countries when compared with export competitors such as Thailand and China. The European Union also imports sizeable quantities of litchi. Litchi are available in India from 15 April (Tripura) onwards up to the 3rd week of June (Gurdaspur, Punjab). Off-season litchi production (December–January) in India offers high potential from higher-altitude (above 1000 m) Southern states.

9.6 Growing Belts

The major litchi-producing belts are Uttaranchal (Dehradun, Pithauragarh, Nainital, Haridwar), West Bengal (Murshidabad, 24-Paraganas), Bihar (Muzzafarpur, East Champaran, Samastipur, Vaisali, Bhagalpur), Assam (Kamrup, Spmotpir, Bongaigaon), Punjab (Gurdaspur, Ropar, Hoshiarpur),Uttar Pradesh (Saharanpur) and Jharkhand and Tripura.

9.7 Varieties

Lychee varieties are highly variable under different climate and soil conditions. Singh in 1954 described 33 varieties classified into 15 group varieties grown in India. Fruit size, shape and taste are also variable, which are influenced by other than genetic factors. Classified the cultivars into five groups with group A (early group) having seven cultivars, and B and C groups (mid-season) and group D (late group). Group E was designated as a very late group as its cultivation is confined to Muzaffarpur (Kotur and Singh 1994). The least colour along with shape and size of the leaf is of importance in varietal identification. A large number of varieties have been identified, grown in different parts of India. Early seedless (Early Bedana), Ross Scented, Dehradun, Gulabi, Calcuttia, Purbi, Kasba, Shahi, Bombai, Late Seedless (Late Bedana), China and Deshi varieties are of great interest for cultivation. The varieties Shahi, Rose Scented and China are commercially cultivated varieties of Muzaffarpur, Kasba and Purbi are the choicest lychee cultivars of the eastern parts of Bihar, whereas the varieties Dehradun and Calcuttia are prominently grown in parts of UP. Bombay and Kalyani selections are extensively grown in areas of West Bengal, and Seedless and Late Bedana are widely grown in Punjab. For commercial cultivation in Chhotanagar an early, noncracking seedless selection, Swaran Roopa, has been identified and selected.

9.7.1 The Popular Varieties of Lychee

9.7.1.1 Shahi

Shahi is the most popular cultivar grown in North Bihar, Jharkhand, Uttaranchal and Uttar Pradesh regions of India. This variety has high-quality fruit with a distinct aroma of rose; hence, it is called Rose Scented. Thus it is known as Shahi in Bihar, Rose Scented in Uttaranchal, and Muzaffarpur in Western Uttar Pradesh. The vegetative flush of this cultivar is light and fruit weight ranges from 20 to 25 g. This cultivar is earliest in maturity, and ripens during the 2nd week of May to the 1st week of June. It matures on 12–15 May in Jharkhand, 25 May in North Bihar, and by the 1st week of June in the Terai region of Uttaranchal. The trees of this cultivar are very vigorous and produce 100–150 kg fruits/plant. Fruits are globous-heart or obtuse in shape. Pulp is greyish white, soft, and moderately juicy and sweet with TSS ranging from 19 to 22 Brix. This cultivar occupies a major part of the area under lychee cultivation in India.

9.7.1.2 China

The origin of this cultivar is not known but the name indicates its superiority and origin from China. This cultivar is tolerant to heat waves and fluctuations in soil moisture which cause fruit cracking. It is commonly known via different names such as Purbi, Calcuttia, Bengalia and Manragi. It is a medium- to late-season cultivar. The fruit ripening process starts from the end of May in West Bengal, and the 1st week of June in Jharkhand and North Bihar. Trees are comparatively dwarf and high yielders. There are 12 to 18 fruit clusters but in some cases 30 are also recorded. Fruits are large in size, medium heavy in weight, oblong in shape and tyrant rose in colour. The aril is creamy white soft, juicy, having 17 Brix total soluble solids (TSS). Seeds are glaucous and medium in size. The flavour is however not so pleasant as Shahi, but because of its high yielding and qualities of not cracking, its cultivation is popular.

9.7.1.3 Early Bedana

This cultivar is also known as Early seedless in Punjab as it ripens having small seeds. It has a distinguishing leaf and flower character. The cultivar is cultivated in the region of Uttar Pradesh and Punjab. Trees are medium, attaining an average height of 5 m, and spread out at 6.2 m. It is a medium yielding cultivar (50–60 kg/tree) but bears fruit regularly. The fruits are medium, oval, or heart shape with red tubercles at maturity. The aril is creamy white, juicy and sweet having 17.2°Brix TSS. The seeds are very small, shrunken, and glabrous. The overall fruit quality of the cultivar is good.

9.7.1.4 Late Bedana

This cultivar is also named Late Seedless because of its late maturing nature, generally ripening in the last week of June in Uttaranchal, the end of May in Jharkhand, and in the 1st week of June in Muzaffarpur. The trees are vigorous, having an average height of 5.5 m and spread of 7 m with a yield ranging from 60 to 80 kg/tree. Although the fruit size is medium, the pulp content is high. The fruits are conical in shape having dark blackish brown tubercles at maturity. The pulp is creamy white, soft, juicy having 18–2°Brix, but acidity is low. Seeds are very small, shrunken and chocolate in colour with fusiform shape. The panicle is compact.

9.7.1.5 Ajhauli

This variety is an early maturing variety selected from Ajhauli village, which yields about 80–100 kg fruit from a 16-year-old tree. The fruits are red in colour, weighing 15–18 g and with big seeds. This variety cannot be distinguished from Shahi on vegetative characteristics as it has many similarities. This variety is largely prone to cracking but this can be minimized with proper irrigation conditions.

9.7.1.6 Bombai

Bombai is an important cultivar in West Bengal. It is a vigorous cultivar attaining a height of 6–7 m and a spread of 7–8 m. This variety also matures early and gives 80–90 kg fruit yield/tree. The fruits are large in size, 3.5 cm long and 3.2 cm in diameter, heart shaped and weighing 15–20 g. The colour of the ripe fruit is an attractive carmine red with uranium green skin background. The pulp is greyish, white soft, juicy, and sweet containing 17°Brix TSS, 11% sugar and 0.5% acidity. The elongated, smooth and shining seed of light chocolate colour is 2.3 cm long, 1.6 cm in diameter, and weighs 3.4 g. This cultivar is similar to China grown in other states.

9.7.1.7 Dehradun

This is an important cultivar grown in the name of Dehra Rose cultivated in regions of Uttar Pradesh (UP) and Punjab. The ripening of fruit starts by the 3rd week of June in UP but in Jharkhand it matures at the time the Shahi variety matures. The trees are vigorous with 5 m height and 7 m spread, producing medium to high yield. The fruits are medium to large in size, measuring about 3.5 cm in length, 3.5 cm in diameter, weighing 15.22 g, and having an oblique heart to conical shape. The pulp of the cultivar is greyish white, soft, moderately juicy with 18°Brix TSS, 10.4% sugar and 0.44% acidity. Seeds remain small, light, shrunken, mostly oblong and dark chocolate in colour. It is mostly prone to cracking, under rainfed conditions.

9.7.1.8 Gulabi

This cultivar is another late-maturing cultivar of North India in which ripening takes place by the 4th week of June. The trees are medium, vigorous, profuse with medium- to large-size fruits. The shape of the fruits is variable from oblong-oval to heart shape.

9.7.1.9 Ellaichi

This cultivar is popular in West Bengal, having prospects for commercialization. The trees are moderately vigorous, attaining an average height of 5–6 m and spread of 6–7 m. The cultivar yields about 50–60 kg fruits annually. The fruits are conical, marigold orange red in colour with an average weight of 1.5–2.0 g. The fruits are less susceptible to sunburn and cracking. This cultivar, however, has not achieved commercial success. The cultivar has 18°Brix TSS, 11.5% total sugars and 0.45% acidity.

9.7.1.10 Longia

This cultivar is well distributed in North Bihar with a preference to late maturity. The trees are medium in size, and leaves are small and light in colour. The fruits are medium in size with the aril having an excellent aroma. Its shy bearing habit has resulted in a declining preference for this cultivar.

9.7.1.11 Kasba

This is a large fruit cultivar selected from Kasba village for its attractive fruit size and colour. The trees are large and compact, having broad and elongated leaves. The fruit weighs between 23 and 27 g, thus being the heaviest fruit among the known varieties, but the fruits are fewer in number.

9.7.1.12 Kaselia

This is a late maturing variety found growing in isolation. The trees are medium in size and mature late. Fruits attain a pink red colour. The pulp content is comparatively low with big seeds. This cultivar is also known as Khatti or pickling. This cultivar has not attained commercial success.

9.7.1.13 Swarna Roopa

The variety is a late maturing selection, a cracking-resistant lychee selected at the Central Horticultural Experiment Station (CHES), Ranchi. The fruits are an attractive red colour with small seeds and high aril percentage. The cultivar is 18–22.5 cm long with compact panicles. Fruits are medium sized with 12–15 g weight and high pulp content. The variety consists of high TSS and low acidity. The cultivar is suitable for extended harvest as it matures after China. This variety is generally priced for its attractive fruit colour and highly recommended for commercial purposes.

9.7.1.14 CHES-2

This is also a late-maturing selection from the population of the China group. It bears fruits both at the outer and inner canopy, which helps in reducing sunburn as well as fruit cracking. The fruits are deep red, conical in shape, and appear in a cluster of about 15 to 20 fruits. The fruit has an average weight of 3.8 g seed and 16.1 g pulp.

9.8 Cultivation of Lychee

Before planting, the land should be cleared and levelled with a gentle slope on one side of the plot, on the opposite direction of the irrigation source. Pits measuring 1 m × 1 m × 1 m in size should be dug at the desired site a few weeks before the actual planting process. These pits are left open for 15–20 days and then refilled with well-rotted farm yard manure, leaf mould and canal silt. Another recommendation could be the use of well-rotted farm yard manure (20–25 kg), bone meal (2 kg) and sulphate of potash (400 g) mixed with soil in a pit containing mycorrhizal fungi. The mycorrhizal fungi is helpful in the establishment and quick growth of newly planted plants. The pits are watered to set the mixture within the earth. Thus, planting is done after the week interval and watering is done immediately after the planting.

Lychee trees are usually planted in a square system which is about 10 cm apart. For disease-free cultivation, quick establishment and lower mortality rate, healthy

6- to 9-month-old true-to-type plants with fine roots should be selected. It is advisable to plant the plants with mycorrhizal fungi.

After planting the land should not be allowed to dry completely; hence, the new plantation is suggested during early monsoon season to avoid the vagaries of situations such as dryness and too much wet. With the increased market base, there is an ample opportunity for increasing the area under litchi in some other areas. The foothills of the Himalayas, that is, the Terai belt, free from frost, offer good scope for plantation of litchi. Litchi can be cultivated up to an altitude of 1000 m above mean sea level. In these foothills, fruits mature late in the season. Extended areas under different situations could be exploited for extended harvest. Based on the fruiting behaviour, quality development and area under cultivation, the litchi-growing districts could be grouped to take full advantage of climatic variability.

However, to increase production and productivity, concerted efforts would be required for technological support and development of infrastructure. Interestingly, the litchi in India matures early in comparison to other litchi-growing countries and offers better domestic and export markets. Accordingly, there is potential to bring an additional 100,000 ha under litchi cultivation.

9.9 Training and Pruning

Training young lychee plants is necessary for achieving a desired shape and a strong framework. In India, this occurs indirectly when a part of the shoot bearing the cluster of fruits is removed during harvesting. Heavy pruning is generally not recommended as it causes profuse vegetative growth, resulting in poor fruiting.

9.10 Manurial Schedule and Fertilization

In India, lychee is cultivated in fertile soils where no manure is given additionally. But the acute shortage of N, P and K seems to limit the forms of litchi growth, including floral initiation. Fertilizer should be applied just after harvesting during rainy seasons. Applying fertilizers late results in more vegetative growth with less fruiting. The whole fertilizer schedule recommended for litchi cultivation in North India and Bihar is summarized in Tables 9.1 and 9.2.

9.10.1 Aftercare Operations for Healthy Lychee Cultivation

To keep the lychee orchard free from disease, good sanitary management conditions are to be followed strictly. Lychee is a deep-rooted tree with most of the roots occupying a region 20–30 cm deep. Remembering that in the deep-rooted system deep tillage is harmful as it directly injures the roots, therefore tillage operations should take place only in the soil layer 7 to 10 cm deep. Because lychee

Table 9.1 The fertilizer schedule recommended for litchi cultivation in North India (Source: *Handbook of Horticulture*)

Number	Manure/ fertilizer	First year	Increasing amount every year (up to 5–6 years)	Fertilizer dose for full-bearing tree
1.	Compost	20 kg	10 kg	60 kg
2.	Castor cake	1 kg	0.5 kg	5 kg
3.	Neem cake	0.5 kg	0.5 kg	3 kg
4.	Single super phosphate	2.5 kg	0.75 kg	5 kg
5.	Muriate of potash	100 g	50 g	0.5 kg
6.	Calcium nitrate	–	0.5 g	0.002 g

Table 9.2 The fertilizer schedule recommended in Bihar

Number	Age of the plant (years)	Fertilizers/plant/year (kg)			
		Farm yard manure	Calcium ammonium nitrate	Superphosphate	Muriate of potash
1.	1–3 years	10–20	0.3–1.00	0.2–0.6	0.05–0.15
2.	4–6 years	25–40	1.0–2.00	0.75–1.25	0.20–0.30
3.	7–10 years	40–50	2.0–3.00	1.50–2.00	0.30–0.50
4.	Above 10 years	60	3.50	2.25	0.60

Source: *Handbook of Horticulture*

stands as a slow-growing fruit that takes about 6 years to come to flowering and fruiting, thus intercropping with vegetables, pulses and barseem is advised; also, fruit plants such as phalsa and papaya can be grown in the early years of plantation.

9.10.2 Protection from Weeds

Common practices such as hand weeding and hoeing are generally time consuming, expensive and laborious. Therefore, the application of the pre-emergence herbicides Diuron or Atrazine at 2 kg/acre at a 1-month interval keeps the weeds under control. The use of black polythene mulch also controls weeds more efficiently than organic mulch.

9.10.3 Irrigation

The critical period in lychee cultivation is the phase of vegetative growth and fruit development, so from the end of January to the onset of the monsoon is the best period for irrigation. Four months before the normal floral initiation period (December–January) in northern India, the plants should not be irrigated. Although lychee is a deep-rooted plant, most of the absorbing roots lie between 20 and 30 cm of depth; thus, this absorption zone should be enriched with 50% soil moisture during the stress period. The basin system should be used for irrigating young trees, whereas furrow and flooding methods are generally incorporated for fully mature plants. The frequency of irrigation generally is related to availability and source of water and soil type. Generally, weekly irrigation is recommended in summer and no irrigation is required during winter in fruiting trees before fruit set.

9.10.4 Harvesting

The colour of a fruit an important criteria for deciding the harvesting stage (Singh and Yadav 1988). Lychee being a non-climacteric fruit requires to be harvested after attaining full maturity on the tree. During maturity acidity decline and TSS increases, deciding the appearance and colour of the fruit. The red pigmentation in lychee is associated with anthocynin pigment (cyanindin-3-glucoside, cyanindin-3-galactoside, and pelargonidin-3-glucoside), which develops better in the direction of good light penetration. Depending upon the cultivars, 65–80 days are required for maturity from the fruit set.

The harvesting of the lychee fruit includes bunch harvesting early in the morning when the temperature and humidity are congenial for longer shelf life of the fruit. For a distant market, fruits are harvested when TSS attains 19°Brix and acidity is 0.3–0.4%. The harvesting period is generally from May to June depending upon cultivar and location.

9.10.5 Yield

The yield of lychee varies from 80 to 150 kg fruits per tree from a tree 14 to 16 years old. However a yield of 160–200 kg per tree has been recorded. The commonly used bee in lychee orchards is *Apis mellifera*, which also provides additional income from honey.

9.10.6 Post-harvest Management

Lychee deteriorates very quickly after harvesting because of pericarp browning which is the major post-harvest problem, rendering the fruit unmarketable (Singh and Yadav 1988). Browning is associated with the perioxidase activity coupled with ascorbic acid oxidation which ultimately leads to anthocyanin degradation. To prevent post-harvest losses in the field of lychee production, sulphur treatment and perforated plastic bag packaging storage under cold conditions are preferred. For sea transportation, 600–650 g sulphur for a duration of 50–60 min is recommended.

The Agricultural Produce Export Development Authority (APEDA) has designed a procedure for quality lychee production.

1. Production
2. Inspection of farms
3. Harvesting
4. Discarding and sorting
5. Receipt at packing house
6. Acceptance of produce
7. Sorting and grading
8. Sulphur treatments
9. Packing and cooling
10. Palletization
11. Storage
12. Container loading
13. Transportation

Aril breakdown is the fault observed in the post-harvest phase wherein fruits become blunt in taste from the loss of turgidity and translucency of aril. Post-harvest decay also occurs from various bacteria, yeast and fungi. A storage temperature of 2–5 °C is considered to extend the shelf life. Controlled atmospheric storage is considered better for the freshness of fruits. On the whole, to have better post-harvest life of fruits, careful harvesting, pre-cooling, and transportation in cool vans, sulphuring and storing at 2–3 °C is essential.

9.10.7 Good Agricultural Practices (GAP) for Cultivation of Lychee

1. Enhance the establishment of air layer plants in the field.
2. Adopt high-density planting.
3. Conserve soil moisture with mulching.
4. Integrate nutrient management.
5. Manage irrigation frequency to avoid fruit cracking.
6. Integrate management of pests and diseases.
7. Lychee lends itself well to polycultures where the trees can be incorporated into a broader production and subsistence system. One or two lychee trees can be more than adequate to supply a family and offer income opportunities.
8. Litchi are being followed by farmers in use of air-layered plants for orchard establishment, rejuvenation to make old litchi orchards productive, high-density planting, conservation of soil moisture with mulching, integrated nutrient management, maintenance of irrigation frequency and spray of zinc/boron chelates to avoid fruit cracking, integrated pest management against fruit borers and post-harvest management through better packing methods with value-added products.

9. Artificial flower induction in litchi is possible by application of potassium per chlorate ($KClO_3$), paclobutrazol and prohexadion-Ca.
10. Technology for controlled atmosphere (CA) and modified atmosphere (MA) storage of litchi are being standardized at ICAR-NRCL and ultimately will be refined to extend the shelf life of litchi and facilitate litchi export to Gulf countries through refrigerated containers and MA cartons.
11. Residue analysis laboratory is needed near production centres. Formidable loss of fruits results from faulty methods of harvesting, lack of storage facility, mismanagement and mishandling of produce, loss during transit, and unsuitable and unscientific packaging.

9.10.8 Criteria and Grades Description for Lychee

(a) Extra class lychee

Lychee must be of superior quality, free from defects, with exception of very slight superficial defects that would not affect the general appearance of the produce or its quality.

(a) Class 1

Lychee must be of good quality, mainly grown for commercial purposes. Slight defects in shape, colouring and skin are allowed provided such do not exceed a total area of 0.25 cm^2.

(b) Class 2

This grade includes lychee which does not quality for inclusion in the higher grades. Defects in skin, colour, and blemishes are permitted if these do not exceed a total area of 0.5 cm^2.

9.10.9 Constraints in the Production Phase of Lychee

The fact that the lychee has a growing demand in the national and international markets supports its superiority as one of the finest fruits. The main constraint in its production lies in its low productivity, widening the gap between potential and existing yield. Reasons for low yield are the narrow genetic base of the crop, nonavailability of suitable superior cultivars, traditional production systems, poor technological innovations, and occurrences of insects, pests and disease. Other limitations and problems witnessed in lychee production are as follows:

1. Long juvenile period of lychee
2. Low female/male flower ratio
3. Premature fruit drop
4. Fruit cracking
5. Production of poor-quality fruits
6. Canopy management

7. Lack of critical stages for flower bud differentiation
8. Short shelf life
9. Paucity of infrastructure of roads, cold storage, reefer vans
10. Low productivity and low proportion of quality fruits
11. Narrow genetic base and lack of natural biodiversity.
12. Limited period of availability of fruits with high post-harvest losses.
13. Lack of quality planting materials with scientific knowledge in rapid multiplication.
14. Lack of complete scientific basis of shoot maturity and flowering
15. Lack of quantified role of microorganism in plant health and yield
16. Lack of standard canopy architecture and its management for different density
17. Unavailability of techniques to manage seed borers
18. Short shelf life and lack of information on proper storage environment
19. Very low level of processing, value addition and export of fresh and processed litchi
20. Insufficient product diversification and dissemination
21. Low level of institutional support, infrastructure and human resource development programme
22. Short production season at a particular locale
23. Absence of post-harvest infrastructure
24. Absence of an efficient and nationwide marketing system
25. Unorganized production by individual farmers
26. Unavailability of trained manpower and entrepreneurs
27. Climatic constraints: absence of sufficient chilling temperature during winter and hot winds and hailstorms during summer months, frost/low temperatures during winter
28. As litchi is a temperature-sensitive fruit, access to market is constrained by the lack of a chain of cooled facilities to transport it to distant markets. It is important that the produce reaches distant locations at ambient temperature within 24–36 h after plucking to retain its desired colour. The supply chain from farm to final consumers outside the state market is not so efficient to maintain the timing. Hence, refrigerated trucks and cold storage facilities are essential for targeting,

9.11 Insect Pests: A Major Threat to Lychee Cultivation

Pest dynamics have been changing over time. Litchi mites and litchi bugs have now become economically less important. Among many factors affecting production and productivity, insect pest losses are a major constraint and have emerged as the most important constraint inflicting high economic losses to growers. By the changing dynamics of pests, *Conopomorpha cramerella*, which was earlier regarded as a minor pest, has became very serious, affecting litchi cultivation not only in Bihar but also in other states such as Uttar Pradesh. Generally, two generations of this insect occur during a litchi season.

Fruit infestations that once occurred earlier are now seen during May. Major insect pests recorded on litchi include the leaf mite (*Acerya litchi* Keifer), leaf miner (*Conopomorpha cramerella*), fruit borers (*Conopomorpha cramerella*, *Platypeplus aprobola* Meyer, and *Dichocrosis* sp.), leaf webber/roller (*Platypepla aprobola* Meyer), litchi bug (*Tessarotoma javanica* Thunb.), bark-eating caterpillar (*Indarbela quadrinotata* and *I. tetraonis*), and shoot borer (*Chlumetia transversa*) (Vinod et al. 2011).

The borer complex of litchi is most important as the borers extensively damage the developing and matured fruits, reducing the marketable yields. Among the diseases occurring before harvest, the anthracnose caused by *Colletotrichum gloeosprioides* Penz. is of economic importance, and the post-harvest losses from fruit rots caused by several pathogens (*Aspergillus*, *Cylindrocarpon*, *Botryodiplodia*, *Colletotrichum*, etc.) are also of concern (Prasad and Bilgrami 1974; Awasthi et al. 2005).

A large number of fruits drop from early infestation by this pest. Overlapping generations during the fruiting period (April to June) have been observed. Indiscriminate use of pesticides to control the fruit borer complex in litchi by the farmers, particularly synthetic pyrethroids, seem to be responsible for the high incidence of *Conopomorpha cramerella* in litchi orchards.

The management of the fruit borer complex hence warrants the integration of alternative methods such as use of pheromone trap, biocontrol agents (*Trichogramma* sp.), removal and destruction of dropped fruits and wild hosts such as *Eugenia jambolana* and *Cassia tora* from orchards, prophylactic spray of neem (*Azadiracta indica*)-based insecticides, and need-based application of chemical insecticides.

9.12 Sustainable Methods for Pest and Disease Management

1. Netting is commonly used to deter fruit-eating birds and bats.
2. Copper fungicides are commonly used to control anthracnose and flower blight.
3. Regular pre-harvest application can minimize the incidence of post-harvest rot.
4. Insecticides are occasionally required for severe outbreaks of insects or mite pests. Thus, pesticides labels should be followed strictly.
5. A cleansing spray of the tree trunk after pruning has been observed to reduce and prevent erinose mite damage to leaf, flower and fruit.
6. Soil moisture should be maintained as the fruit moisture.

9.13 New Emerging Diseases

Among the emerging disease problems in lychee production, 'twig blight' and litchi sudden death disease are very important. The symptoms appear as the death of leaves on new shoots and a foliar blight and tip dieback, which are difficult to separate. The leaf blight appears as tan spots on the leaves. The afflicted leaves look as though they were scorched by the sun.

9.13.1 Problems Related to Lychee Cultivation in India

There has been ever-increasing demand for litchi in domestic and export markets. Owing to specific climatic requirements, successful litchi cultivation has been restricted in certain areas but now, with the development of improved cultivation technologies, it is spreading to many other parts of India.

Saturation of areas in the traditional belt, and the lack of post-harvest management facilities and marketing network for the highly perishable litchi fruit, have been responsible for the slow increase in growing area during past decades. An unstable market trend has resulted in low and declining litchi prices, which has further decreased the farmers' profits. India is the second largest producer of litchi, and its productivity is higher than that of China. The existing gap in productivity and potential yield needs to be addressed. In spite of its vast export potential, the quantity of exported litchi fruits is negligible. Therefore, there is a need for not only increasing the production but also the proportion of export of quality litchi fruit. Processing and value addition, and human resource development in the selected sectors. will further boost the litchi industry.

Attention to research and development for overall improvement in the production and productivity of litchi in India has not been adequate. The need for systematic research for identification of additional potential areas, development and refinement of technologies for enhancing productivity and quality along with post-harvest handling, processing and value addition, and the development of high-temperature-tolerant varieties and high yield and processing has long been felt to require suitable focussing.

9.13.2 Pericarp Browning: A Threat to Lychee Production

The availability of lychee fruit has been limited by variability in yield, alternate bearing, seasonal differences, and most commonly post-harvest problems. Lychee has a short life span during which the red colour turns brown, which greatly affects the appeal to the consumer. The pericarp of lychee is also sensitive to desiccation and turns brown and brittle if moisture is reduced to half.

9.14 Genetic Relationship Studies on Lychee

The studies related to genetic relationships and population structures of litchi germplasm are crucial for litchi breeding projects by facilitating the selection criteria of optimal parental combinations and management of germplasm to avoid genetic redundancy. Traditionally the most commonly used criteria for lychee accession were the shape of the skin segments and protuberances, based on which the germplasm is divided into three main types:

(a) Smooth protuberances
(b) Protruding and hard protuberances
(c) Hair-like protuberances

In contrast to the traditional method, an UPGMA dendrogram reveals four main groups of litchi germplasm congruent with fruit maturation time (Liu and Mei 2006). Therefore, 2010 marked an era of a litchi genome sequencing and resequencing project, initiated by researchers from the China Lychee and Longan Industry Technology Research System, which will ensure high throughput tools for standardization and genetic relationship assessment of lychee collection. Moreover, molecular genetic marker technology provides the most direct means for cultivar identification and genetic relationship analysis with a number of systems including random amplified polymorphic DNA (RAPD) (Wang et al. 2006), amplified fragment length polymorphism (AFLP), sequence-related amplified polymorphism (SRAP), and simple sequence repeat polymorphism (SSR). However, to date the studies are not based on a broad genomic infrastructure and the market numbers.

9.15 Government Policies and Plans Framed for Research and Development of Lychee

- The All India coordinated research project on subtropical fruits is providing research support for varietal and production technology improvement.
- The Central Horticultural Experimental Station, Ranchi, Jharkhand RAU, PUSA, Samastipur, Bihar, G.B. Pant University of Agriculture and Technology, Patanagar, West Bengal are engaged in the research projects. The main thrust of research is on these topics:

1. Augmentation of germplasm
2. Propagation studies
3. Technologies for development of fruit production
4. Improved shelf life

A network project for improving productivity of lychee has also been initiated. For the strategic and basic research on lychee, a National Research Centre on Litchi has been started.

9.15.1 The Centre Has Delineated Its Functions

1. To undertake basic, strategic, and applied research for enhancing productivity, improving quality and utilization of litchi fruits.

2. To act as a repository of genetic resources and scientific information on litchi.
3. To demonstrate improved technologies and impart training to stakeholders for upgrading their knowledge base.
4. The Centre has established a National Active Germplasm site of litchi, and total of 52 germplasms including 20 superior clones have been planted for detailed characterization. The Centre has developed a comprehensive plan to identify and collect superior genotypes and clones available in the country by 2017.
5. Development of ideal litchi cultivars is the main focus of the centre for which precocious gene harnessing through natural cross-pollination, provincial population, traditional breeding techniques and understanding the behaviour of seedlessness including molecular characterization and barcoding of genotypes are priority areas of research during the coming years.
6. Widening of the genetic base through creation of variability, conservation of germplasm, description of elite/ideal type, and selecting superior activities.
7. A systematic approach for a litchi descriptor is needed.
8. Faster multiplication techniques for the production of quality planting material are needed (exploitation of grafting and tissue culture techniques).
9. The development of nutrition management to maintain tree health and encourage successful flowering, fruiting and quality in a sustainable manner requires attention.
10. Monitoring of nutrition in litchi through leaf analysis and establishment of correlation between deficiency symptoms and micro-nutrient imbalances would be an approach for efficient fertilizer use.
11. Through effective recycling of residues coupled with organic manure, it is possible to improve soil health (immense potential for organic production of litchi).
12. Infrastructure for post-harvest management requires emphasis to reduce risk of pesticide residue.
13. Starting of a network programme on litchi to boost the production and ensure livelihood security of the people.

In the past, the Centre has developed quite a good number of technologies and refined the package of practices. The technology developed for cropping sequence and crop combinations for interspaces in young orchards to sustain the soil fertility and enhance income in initial years has been widely adopted and appreciated by growers. The protocol for rejuvenation of old senile litchi orchards developed by the Centre has been adopted by the National Horticulture Mission (Bihar) for the large-scale conversion of unproductive orchards to productive ones.

Further refinement of this technology is required for sustaining productivity for a longer period. Some positive indications for management of fruit borers and litchi mites, use of bio-fertilizers and mycorrhizae, conservation horticulture and canopy architecture management have been visualized that need further research intensification for providing a total solution to litchi growers by 2050.

9.15.2 Goals for Enhancing Lychee Production

1. Enhancing litchi production and productivity
2. Reducing cost of litchi production and processing (pre- and post-harvest stages)
3. Exploitation of marginal and degraded lands in traditional and nontraditional areas for commercial cultivation, maintaining the ecosystem
4. Integrating litchi in a farming system approach including agro-forestry or social forestry.
5. Technology assessment, release, refinement and transfer to beneficiaries.
6. Post-harvest technology for litchi fresh fruits and market intelligence.
7. Human resource development.
8. Developing ICAR-NRCL as a Center of Excellence for R&D on all aspects of litchi.

9.15.3 Targets for Improving Lychee Production

1. To improve the competitiveness of domestically produced fruits for international markets.
2. To adjust regional distribution of cultivars and improve the quality, unit yield and economic benefit based on the existing orchard area.
3. Rationalization of cultivar structure that extends the production season of litchi through selection, breeding and biotechnological methods to have high-quality early litchi cultivars and extreme late-season ones.
4. To clarify mechanisms of flower induction, fruit setting, cracking susceptibility, and quality so as to formulate relevant technical strategies to ensure flower induction, improve fruit set and quality, and to prevent fruit cracking

9.15.4 Strategies to Achieve Targets

1. Germplasm collection, conservation, characterization, evaluation and documentation.
2. Selection/improvement of cultivars through traditional and modern tools (biotechnology, bio-informatics, genetic engineering, etc.) for higher quality production and productivity.
3. Litchi cultivation integrated with social and agro-forestry and a farming system approach.
4. Extending intensive cultivation in nontraditional areas.
5. Improved integrated management system for nutrients, water, pests and diseases, and climate resilience.
6. Sustainable and economical production/protection/management system through natural resource management.
7. Development of models for forewarning against natural/biological disaster.
8. Developing models for forecasting litchi production.

9. Mechanized farming system in litchi.
10. Multiplication of quality certified planting materials for timely availability.
11. Litchi marketing research and infrastructure.
12. Simple, low-cost and high-recovery technology for processing of litchi along with packaging.
13. Linkages with litchi-related national and international organizations through networking and collaboration.
14. Creating and maintaining a database on all aspects of litchi.

9.16 Role of PGPRs in Lychee Production

Plant growth-promoting regulators (PGPRs) are a class of organic compounds other than nutrients supplying either energy or mineral elements that in small amounts promote, inhibit or otherwise modify any physiological processes in plants.

Five well-established classes of phytohormones are necessary at different stages of litchi cultivation.

1. Auxins
2. Gibberellins
3. Cytokinins
4. Abscisic acid
5. Ethylene

Role of auxins:

(a) Propagation
(b) Stimulation of fruit set
(c) Chemical thinning
(d) Prevention of fruit drop
(e) Herbicidal action

Role of gibberellins:

(a) Increasing fruit size
(b) Stimulating fruit set
(c) Ripening of fruit
(d) Controlling flower bud production
(e) Controlling environmental stress

Role of cytokinins:

(a) Cell division
(b) Increase of internal bud differentiation

(c) Morphogenesis

Role of abscisic acid:

(a) Stomata closure
(b) Cold hardiness
(c) Delayed flowering
(d) Induced storage proteins in seeds

Role of ethylene:

(a) Promotes fruit ripening
(b) Increases latex flow
(c) Delays flowering

Role of jasmonates:

(a) Inhibits growth and seed germination
(b) Promotes abscission, tuber formation and pigment formation

Role of brassinosteroids:

(a) Promotes stem elongation
(b) Promotes ethylene biosynthesis and epinasty

9.17 Challenges Faced During Lychee Cultivation

Over a period of time, a large number of ecotypes in various countries including India have been identified; however, a cultivar having all the required characters in one has not been possible. Litchi plants are very good yielders and generally not affected by most serious biotic stresses, but abiotic stress, particularly that of climatic aberrations, affects the fruit yield and quality. Numerous post-harvest diseases have their origins from pre-harvest management.

(a) Fruit drop problems:

The most problematic phase in lychee cultivation is the phase of fruit drop which occurs because of the failure of fertilization of the ovule immediately after fruit set. The maximum fruit set has been reported in Muzzaffurpur and Calcutta.

(b) Healthy hygienic conditions

Good field sanitation, pre-harvest disease management, and discarding fruit with symptoms of fruit-piercing insects, cracks and sun scorch are effective in minimizing losses.

(c) Insect–pest management

Disease-causing organisms include *Aspergillus* spp., *Pestalotiopsis* spp., *Peronophytophthora* spp., sour rot caused by *Geotrichum candidum*, and yeasty rots caused by *Botryodiplodia theobromae*, *Colletotrichum gleosporioides*, and *Rhizopus oryzae*. Insect and pest management is the need of the hour.

(d) Production challenges

Among production challenges, optimum shoot maturity for flowering and encouraging a sufficient number of mature shoots in a litchi tree needs the most focused attention. The number, opening time, duration of flowering and strength of female and functional male flowers coupled with a pollinator are the main deciding factors for litchi yield and need much critical attention. The hormonal status and balance of different phytohormones in the developing fruit and even in the flowering twig also need observation. The sink–source relationship, the nutrients reserve in different plant parts, is essential to determine plant performance. Major nutrient and micronutrient availability via sustainable supplementation through various inputs and means are the focused areas to be addressed. Optimization of orchard environment above the soil and subsurface during fruit development and maturity through advance tools and their standardization is also a major challenge for litchi.

9.18 Integration of Plans (Vision 2050) to Promote Lychee Production and Cultivation by NRCL (National Research Centre on Litchi)

1. Production of litchi to be enhanced by 33% and also to increase the sale of Indian litchi in export markets is also included in our priority list. Litchi remain a 'special' fruit in the minds of consumers and that niche enables a premium price to be delivered; there is much potential for sales growth in the domestic market. Research reveals a high proportion of consumers (of non-Asian background) who are unfamiliar with or have not tried litchi, which can also enhance demand. With adequate marketing effort (and at the same time addressing the retail quality issues) the industry should be able to capit0alize on this market growth. Considering the present constraints and opportunities the following actions may be undertaken immediately: introduction/selection of high-quality litchi varieties of early, mid, and late maturity.

2. Intensive training of farmers and the extension agent on the modern methods of litchi cultivation and management (i.e., high-density orcharding with precise application of inputs).
3. Production and distribution of quality planting materials.
4. Development of complete protocol of post-harvest management from harvest to wholesale mandis of a distant market with inhibited pericarp browning.
5. Ethnic advice for establishment of market infrastructure and information system.
6. Setting up of cooperative marketing units for marketing of fresh litchi produce.
7. Establishment of processing units in litchi production zone.

9.19 Litchi Marketing

National and international marketing of litchi is expanding very fast. The Indian litchi has a period advantage of fruit availability (May–July) when premium quality litchi is not available elsewhere except Thailand (May–June) and Israel (July), but the quantum of produce available from these (two) countries is very minor or negligible. Despite the period advantage, India has negligible share in the world litchi trade. This happens because of strong interstate trade of litchi and more demand of produce at higher price in domestic market. Besides meeting domestic demands efforts are needed to promote export of litchi during main season. Before exporting litchi abroad, certain requirements of international marketing must be fulfilled.

(a) The fruits should be of extra grade: the extra-grade fruits measuring 33 mm in diameter and coinciding with nearly 95% to the specified characters of variety with respect to fruit colour and shape. Only 15–20% of fruits have this character and can be graded thus for export.
(b) The fruits should be properly packed and stored: sulphited litchi fruits are banned in the international market. Therefore, to keep the fruits in proper condition on arrival at a terminal market, modified atmospheric packaging (MAP) along with control atmospheric storage (CAS) condition are required.
(c) An effective communication and networking between fruit growers and various stakeholders in the value chain is needed for organized export.
(d) Most of the litchi produced in Indi, particularly in Assam, Tripura, and Bihar, is organic in nature and should be exported as organic litchi to the foreign market. Even for normal produce, "brand promotion" is needed to catch the market.

9.19.1 Reasons for Declining Status of Litchi Export

The export share of litchi is declining for the following reasons.

1. The domestic demand for the fruit is very high.

2. Limited produce is available for export purposes and the U.S. market allows only produce free from the oriental fruit fly.
3. Lack of initiative for export facilities.
4. Fresh fruits are highly perishable. As the sulphuration process is prohibited in the international market; the shelf life and appearance of the fruit are reduced.
5. Lack of initiative for export facilities.
6. High freight charges mean only exporting litchi by ship to nearby markets may be profitable. In comparison to long-distance export, profit is better in the domestic market.

The quality of the produce as per the cultivar codex, infrastructure support for transport and market information system via the government polices are vital in marketing of highly perishable litchi fruits.

The pre-harvest contract system is the most prevailed and acceptable method of litchi marketing as most of the growers prefer it for obvious reasons; however, some of them undertake self-marketing. Growers, who do not intend to market their litchi harvest give their litchi orchards to pre-harvest contractors (PHC) during the month of January. PHC sells the orchard to the wholesale or commission agent, who undertakes the harvesting, packing and transportation of the produce to the market. Self-marketing in the distant domestic market is considered to be more profitable with higher net gain but it involves a high marketing risk. Absence of price information and poor leadership are the constraints in self-marketing of litchi fruits. The short span of fruit availability in a particular orchard, high marketing expenditure, long distance to main market centres, remote production areas, absence of market regulation and exploitative marketing practices by market intermediaries are some of the important factors impeding self-marketing. The major marketing difficulty with litchi involves preserving its attractive red colour.

9.19.2 Steps for Enhancing Competitiveness for Exports

1. To enhance the competitiveness of India in exporting litchi, the following measures can be suitably undertaken.
2. To exploit export of organic litchi in foreign markets, its cultivation in Tripura and Assam needs to be encouraged. and to facilitate it, packing houses need to be established in a phased manner. Side-by-side markets for organic litchi need to be identified.
3. Litchi-producing areas in the Gurdaspur and Hoshiarpur districts of Punjab are near the Amritsar international airport. Export of litchi from this area can be enhanced by setting up a packing house in the area.
4. Technology for CA and MA storage of litchi needs to be standardized so that the shelf life of litchi can be extended, and then litchi can be sent to nearer areas such as the Gulf countries by reefer containers through MA cartons.
5. A residue analysis laboratory need to be set up appropriately to minimize the risk of rejection of a consignment.

9.20 Future Recommendations for Lychee Cultivation

1. The quality of some of the cultivars is inferior compared to the traditional types. New cultivars need to be developed having better fruit quality.
2. More research is required to define the optimal temperatures for flowering in the major commercial cultivars.
3. Much work is required to raise productivity as a large gap exists between actual and potential yields.
4. Prospects for increasing production and marketing of this crop are high if some of the growing, postharvest and marketing issues are resolved.
5. Intraregional cooperation would assist industry development and hence amplify the importance of the crop to local economies.
6. Training sessions for extension and scientific staff are also a priority.
7. New protocols and techniques need to be designed for different microclimatic conditions.
8. A long-term breeding programme is needed to develop better cultivars, where the best is implemented with a regional focus.
9. The current gene pool needs to be systemically evaluated.
10. A much stronger varietal improvement programme is needed to be initiated throughout the countries.
11. More research needs to be focused on standardizing nursery techniques.
12. Education of nursery workers and growers in tree care and the provision of adequate irrigation need serious attention.
13. Organic farming needs to be implemented for better nutrient cycling and sustainable cultivation.
14. The preferred cultivars, packaging, etc., for each market need to be developed.

9.21 Conclusion and Prospects of Lychee Cultivation

Lychee, a climatic evergreen fruit plant introduced to this country in the nineteenth century, has adapted well to the climate in Eastern India. Because of its increasing demand, the area under lychee cultivation has constantly increased. However, there is a need for improving productivity and also widening the genetic base.

To progress in this direction, a concerted research effort and effective linkages are essential. On the other hand, suitable cultivars are needed for different climatic conditions. It is also essential to develop promising hybrids, having larger fruit size, small/chicken-tongued seeds, and tolerance to pericarp splitting and having various maturity groupings. Agro-techniques, particularly for source and sink management, micronutrients, post-harvest technology, and effective marketing, need due concern.

9.21.1 The Following Points Need Due Considerations

1. Lychee depends on a narrow genetic base, which needs to be widened by proper selection of genotypes from the existing population.
2. Target-oriented projects must be projected for germplasm conservation and use.
3. A lychee descriptor needs to be developed for the description of cultivars.
4. Propagation technology needs to be developed to allow faster multiplication of quality plants.
5. Sustainable nutrient management would encourage successful flowering, fruiting and quality.
6. Lychee nutrition monitoring through leaf analysis would be a step for efficient fertilizer use.

To grow and market lychee successfully, careful planning and management technologies need to be implemented:

1. Routine operations should be followed during lychee production and the cultivation process which should include leaf and soil analysis, fertilizing, irrigation scheduling, weed management, pest monitoring and bird and flying fox control.
2. A quality assurance system is needed that addresses food safety requirements and incorporates hazard analysis and critical control point (HACCP) principles.
3. Organize labour for harvesting and packing lychees.
4. Correct picking and grading of lychee fruits.
5. Distance from main metropolitan markets for some growers results in high freight cost.
6. A shifting market preference to small-seeded varieties at the expense of some established large-seeded varieties.
7. Quarantine issues slow the development of export markets.
8. The short intense harvest period coinciding with the summer wet season and often around Christmas time can make the labour and transport difficult.

References

Awasthi DP, Sarkar S, Mishra NK, Kaiser SAKM (2005) Disease situation of some major fruit crops in new alluvial plains of West Bengal. Environ Ecol 23:497–499

Chadha KL (1968) Litchi cultivation in India. Indian Horticult 12:13–16

Duan X, Jiang Y, Su X, Zhang ZY, Shi J (2007a) Antioxidant properties of anthocyanins extracted from litchi (*Litchi chinenesis* Sonn.) fruit pericarp tissues in relation to their role in the pericarp browning. Food Chem 101:1365–1371

Duan X, Wu G, Jiang Y (2007b) Evaluation of the antioxidant properties of litchi fruit phenolics in relation to pericarp browning prevention. Molecules 12:759–771

Jiang G, Lin S, Wen L, Jiang Y, Zhao M, Chen F, Prasad KN, Duan X, Yang B (2013) Identification of a novel phenolic compound in litchi (*Litchi chinensis* Sonn.) pericarp and bioactivity evaluation. Food Chem 136:563–568

Kotur SC, Singh HP (1994) Varietal differences in leaf nutrient composition of litchi. Indian J Horticult 51:59–62

Li W, Liang H, Zhang MW, Zhang RF, Deng YY, Wei ZC, Zhang Y, Tang XJ (2012) Phenolic profiles and antioxidant activity of litchi (*Litchi chinensis* Sonn.) fruit pericarp from different commercially available cultivars. Molecules 17:14954–14967

Liu CM, Mei MT (2006) Classification of lychee cultivars with RAPD analysis. Acta Hortic 665:149–159

Pandey RM, Sharma HC (1989) The litchi. ICAR Publication, New Delhi

Prasad SS, Bilgrami RS (1974) Investigations on diseases of litchi: VI. Post harvest diseases of fruit in India. Plant Dis Rep 58(12):1134–1136

Prasad NK, Yang B, Zhao M, Wang BS, Chen F, Jiang Y (2009) Effects of high-pressure treatment on the extraction yield, phenolic content and antioxidant activity of litchi (*Litchi chinensis* Sonn.) fruit pericarp. Int J Food Sci Technol 44:960–966

Sakurai T, Hiroshi N, Hajime F, Hideki O (2008) Antioxidative effects of a new lychee fruit derived polyphenol mixture, oligonol, converted in to a low molecular form in adipocytes. Biotechnol Bioeng 72(2):463–476

Singh HP, Yadav IS (1988) Physio-chemical changes during fruit development in litchi cultivar. Ind J Horticult 45:212–218

Singh HP, Yadav IS (1992) CHES-1, a promising selection of litchi. In: National seminar on recent development in litchi production, RAU, PUSA, 30–31 May 1992

Vinod K, Amrendra K, Vishal N (2011) Emerging pests and diseases of litchi (*Litchi chinensis* Sonn.) Pest Manag Horticult Ecosyst 17(1):11–13

Wang JB, Deng SS, Liu ZY, Liu LZ, Du ZJ, Xu BY (2006) RAPD analysis on main cultivars of litchi (*Litchi chinensis Sonn.*) in Hainan. J Agric Biotechnol 14:391–396

Zhao M, Yang B, Wang J, Li B, Jiang Y (2006) Identification of the major flavonoids from pericarp tissues of lychee fruit in relation to their antioxidant activities. Food Chem 98:539–544

Lychee Diseases and its Viable Control 10

Shanker Kumar Pandey, Awadh Kishore Roy, Shyam Chandra Roy Chandel, Anuj Rani, Rashmi Singh, and Ravi Shankar Mishra

Abstract

Litchi chinensis Sonn. is a common fruit of India especially grown in the state of Bihar and surrounding states. Common forest regions, botanical gardens, and plant conservatories were surveyed at a regular interval of 3 months during the period April 2002 to March 2012 for the collection of diseased samples under the AICOPTAX program. During the course of the survey a number of diseases such as tar leaf spot were observed. Older leaves show a higher percentage of infections than younger leaves. The diseased leaves serve as the source of primary infection. The leaf spots spread rapidly during humid and rainy months. The lesions coalesced at the later stages. The diseased leaves show yellow brown to brick red areas, mostly towards the margins. The percentage of disease incidence varied in different localities. The coloured areas gradually become light brown and show black pinheads resembling pycnidia. *Pestalotia pauciseta* Sacc., *Botryodiplodia theobromae* Pat., and *Colletotrichum gloeosporoides* occur very commonly in litchi orchards, causing tar disease. Another serious leaf spot disease is incited by *Microdiplodia litchi* and *Clitocybe tabescence*. Sudden wilt caused by *Fusarium* sp., dieback from *Phomopsis*, and foliar necrosis caused by *Gloeosporium* spp. are reported. Some of the post-harvest diseases are carried along with the fruits during transit and storage. Among these pathogens, such as *Botryodiplodia theobromae*, *Colletotrichum gloeosporioides*, *Cylindrocorpon forkinense*, *Phomopsis* spp. and species of *Pestalotia* are the most important and need attention as they also cause diseases on the plants.

S.K. Pandey (✉) • S.C.R. Chandel
Department of Botany, Koshi College, Bhagalpur University, Bhagalpur 851205, India
e-mail: spfungi@gmail.com

A.K. Roy • A. Rani • R. Singh • R.S. Mishra
Department of Botany, TM Bhagalpur University, Bhagalpur 812007, India

M. Kumar et al. (eds.), *Lychee Disease Management*,
DOI 10.1007/978-981-10-4247-8_10

Spraying of leaf extracts of *Andrographis paniculata*, *Azadirachta indica*, *Plumbago zeylanica*, and *Ocimum sactum* reduced the infection; however, fungicides such as bavistin, topsin M, baycor, and dithiocarbmates were found to be very effective to control the disease.

Keywords
Litchi chinensis • Infection • Disease • Percentage disease incidence • Biocontrol

10.1 Introduction

The medicinal value of *Litchi chinensis* Sonn. (Sapindaceae) is well known as it has been widely used in many cultures for the treatment of cough, flatulence, stomach ulcers, diabetes, obesity, testicular swelling, hernia-like conditions, and epigastric and neuralgic pains (Butani 1977; Chadha 1968). The ethnopharmacologial history of *L. chinensis* indicates that it possesses hypoglycemic, anticancerous, antibacterial, antihyperlipidemic, antiplatelet, antitussive, analgesic, antipyretic, hemostatic, diuretic, and antiviral activities (Kotur and Singh 1994; Kunwar and Singh 1993; Mallik and Singh 1965; Pandey and Sharma 1989; Rao et al. 1985; Singh 1992). The lychee (*Litchi chinensis* Sonn.), an important subtropical evergreen fruit crop belonging to family Sapindaceae, is believed to have originated in China, where it has been grown in Southern Guangdong State for thousands of years. It is highly specific to climatic requirements, and probably for this reason its cultivation is restricted to a few countries in the world. In India, lychee was introduced in the eighteenth century through Burma, and from there, it spread to many countries. India and China account for 91% of the world lychee production but it is mainly marketed locally. In India, 4,28,900 metric tonnes of lychee is produced annually from 56,200 ha. Lychee, being exacting in climatic requirements, is confined to a few states with 74% of production recorded in Bihar. In this state, lychee is the livelihood for millions of people as it provides both on-farm and off-farm employment. Small and marginal farmers obtain additional income from lychee plants in their homesteads. Thus, lychee cultivation is the livelihood security for a large population, especially in the state of Bihar. The lychee tree is handsome, dense, round topped and slow growing with six to nine elliptical oblong and lanceolate, abruptly pointed, evergreen leaves. Leaf colour varies from light green to dark green. Greenish white or yellowish flowers are borne in clusters. Fruits are round or heart shaped having thin, leathery skin. The colour of fruits varies with the cultivar but is red or rose or pinkish. The edible portion or fruit is the aril, which is immediately beneath the skin. The flavour of the aril, which is distinctive, varies with the cultivar. Seeds are bold but in some cultivars the seeds are partially developed from failure of pollination, referred to as 'chicken-tongue' seed. The trees with small-seeded fruits are prized because of the greater portion of pulp. Considering the importance of this fruit crop in the region, efforts are made to provide technological support through research and promoting production, postharvest management and marketing, including export, through development programmes. Lychee has also been identified as an important crop for export. Currently, Indian export of lychee remains quite small because of the expanded domestic market. The growing of lychee in different states under various climatic conditions has advantages in terms of earliness and extended harvest. With a narrow genetic base, under given climatic conditions, fruits are available only for 3–4 weeks. However, the spread of cultivation over a wide range of climates offers the possibility for extending the cropping period from the 1st week of May to the 1st week of July (Dass and Choudhary 1958; Jamaluddin and Tandon 1971, 1973, 1975, 1976, 2004). Evidently, with an expanding market, there is ample potential for increasing area and production with improved production technology and efficient post-harvest management and storage (Kotur and Singh 1993; Pandey and Misra 1975). This chapter is concerned with the current status and identifies the constraints that need to be addressed.

A few diseases affect the leaves, flowers, fruits, and some other parts, causing tree deaths or decline. Lychee plants and fruits are affected by microbes, insects, and pests which cause considerable losses if not managed (Verma et al. 1980; Yadav and Singh 1992). Lychee plants as compared to many fruit-bearing species are least affected by diseases. A few leaf spot diseases have come to light that are caused by fungal pathogens. No bacterial or viral infections have been reported so far. Powdery mildew (*Oidium* spp.), anthracnose or leaf spot (*Botryodiplodia theobormae* Pat, *Colletotrichum gloeosporioides* Penz) and red rust (*Cephaleuros mycoides*) are diseases which cause some damage to the lychee crop, but severity varies from season to season, even in the same locality. Control measures consist of one or two applications of the proper fungicides; for red rust, sulphur washes in September–October and February–March are sufficient (Ray et al. 1984; Singh and Yadav 1988, 1992a, b). About 40 insects, mites, and pests are reported to affect lychee trees and fruits at different stages of growth.

10.2 Sampling

Different forest regions, botanical gardens, and plant conservatories of Bihar were surveyed at regular intervals of 3 months during the past 3 years for the collection of diseased samples of litchi plants under the AICOPTAX program. Collections were placed separately in polythene bags, and a paper slip giving the name of the host, place and date of collection, and infected plant parts tagged each collection. The collected materials were brought to the laboratory for pathological and taxonomic studies Singh HP (1998).

10.3 Isolation, Purification and Systematization of Pathogens

Isolation was made from the junction of healthy and diseased regions after surface sterilization with 90% alcohol or 2% NaOCl and allowed to grow on potato dextrose agar (PDA) medium in culture tubes. Fungal colonies were developed within 3–7 days and were purified by the single-spore isolation method. Media were also changed to Asthana and Hawkers and Czapex Dox agar as per requirements of fungi. The stock culture of each fungus was determined by regular subculturing after 6 weeks.

10.4 Identification

Identification keys of different genera and species of fungal pathogens proposed by were followed in determining the identification of fungi, which were further authenticated by comparing with culture and herbarium specimens of IARI, New Delhi. Algal members were identified with the help of recent and relevant manuals.

10.5 Preservation of Diseased Specimens and Cultures

The cultures were preserved by following standard techniques. The diseased specimens were pressed in blotter paper repeatedly to remove moisture (Singh and Singh 1954; Singh and Kumar 1988). Each specimen was enumerated, documented and also deposited to the Mycology and Plant Pathology Division of IARI, New Delhi, for confirmation of identity of the pathogen and also the HCIO numbers.

10.6 Percentage Disease Incidence (PDI)

During the course of collection of diseased leaves the percentage disease incidence (PDI) was also determined by adopting the following formula:

$$\frac{\text{Total no.of diseased leaves in a plant/twig}}{\text{Total no.of healthy and diseased leaves in a plant/twig}} \times 100$$

10.7 Pathogenicity Tests

Tests were conducted by following the usual methods:

- By putting inoculum directly on the leaf surface
- By spraying spore suspension on the injured and uninjured leaf surface

Injuries were made with sterilized needle or sterilized sand, and care was taken that the injury was only superficial. Whenever more than one fungal organism was isolated from a particular leaf spot, the pathogenicity of each organism was tested separately. Pathogenicity was confirmed only when Koch's postulates were fully satisfied. The host range of different isolates was determined by usual inoculation methods. During reisolation and stocking of cultures, different culture media were used. The stock cultures were maintained either on PDA or Asthana & Hawkers media when required.

10.8 Application of Botanicals

Leaf extracts of five medicinal plants, viz. *Azadirachta indica*, *Andrographis paniculata*, *Cymbopogon citratus*, *Ocimum sanctum*, and *Plumbago zeylanica*, were prepared by the following methods of Kumar and Sachan (1979). For extracts, 50 g fresh leaves of plants were thoroughly washed in tap water and finally in distilled water. The leaves were mixed with 100 ml sterilized water and then crushed in a mixer and squeezed. The extracts were filtered through two layers of muslin cloth. The extract was then filtered through Whatman no. 1 filter paper. The filtrate was centrifuged at 5000 rpm for 10 min. The supernatant was then passed through a Millipore filter and was also used against radial growth, spore germination and dry biomass of test fungi.

10.9 Radial Growth of the Fungi

PDA and Asthana & Hawkers medium were prepared by adding leaf extracts and aqueous solution of chemicals in equal volume (10:10 v/v) in such a way to maintain the concentration as was the control medium. In the control, only 20 ml of medium was taken. The sterilized medium was poured in Petri plates, allowed to solidify, and then inoculated with a 8- to 10-day-old culture of test fungi (*Colletotrichum gloeosporioides*). The plates were incubated at 28 ± 2 °C in a biological oxygen demand (BOD) incubator for 7 days. The radius of the developed colonies was measured to estimate the fungal growth.

10.10 Mycelial Dry Weight

The dry weight was determined by the following method of Dwivedi and Pathak (1978) for which liquid medium and leaf extracts in equal ratio (20:20 V/V) were placed in a flask. A piece of test fungus (*Colletotrichum gloeosporioides*) was placed in each flask and left for 10 days at 28 ± 2 °C for growth. After 7 days of incubation, the contents of each flask were stirred vigorously and then filtered through filter paper. The mycelial mat obtained as residue was dried at 60 °C for two successive days and mycelial dry weight was then determined.

10.11 Spore Germination

Standard spore suspension of the test pathogen was prepared with fresh culture in 10 ml sterilized water. The suspension was centrifuged at 3000 rpm for 15 min and washed twice by distilled water. Finally, the concentration of spores in each case was maintained 100–150/0.5 ml water by the dilution method: 1 ml spore suspension was mixed with equal volume of leaf extract, and spore germination was studied following the hanging drop method. First reading was taken after 6 h and subsequent readings were taken up to 22 h at intervals of 2 h. Readings were taken to determine the percentage of spore germination. The entire study was conducted in triplicate. Fungal toxicity of leaf extracts chemicals was expressed in terms of percentage of spore germination inhibition and was calculated by the following formula (Dixit et al. 1978).

$$\text{Percentage inhibition of spore germination} = \text{Gc} - \text{Gt} \times 100$$

Whereas

Gc = Average spore germination in control and
Gt = Average spore germination in treatment

10.12 Pathogenic Invasion

In nature, the aerial part of the plants in general and leaves in particular harbour a large number of pathogenic or nonpathogenic airborne fungi, but so far as disease development is concerned, the event starts from spore germination, penetration of the appressorium inside the host, and there its further development derives nutrition from the host. Thus, the host–pathogen interaction is an intricate process which involves biochemical warfare with the production of enzymes and toxins as the armory. A regular survey of study sites were made during the past 5 years for the study of plant diseases. A number of diseased leaves manifested perceptible symptoms such as leaf blight and leaf spots, shot holes, and anthracnose. Leaf spot diseases are caused by different fungi. During the survey, symptoms were recorded and the infected leaves and other parts were collected for the identification of fungal pathogens. Litchi plants as compared to many fruit-bearing species are least affected by diseases. Diseases are more important before harvesting and after harvest, although undoubtedly many of the fruit are infected before picking. There are a few organisms that infect the leaves, flowers and fruits, and a few others that are associated with tree decline and tree death. A few leaf spot diseases have also been reported which are caused by fungal pathogens. No bacterial or viral infections have been reported so far. A few reports of algal infections are also available. The details of common diseases observed are as follows.

10.12.1 Brown Blight

Brown blight, known as blossom blight or downy mildew, is caused by *Peronophythora litchi.* It attacks both young and ripe fruit, and the pedicels and leaves of litchi. The infection reduces production and shelf life. Flower panicles are particularly susceptible. Browning of flowers and desiccation of panicles, attacking young and ripe fruits causing brown lesions and white downy growth or premature fruit drop, are the symptoms. Immature fruit turn brown, and those infected before harvest have a white mildew growing on the skin. The pathogen probably persists as zoospores or dormant mycelium in the soil or in plant debris. Higher temperatures during the day are suitable for sporulation, germination and infection by the pathogen, and lower temperatures and high humidity at night facilitate zoospore release and distribution. Continuous rain and reinfection are the most important factors leading to the wide distribution of this disease.

10.12.2 Anthracnose

In anthracnose, caused by *Botryodiplodia theobromae*, the spots usually start from the tip or the margin of the lamina. These spots are deep chocolate in colour. The limiting margins of the spots with irregular outline are Vandyke brown. Black pycnidia appear on both surfaces of the leaves but more often on the upper one. These spots are irregular in outline and are brick brown in colour with prominent marsh brown margins encircling them.

10.12.3 Tar Disease (Anthracnose)

Mummy brown, waxy subepidermal acervuli appear on both surfaces of the infected leaves but especially so on the upper one. Anthracnose (*Colletotrichum gloeosporioides*) attacks leaves and branches as well as flowers and flower stalks and fruit. Lesions on the leaves may appear as small round light grey areas, or irregular brown marks at the tips. In contrast, infections are much more obvious on the flowers and fruit. Outbreaks are common after warm wet weather. The infection remains restricted to the lower leaf surface. Under no circumstances does the upper surface of healthy leaves become infected. During the rainy season, rain water trickling through the diseased leaves is richly laden with the spores of the pathogens to cause further infections. The fungus may not always cause immediate disease, which sometimes only becomes apparent after harvest. Fungicides are used during an initial outbreak but are not always effective. Prevailing climatic conditions (high temperature and high relative humidity) are highly conducive for the development and spread of anthracnose. Higher latent infection rates of anthracnose of the fruits cause more serious post-harvest decay and browning of the fruits.

10.12.4 Tree Decline and Root Rot

A slow decline and sudden death have been recorded in litchi, which can affect the whole tree or just one or two branches. The symptoms include a sudden branch wilt that is followed by the decline of new growth on the affected branch over a period. In other situations, the tips die without wilting. The tree or branch may recover temporarily but subsequently dies. Parts of the tree flush and grow while other sections die. A number of organisms including *Phytophthora*, *Pythium*, and *Fusarium* have been isolated from the roots of trees, but it is not known where they cause the disease. The fungus may survive in the soil or on stumps and roots of various trees for many years. In some parts of the litchi belt, trees are killed by the root rot.

10.12.5 Root Collar Rot

This disease, caused by *Botryodiplodia theobromae* Patr., can quickly kill the trees, and no satisfactory control method has been established. A species of *Fusarium* has been isolated that is associated with the sudden death of the plant. One side of the tree's crown may be perfectly sound and the other totally necrosed. Leaf shed never occurs (it does in the case of a nematode attack), and the internal parts of the roots are characteristically red in colour. As with the previously mentioned cases, no control method has been reported. No method has been found to save the tree once it has become infected.

10.12.6 Red Rust

Red rust is caused by *Cephaleuros mycoides*, a parasitic alga which occasionally attacks trees, causing loss of vigour. The alga starts its appearance as small dark isolated patches which spread very fast and ultimately develop into a velvet reddish brown to orange cushion-like growth. Algal filaments associated with both asexual and sexual reproductive organs grow on both surfaces of the leaves and penetrate deep between cuticle and epidermis, sometimes extending between adjacent epidermal cells into the layer of parenchyma below the epidermis. The growth of this alga initiates the development of cork tissue in a few upper layers of leaves, thus causing their death. Severely infected leaves exhibit curling inward towards dorsal side. The disease first appears on the young unfolded tender branches. On the infected young leaves, small lesions of velvety white growth appeared on the lower surface. On the upper surface, just at the opposite site of the lesion, chlorotic patches occur. As the leaves unfold and increase in size, the velvety growth becomes more prominent and dense. Later, larger areas of leaves are affected with this growth. Old and thick leaves show various types of malformation. The velvety growth turns light brown to brick red.

10.12.7 Fruit Rot

Fruit rot of litchi has been a serious problem. Litchi is host to a range of post-harvest pathogens, often with quite different modes of infection. Several fungi have been found to be associated, as reported by several workers. Usually large water-soaked lesions appear on the surface of fruits. Initially the disease symptoms are perceptible on the injured portion of the fruits. With the advance of the disease, the decayed areas become depressed. The rot gradually penetrates deep into the pulp. Ultimately the rind of infected fruits cracks off, exposing the pulp which subsequently is covered with thick cottony mycelium. Such affected fruits emit an odour of fermentation. This disease is caused by *Pythium* spp.

10.12.8 Powdery Mildew

In winter season a white powdery mass is observed on the surface of the leaves. Initially it is small but gradually develops to cover the maximum photosynthetic area. On isolation, *Oidium* spp. were observed.

10.12.9 Leaf Spot

In summer season, reddish to dark black spherical spots are irregularly distributed on the leaf. Black pinhead-like structures were noticed on both surfaces of the leaf. In the early stage the the spots were minute but these spread throughout the leaf gradually. *Phomopsis* spp. has been isolated from such leaf spots.

10.12.10 Leaf Spot

Light brown irregular spots starting from the margin were recorded. No fruiting body was noticed. *Nigrospora sphaerica* (Sacc.) Mason was isolated from this symptom, a dark brown irregular big spot on the leaf. Blackish brown wet points were present on the dorsal side of the leaf. On pathological examination *Pestalotia* spp. was isolated from these spots.

10.12.11 Leaf Spot

Reddish-brown irregular spots were recorded on the surfaces of the leaf starting from the margin. Light- to dark-brown fruiting bodies were found on the leaf surfaces. *Pestalotia pauciseta* was identified as the causal organism.

A large number of other symptoms were observed on the leaves, but the pathogens were the same.

10.13 Disease Control by Botanicals

In view of known antifungal properties, the effects of leaf extracts of five medicinal plants (*Andrographis paniculata*, *Azadirachta indica*, *Cymbopogon citratus*, *Ocimum sanctum*, and *Plumbago zeylanica*) on growth behaviour and germinability of the test fungus tar pathogens has been studied to determine the effectiveness so that the positive results may be applied to control leaf spot disease. The results are depicted in Table 10.1.

The findings as reported Table 10.1 reveal that the fungus attained 70-mm colony diameter at 370 mg mycelial dry weight (wt.) and 70% of spore germinability under control. The growth behaviour and germinability of the test fungus (*C. gloeosporioides*) under the influence of the plant extracts varied considerably. Leaf extracts of *A. paniculata* showed minimum colony diameter (8 mm), followed by *C. citratus* (14 mm), *P. zeylanica* (16 mm), *A. indica* (17 mm), and *O. sanctum* (20 mm). The maximum inhibition in colony diameter was observed in *A. paniculata* (89%), followed by *C. citratus* (80%), *P. zeylanica* (77%), *A. indica* (76%), and *O. sanctum* (76%) over control. The mycelial dry weight was found minimum by the application of *P. zeylanica* (43 mg), followed by *O. sanctum* (53 mg), *C. citratus* (65 mg), *A. paniculata* (146 mg), and *A. indica* (172 mg). The maximum percentage inhibition was noticed in *P. zeylanica* (88%), followed by *O. sanctum* (85%), *C. citratus* (82%), *A. paniculata* (61%), and *A. indica* (54%). Percentage germination of the test fungi was found minimum in *O. sanctum* (06%), followed by *A. indica* (08%), *A. paniculata* (10%), *C. citratus* (12%), and *P. zeylanica* (15%). The maximum percentage inhibition was found in *O. sanctum* (91%), followed by *A. indica* (89%), *A. paniculata* (86%), *C. citratus* (82%), and *P. zeylanica* (79%).

Table 10.1 Effects of leaf extracts on growth behaviour and germinability on tar fungi

Number	Plant extracts	Colony diameter (mm)		Mycelial dry wt (mg)		Germinability	
1	Control	70	Percent (%) inhibition	370	Percent (%) inhibition	70	Percent (%) inhibition
2	*Andrographis paniculata*	08	89	146	61	10	86
3	*Azadirachta indica*	17	76	172	54	08	89
4	*Cymbopogon citratus*	14	80	65	82	12	82
5	*Ocimum sanctum*	20	71	55	85	06	91
6	*Plumbago zeylanica*	16	77	43	88	15	79

10.14 Disease Control by Chemicals

Cleaning up the orchard by removing shaded, infected and dead branches after harvest will control disease. Copper oxychloride spray during winter and copper sulphate in spring also help to reduce inoculum levels. Metalaxyl spray during flowering and fruit development reduces disease occurrence. Overcrowding of trees and branches in the orchard should be avoided. Pruning of affected plants and burning has been suggested to minimize the chances of fresh infections. Copper oxychloride spray has been found effective but calendar sprays of copper are costly and could lead to unacceptable residues if used close to harvest. Spray of 3:3:50 Bordeaux mixture in February, April, and September–October or application of Caption 50 WP at 0.2% was effective. The rates of latent infection of the fruits could be evidently controlled by integrated management of the disease in the growing season, and the post-harvest decay and browning of the fruits are effectively reduced. The effect of storage can be improved when the measure is applied to control the latent infection of anthracnose on litchi fruit before harvest. Growers are recommended to remove all roots from other host trees before planting litchis. Setting up a drainage system before planting the litchi ensures that there is no standing water around the base of the trees. Pruning the tree to reduce evaporation and encourage root growth helps the tree to recover quickly. Adding organic manure improves the biological resilience of the soil. Do not propagate trees from cuttings in soil from diseased areas. Foliar application of copper oxychloride, 0.3%, should be done in the months of July and October. Spray 5:5:50 Bordeaux mixture during autumn (September–October) and spring (February–March) at 15-day intervals depending upon degree of infestation. Spray of 0.25% ziram also reduces the disease. Fruit rot of litchi has been a serious problem. Litchi is host to a range of post-harvest pathogens, often with quite different modes of infection. Several fungi have been found to be associated as reported by several workers.

Usually large water-soaked lesions appear on the surface of fruits. Initially the disease symptoms are perceptible on injured portion of the fruits, as prescribed by National Research Centre, India on litchi (Fig. 10.1).

With the advance of the disease the decayed areas become depressed as the rot gradually penetrates deep into the pulp. Ultimately the rind of infected fruits cracks off, exposing the pulp which subsequently is covered with thick cottony mycelium. Such affected fruits emit an odour of fermentation. Low-temperature storage is the most successful means of slowing rot development. For instance, fruit stored at 22 °C rotted three times more quickly than fruit stored at 5 °C. Fungicides are also effective. Hot benomyl dips at 48–52 °C slow rot development compared with undipped fruit. Rots still affect the dipped fruit, although the fungicide slows the spread of the diseases. This technology has not been used because of health concerns surrounding fungicides. Spraying with a baking soda solution (1 tablespoon of baking soda, 2.5 tablespoons of vegetable oil, 1 teaspoon of liquid soap, not detergent, to 1 gal water) or neem oil (do not use when pollinating insects including bees or other beneficial insects are present) are helpful. Baking soda may burn some plant leaves. Spray only a few and then check for a reaction before

Fig. 10.1 Infection with soaked lesions appearing on the surface of litchi fruits (From National Research Centre on Litchi, India)

applying applications every 2 weeks. Apply copper-based fungicides weekly at first sign of disease to prevent its spread. This organic fungicide will not kill leaf spot, but prevents the fungus spores from germinating. Applications of Actinovate Lawn & Garden Fungicide control the bacterial form of the disease when used as a preventive treatment or applied at the first sign of water-soaked leaves.

10.15 Conclusion

Selective diseases have been reported from any lychee-growing locality. The glossy leaves are very resistant to fungi. In India lychee trees are occasionally subject to green scurf, or algal leaf spot (*Cephaleuros mycoides*). In India, leaf spot caused by *Pestalotia pauciseta*, *Pestalotia* spp. and *Pestalotia versicolor*, and *Phomopsis* spp. may be prevalent in December and can be controlled by lime-sulphur sprays. Leaf spots are caused by *Botryodiplodia theobromae* and *Colletotrichum gloeosporioides,* which begin at the tip of the leaflet. The main post-harvest problem is spoilage by the yeast-like organism, which is quick to attack warm, moist fruits. It is important to keep the fruits dry and cool, with good circulation of air. When conditions favor rotting, dusting with fungicide will be necessary. The maximum inhibition in colony diameter was observed in case of *Andrographis paniculata* (89%), followed by *Cymbopogon citratus* (80%), *Plumbago zeylanica* (77%), and *Azadirachta indica* (76%), whereas *Ocimum sanctum* showed minimum inhibition (71%) among all. The leaf extract of *P. zeylanica* showed maximum inhibition (88%) in mycelial dry wt. followed by *O. sanctum* (85%), *C. citratus* and *A. paniculata*, whereas *A. indica* showed minimum inhibition (61%). Spore germination was inhibited maximally by *O. sanctum* (91%); however, the least inhibition was observed in *P. zeylanica*.

References

Butani DK (1977) Pests of litchi in India and their control. Fruits 32:269–273

Chadha KL (1968) Litchi cultivation in India. Indian Hortic 12:13–16

Dass CS, Choudhary KR (1958) Floral biology of litchi (*Litchi chinensis* Sonn.) South Indian Hortic 6:17–22

Dwivedi RS, Pathak SP (1978) Proc Natl Acad Sci. India 48(B):287–288

Dixit SN, Tripathi NN, Tripathi SC (1978) Fungitoxicity of some seed extracts. Natl Acad Sci Lett 1(8):287–288

Jamaluddin, Tandon MP (1973) Some new leaf spot diseases. Curr Sci 42:103

Jamaluddin, Tandon MP (1976) Some new market diseases of vegetables and fruits. Indian Phytopathol 29:74–75

Jamaluddin, Tandon MP, Tandon RN (1971) A storage rot of tomato fruits (*Lycopersicon esculentum*) caused by *Geotrichum candidum* Link ex Pers. Curr Sci 40:415–416

Jamaluddin, Tandon MP, Tandon RN (1975) Rot of fruits of litchi (*Litchi chinensis*) in marketing process. Indian Phytopathol 28:530–531

Jamalulddin, Goswami MG, Ojha BM (2004) Fungi of India (1989-2001). Scientific Publishers, Jodhpur

Kotur SC, Singh HP (1993) Leaf sampling technique in litchi (*Litchi chinesis* Sonn.). Indian J. Agric Sci 63:632–638

Kotur SC, Singh HP (1994) Varietal differences in leaf nutrient composition of litchi (*Litchi chinesis* Sonn.) Indian J Hortic 51:59–62

Kumar R, Sachan SN (1979) Effect of some plant extracts on the conidial germination of *Curvularia pallescens*. Indian Phytopathol 32(3):489

Kunwar R, Singh R (1993) Note on nutritional survey of litchi orchards in Doon Valley of Garhwal Hills of Uttar Pradesh. Prog Hortic 25:164–165

Mallik PC, Singh DL (1965) Hunger sign in litchi. Indian Agriculturist 91:127–138

Pandey RM, Sharma HC (1989) The litchi. ICAR Publication, New Delhi

Pandey S, Misra AP (1975) Mycorrhiza in relation to growth and fruiting of *Litchi chinesis* Sonn. J Indian Bot. Soc 54:280–293

Rao DP, Mukherjee SK, Ray RN (1985) Nutritional studies in litchi orchards in different districts of West Bengal. Indian Hortic 42:1–7

Ray PK, Sharma SB, Mishra KA (1984) Important litchi cultivars of Bihar. Indian Hortic 30 (1):9–13

Singh HP 1992 Physico-chemical evaluation of litchi cultivars in South Bihar. In: National seminar on recent developments in litchi production, 30–31 May, RAU, Pusa

Singh HP, Yadav IS (1988) Physico-chemical changes during fruit development in litchi cultivar. India J Hortic 45:212–218

Singh HP, Yadav IS (1992a) CHES-1, a promising selection of litchi. In: National Seminar on recent development in litchi production, RAU, Pusa, 30–31 May 1992

Singh HP, Yadav IS (1992b) Studies on seed germination in litchi cultivars affected by physiological maturity. In: National seminar on recent developments in litchi production, RAU, Pusa, 30–31 May, p 16

Singh LB, Singh OP (1954) The litchi. Superintendent Printing Press, Allahabad, p 110

Singh HP (1998) Genetic diversity, breeding and utilisation of the genepool of litchi. In: Arora RK, Roa R (eds) Tropical fruits in Asia, IPGRI, pp 171–184
Singh HP, Kumar KK (1988) Litchi growing in India. Chona Hortic 5(1):22–32
Verma SK, Jain BP, Das SR (1980) Preliminary studies on the evaluation of the effect of growth substances with minor elements in controlling fruit drop in litchi (*Litchi chinesis* Sonn.) Haryana J Hortic 10:4–10
Yadav IS Singh HP (1992) Genetic resources of litchi in India. In: National seminar on recent developments in litchi production, RAU, Pusa, 30–31 May 1992

Lychee Juice and Pulp Extracts as Potential Components for Production of Extracellular Phosphate-Binding Biopolymer from *Acinetobacter haemolyticus* 11

Moushumi Ghosh

Abstract

This study evaluated extracellular polymer production by *Acinetobacter haemolyticus* with lychee juice and pulp in cultivation medium. Our earlier studies have established the biotechnological importance (phosphate binding) of polymer produced by this strain. Waste/spoilt lychee juice and homogenised pulp were incorporated in culture medium at concentrations of 5%, 10%, 15%, and 20%, respectively, and extracellular biopolymer was produced, evaluated and compared with biopolymer produced under normal cultivation media for phosphate-binding function, characteristics and toxicological properties. Concentrations of 15% lychee extracts/juice led to comparable biopolymer yield; no significant ($p < 0.05$) changes were observed in either functionality, characteristics, or yield. Moreover, the biopolymer did not elicit either haemolytic activity or toxicity in RAW cell lines. Haematological, histopathological and general examinations indicated no adverse effects in Swiss albino mice fed with the biopolymer (120 mg kg^{-1} body $weight^{-1}$ day^{1}) during a period of 30 days. These results suggested that spoilt lychee can be utilised for production of the biopolymer following the industrial process parameters. Optimisation of scale-up of conditions is, however, required for commercial applicability of utilising the agro-waste to generate this value-added product.

Keywords

Phosphate-binding toxicity • Polymers • Lychee juice/extract • *Acinetobacter haemolyticus*

M. Ghosh (✉)
Department of Biotechnology, Thapar University, Patiala, Punjab, India
e-mail: mghosh@thapar.edu

M. Kumar et al. (eds.), *Lychee Disease Management*,
DOI 10.1007/978-981-10-4247-8_11

11.1 Introduction

India is the second biggest lychee (*Litchi chinensis* (Gaertn.) Sonn.) producer with an estimated 80% export worth Rs. 5 crore and more. However, postharvest losses have consistently accounted for reduction in foreign exchange. Although various approaches are being used for preventing postharvest losses, pericarp browning of lychees following harvest has not been successfully resolved. A substantial quantity of spoilt lychees from browning of the pericarp is usually discarded, necessitating suitable processes to reutilise these to obtain value-added products. Biotechnological processes designed to valorise the wastes, especially spoilt lychees, for bioproducts may be important according to economical benefits accrued upon losses incurred to the lychee industry.

In the past three decades, an increasing number of reports of novel structures and diverse biological activities of microbial exopolysaccharides, lacking in plant polysaccharides, have prompted the application of microbial exopolymers as bioproducts of prominence in potentially important industrial and biomedical areas. The significantly expanded scope of exopolysaccharide for commercial applications over the current years is based on strong insights on their diverse technological and putative environmental functionalities (Horn et al. 2013; Jing et al. 2013). Recently, Kaur and Ghosh (2015a, b) reported high phosphate-binding biopolymer production (220 mg/l) by *Acinetobacter haemolyticus*. Phosphate is a persistent problem in surface water and requires effective, economical remediation. A potential role of green polymers as described may be appropriate; however, the production economics need to be reduced. Agro-wastes may be viewed in this regard as an important sustainable component for reshaping production strategy. We envisioned that spoilt lychee could be utilised for cultivating the strain and attempted to study whether incorporation of spoilt lychee juice or pulp in the cultivation medium would still support exopolymer production. The phosphate-binding capability structure was compared with biopolymer obtained by growing the culture in media lacking lychee juice or extract.

A key aspect for viability of biopolymers relies on their toxicity profile and requires to be addressed through well-designed studies. The in vitro cytotoxicity of the exopolymer determined using the RAW cell lines and its oral toxicity investigated in murine models are reported in this study.

11.2 Biopolymer Source

Acinetobacter haemolyticus TK15 was cultivated as described by Kaur and Ghosh (2015a, b), in 2 l FIB medium on a rotary shaker (120 rpm/min) at 30 °C for 48 h. Cells were removed from the culture medium by centrifugation at 12,000 *g* for 30 min at 4 °C, and polysaccharide was separated from the supernatant by the addition of two volumes of ethanol and precipitation at 4 °C for 24 h. The precipitated polysaccharide was collected by filtration (Whatman GF filter), dissolved in deionised water, dialysed extensively against deionised water and lyophilised. Crude exopolysaccharide was dissolved in deionised water and reprecipitated by adding a 10% solution of cetylpyridinium chloride (CPC). The precipitated polysaccharide complex was collected by centrifugation at 10,000 *g* for 20 min at 4 °C and redissolved in a 10% NaCl solution. The precipitated polysaccharide was recovered by addition of three volumes of ethanol. The extracted polysaccharide was dissolved in deionised water, dialysed twice against deionised water and lyophilised. Spoilt lychee pulp was homogenised for 30 min in sterile distilled water, whereas lychee juice was extracted using a juice extractor (Philips, UK). Concentrations added to FIB for cultivating the strain were 5%, 10%, 15%, and 20%, respectively. The biopolymer from each combination was purified as described.

11.3 Phosphate-Binding Activity, Structure and Cytotoxicity of the Biopolymer

Phosphate sorption by the biopolymers was determined as described by Kaur and Ghosh (2015a, b). Total protein, carbohydrate, pyruvic acid and uronic acids were determined by the methods of Lowry et al. (1946), Fredermann and Haugen (1945). Monomeric constituents were analysed by gas chromatography–mass spectrometry (GC-MS) as reported by Kaur and Ghosh (2015a, b). Exobiopolymer (EPS) cellular toxicity was evaluated by a haemolytic assay as described by Niidome et al. (2000), using sheep blood cells (BioMérieux, France). Control experiments comprised similar combinations but lacked biopolymer. Percentage haemolysis was calculated thus:

$$\text{Haemolysis } \% = \frac{\text{O.Ds} - \text{O.Dnc}}{\text{O.Dpc} - \text{O.Dnc}}$$

where O.Ds, O.Dnc and O.Dpc are optical densities of samples, negative control and positive control, respectively.

To determine the cytotoxic activity of the microbial exopolymer on macrophages, RAW 264.7 cells (1×10^4 cells/ml) were cultured in Dulbecco's modified Eagle's medium (DMEM; HiMedia, Mumbai, India) supplemented with 10% foetal calf serum, 1% penicillin-streptomycin solution, 1% L-glutamine and HEPES (4-(2-hydroxyethyl)-1-piperazineethanesulfonic acid) in a 96-well plate at 37 °C in an atmosphere of 5% carbon dioxide (CO_2) for 24 h, followed by treatment of the cells with different concentrations of biopolymers for another 24 h. To determine cell viability, MTT (at a concentration of 0.1 mg/ml) was added to the wells and incubated for 4 h at 37 °C in an atmosphere of 5% CO_2 in the dark. In metabolically active cells, MTT was reduced to an insoluble, dark purple formazan. The formazan crystals were dissolved in a dissolving buffer [sodium dodecyl sulphate (SDS) (11 g) in 0.02 M hydrochloric acid (HCl) (50 ml) and isopropanol (50 ml)]. The absorbance was read at 570 nm.

11.4 Oral Toxicity

Biopolymer solutions were prepared with sterile distilled water and kept under refrigeration until use. The toxicity studies were carried out in accordance with the OECD guidelines 423 (OECD) and ethical guidelines. Twenty-five pathogen-free male Swiss mice, 8 weeks old, were acclimatised to a 12:12 h light:dark cycle for 1 week. They were housed in a polystyrene cage, allowed free access to feed and sterile tap water, divided into five groups of five animals per cage and identified. The biopolymer was administered orally at a dose of 25, 75, 120 and 140 mg kg^{-1} in distilled water for 30 days. The control group received sterile distilled water. The animals were observed for mortality and morbidity such as convulsions, tremors, grip strength and pupil dilation. All animals were weighed before treatment, weekly during treatment and at the end of the study. Feed consumption was recorded weekly.

11.5 Haematology

At the end of the experiment, animals were anaesthetised in an ether chamber, and blood was collected by cardiac puncture. Analyses of red blood cell (RBC) count, haematocrit (Hct), haemoglobin (Hb) concentration, mean corpuscular volume (MCV), mean corpuscular haemoglobin (MCH), mean corpuscular haemoglobin concentration (MCHC) and differential cell count were conducted.

11.6 Histology

Necropsies were performed on all study animals; the liver and kidney were analysed macroscopically. To microscopically examine the tissues, the latter were fixed in aqueous Bouin, processed, embedded in Paraplast, sectioned to a thickness of 7 μm and stained with haematoxylin and eosin. Histological analysis of organs was done as described by Gauthier et al. (2003).

11.7 Statistical Analysis

Data are presented as mean ± SD of three independent experiments. During the 30-day oral toxicity test, treated and control groups were compared by analysis of variance (ANOVA) with significance $p = 0.05$, using STATISTICA 11.01 (2012).

11.8 Biopolymer Production

All concentrations of lychee juice and extracts supported growth and biopolymer production; however, yields were notable at concentrations of 15%, implying a maximum amount of required nutrients desirable for the cells to initiate biopolymer synthesis (Table 11.1 and 11.2).

The monomer composition of the polysaccharide was determined by GC-MS and found to be a heteropolysaccharide, composed mainly of pentose and hexose sugars in approximately equal proportions. The compositional analysis of the phosphate-binding biopolymer was determined in our previous study.

The relative ratios of sugars (depicted in Fig. 11.1) by GC-MS were determined as glucose/xylose/arabinose/ribose/galactose/allose/lyxose/mannose/fructose (3:2:2:2:2:2:1:0.2:0.1). No change in monomeric composition was noted in biopolymers obtained from cultures grown in lychee juices or extracts, indicating complete chemical identity.

The biopolymers obtained from cultures grown in lychee juice or extracts demonstrated phosphate binding. Kaur and Ghosh (2015a, b) reported a contact time of 240 min and biopolymer concentration of 100 mg/l as maximum phosphate adsorption (Fig. 11.2). The values observed in this study did not differ significantly ($p > 0.05$) from those observed earlier, suggesting a complete congruency in biopolymer functionality.

Table 11.1 Exobiopolymer (EBP) yields and phosphate removal trends observed with *Acinetobacter haemolyticus* grown in lychee extracts/juices

A. haemolyticus TK15 medium conditions	Biopolymer yield (mg/l)	Phosphate removal (%)
TK15 grown in FIB	250	21
TK15 grown in FIB with 15% lychee juice	253	20.8
TK15 grown in FIB with lychee pulp homogenate (15%)	250	22

Table 11.2 Chemical composition of biopolymer TK15 obtained upon cultivation in medium containing 15% lychee juice

Constituent	Percent (%)
Sugars	74.2
Proteins	8
Amino sugars	2.5
Pyruvic acid	0.98
Uronic acids	2.1
DNA	ND
RNA	ND
Carbon	18.7
Hydrogen	4.35
Nitrogen	6.14
Sulphur'	0.2

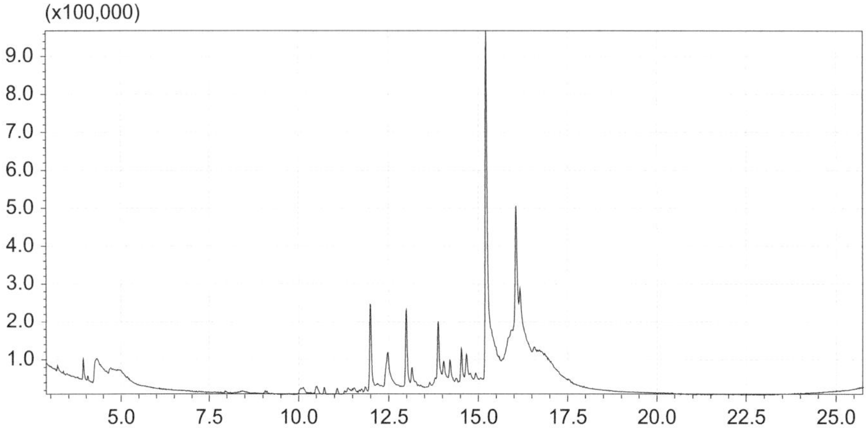

Fig. 11.1 Gas chromatography–mass spectrometry (GC-MS) profile and monomeric composition of exobiopolymer (EBP) of *Acinetobacter haemolyticus* obtained upon growth in lychee juice/extracts

Fig. 11.2 Phosphate removal with different concentrations of exobiopolymer (EBP) obtained from *Acinetobacter haemolyticus* grown in lychee juice/extract. *Qe* denotes adsorption capacity. Values are average of three independent experiments

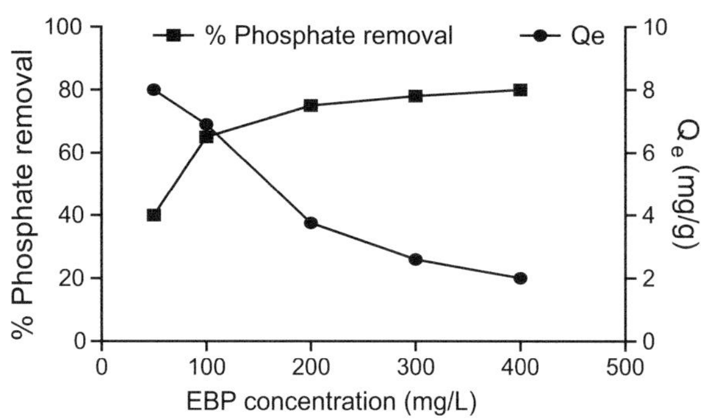

ASTM F 756-00 (2000) classifies materials into three categories based on their haemolytic index. Those exhibiting >5% haemolysis are considered haemolytic, between 2 and 5% are considered slightly haemolytic, and those having <2% haemolysis are nonhaemolytic materials. Haemolysis observed by TK15 was less than 2%, indicating a strong possibility for being safe. The risk of haemolytic character is evaluated against the clinical benefits. To further substantiate the finding, RAW cells were challenged with various concentrations of this biopolymer. Microscopic observations of the treated macrophages were also carried out (results not shown), indicating no alterations with respect to control (Table 11.3).

Table 11.3 Cytotoxic activity of the mucoadhesive, antimicrobial biopolymer TK15 on mouse macrophage RAW 264.7 cells

Biopolymer dose	Cell viability
10	98.2% ± 5.30
20	98% ± 4.80
40	98.3% ± 5.20
60	99% ± 8.80
80	98.6% ± 7.50
Control	100% ± 9.60

Macrophages were treated with different concentrations of biopolymer TK15 for 24 h; cell viability was determined by MTT (3-(4,5-dimethylthiazol-2-yl)-2,5-diphenyltetrazolium bromide) assay. The experiments were performed in triplicate; results are shown as mean plus or minus standard deviation

11.8.1 Oral Toxicity Studies

Mortality of animals was not observed over the period of experiment. Body weight gains (Table 11.4), rectal temperature profile (results not shown) and feed consumption in all concentrations of the biopolymer tested were comparable to control group values. Behavioural changes or changes in body weight of TK15-exposed mice were not observed. Gross and microscopic examination revealed no changes attributable to the administration of either of the biopolymers. Haematology indicated no significant ($p > 0.05$) treatment-related changes that were evident during the experimental period (Tables 11.5 and 11.6) upon comparison with the control group. Liver sections from mice fed with biopolymers exhibited a similar profile as the blank groups in liver tissues, suggesting the absence of any untoward response.

11.9 Bioprocess

To the best of our knowledge, a valorisation process for spoilt lychees has not been reported. Considering the enormous potential for this produce as a foreign exchange earner, few efforts in this direction have actually been done. Valorisation of agriproduct wastes represents an important area of bioprocess/bioproduct generation through biotechnological applications in India. Lychee juice and pulp extracts contain 82.5% fermentable sugars with predominance of glucose, in addition to nitrogen sources and degradable fibres (Singh and Kaur 2009), making it a natural media for *Acinetobacter*, glucose being a preferred source for growth and biopolymer synthesis (Kaur and Ghosh 2015a, b). Based on results of our previous studies on microbially produced exopolymers, we anticipated that juice or pulp from spoilt pulp from spoilt lychees may support growth and biopolymer production by *Acinetobacter haemolyticus*, an industrially important nonvirulent strain. An important aspect for any molecule to be used as a therapeutic agent is that the molecule should eliminate the target without affecting the viability of mammalian

Table 11.4 Mean body weight profile of mice after every fifth day of administration of biopolymer TK

Group	Dose (mg/kg bw)	Days: 0	5	10	15	20	25	30
G1	60 ± 0	23.5±	23±	24±	24±	23.5±	23.4±	23.3±
G2	180±	23.0±	22±	22.5±	23.5±	23.0±	22.8±	22.8±
G3	220±	23.0±	22.5±	23.4±	22.2±	22.6±	23.5±	23.2±

Table 11.5 Significant mean value of red blood cell parameters in mice administered with the biopolymer TK15 by gavage for 30 days

	Dose				
Parameter	0	25	80	120	140
Red blood cells (millions/mm^3)	9.0 ± 0.2	7.4 ± 0.2	6.5 ± 0.9	7.0 ± 1.5	8.3 ± 0.2
Haemoglobin (g dl 1)	13.1 ± 2.0	13.3 ± 1.2	46 ± 1.2	15.4 ± 2.4	18.5 ± 3.5
Haematocrit (%)	49.3 ± 5.3	53 ± 3.8	51 ± 3.4	50 ± 5.1	48.7 ± 2.0
Mean corpuscular volume (μm^3)	50.8 ± 2.2	60 ± 10.5	464 ± 7.3	63.2 ± 10.5	50 ± 2.3
Mean corpuscular haemoglobin (pg)	17.2 ± 0.7	18.2 ± 1.0	21 ± 3.0	20.5 ± 2.5	21.9 ± 4.0
Mean corpuscular haemoglobin concentration (%)	28.2 ± 4.3	25 ± 4.8	26 ± 4.2	30 ± 2.5	39 ± 9.3

Table 11.6 Means of white blood cell parameters of mice administered with the biopolymer TK15 by gavage for 30 days

Parameter (%)	0	5	15	25	35
Eosinophils	0.8 ± 0.9	1.3 ± 0.7	1.6 ± 0.6	1.4 ± 1.0	1.34 ± 0.6
Monocytes	4.4 ± 1.0	4.8 ± 1.5	4.8 ± 1.2	4.2 ± 1.5	4.2 ± 1.6
Lymphocytes	.07 ± 2.5	83 ± 6.4	8.8 ± 3.0	80.3 ± 7.2	80.1 ± 8.5
Neutrophils	1.2 ± 2.0	13 ± 4.2	1.4 ± 4.1	12.3 ± 8.2	13.02 ± 4.2

cells. In this study, the cytotoxic effect of the biopolymer was shown on mouse macrophage (i.e RAW 264.7 cell line) and analysed with zero cytotoxic effect when exposed to the bactericidal dose at 24 hrs interval. Microscopic observations of macrophages treated with the bactericidal dose of TK15 did not show distinct changes when compared with control cells.

The safety of polysaccharide biopolymers for chitin derived from crustacean shells and similarily β-glucan from barley and yeast were well documented. A general interpretation from these studies suggests that differences in the molecular structures of polysaccharide chains (e.g., chain length, type of linkage) in biopolymers influence physicochemical properties and, consequently, biological activity as well as a differential response in mice under in vivo conditions (intestinal behaviour); however, these biopolymers were intended as dietary supplements. Such detailed examination, including caecum, urine and faeces, was not deemed important because the present study was designed to judge the toxicity of the antimicrobial polymers. The absence of lipids, and a primarily polysaccharide composition of TK15, eliminated the possibility of a cytotoxic effect as indicated in toxicity studies with cell lines. A critical comparison of the results obtained from our studies was not possible because of the lack of similar studies in which microbially derived biopolymers have been screened for antimicrobial property. The absence of negative effects of the biopolymers on behaviour, body weight, haematology and organ histology at higher concentrations representing 1000 fold more than the anticipated application suggests the safety and, therefore, a potential applicability of the biopolymer.

11.10 Conclusion

TK15, an extracellular polymer with high phosphate-binding capability produced by *Acinetobacter haemolyticus*, could be successfully produced in media containing lychee juices as well as lychee pulp extracts obtained from discarded spoilt lychee. Yield values, chemical composition and phosphate-binding efficacy of the biopolymer produced by *A. haemolyticus* remained unaltered upon cultivation with lychee juice/extract. The biopolymer was non-haemolytical and non-cytotoxic, as evident from RAW cell lines. Murine models suggests the oral toxicity of the biopolymers and its safety up to a working concentration, the results of our study imply an interesting commercial prospect for further exploitation of spoilt lychee in scaling up biopolymer production.

Acknowledgments The work was supported by a grant from the University Grants Commission (UGC), Government of India. We are thankful to Venus Remedies, Baddi, for carrying out the animal studies.

Conflict of Interest None.

References

Avril JL, Mesnard R (1991) Factors influencing the virulence of *Acinetobacter*. In: Towner KJ, Bergogne-Bérézin E, Fewson CA (eds) The biology of *Acinetobacter*, vol 57. Springer, New York, pp 77–82.

Boels IC, Kleerebezem M et al (2003) Engineering of carbon distribution between glycolysis and sugar nucleotide biosynthesis in *Lactococcus lactis*. Appl Environ Microbiol 69(2):1129–1135

Breslow JM, Meissler JJ et al (2011) Innate immune responses to systemic *Acinetobacter baumannii* infection in mice: neutrophils, but not interleukin-17, mediate host resistance. Infect Immun 79(8):3317–3327

Castellanos Martinez E, Telenti Asensio M et al (1995) Infective endocarditis of an interventricular patch caused by *Acinetobacter haemolyticus*. Infection 23(4):243–245

Choi SH, E. J. Choo EJ , et al (2006) Clinical characteristics and outcomes of bacteremia caused by *Acinetobacter* species other than *A. baumannii*: comparison with *A. baumannii* bacteremia. J Infect Chemother 12(6):380–386

Fredermann TE, Haugen GE et al (1945) The level of pyruvtc and lactic acids and the lactic-pyruvic ratio, in the blood of human subjects. The effect of food, light muscular activity, and anoxia at high altitude. J Biol Chem 157:673–689

Grotiuz G, Sirok A et al (2006) Shiga toxin 2-producing *Acinetobacter haemolyticus* associated with a case of bloody diarrhea. J Clin Microbiol 44(10):3838–3841

Gauthier DT, Rhodes MW et al (2003) Experimental mycobacteriosis in striped bass Morone saxatilis. Dis Aquat Org 54:105–117

Hong HA, Huang JM et al (2008) The safety of *Bacillus subtilis* and *Bacillus indicus* as food probiotics. J Appl Microbiol 105(2):510–520

Horn N, Wegmann U et al (2013) Spontaneous mutation reveals influence of exopolysaccharide on *Lactobacillus johnsonii* surface characteristics. PLoS One 8(3):e59957

Jagtap S, Yavankar S et al (2010) Production of bioemulsifier by *Acinetobacter* species isolated from healthy human skin. Indian J Exp Biol 48:70–76

Jing X, Mao D et al (2013) Medium optimization, molecular characterization, and bioactivity of exopolysaccharides from *Pleurotus eryngii*. Arch Microbiol 195(10–11):749–757

Kaur T, Ghosh M (2015a) *Acinetobacter haemolyticus* MG606 produces a novel phosphate. Carbohydr Polym 132:72–79

Kaur T, Ghosh M (2015b) Characterization and upregulation of bifunctional phosphoglucomutase/phosphomannomutase enzyme in an exobiopolymer overproducing strain of *Acinetobacter haemolyticus*. Microbiol Res 181:8–14

Levander F, Radstrom P (2001) Requirement for phosphoglucomutase in exopolysaccharide biosynthesis in glucose- and lactose-utilizing *Streptococcus thermophilus*. Appl Environ Microbiol 67(6):2734–2738

Lowry OH, Bessey OA (1946) J Biol Chem 183:633

Panagiotou G, Christakopoulos P et al (2005) Simultaneous saccharification and fermentation of cellulose by *Fusarium oxysporum* F3: growth characteristics and metabolite profiling. Enzym Microb Technol 36(5-6):693–699

Niidome T, Tsuiki M et al (2000) Antibacterial activity of Arg/Pro-rich bactenecin 5 model peptides and their interaction with phospholipid membranes. Bull Chem Soc Jpn 73:1397–1402. doi: 10.1246/bcsj.73.1397

Patil JR, Chopade BA (2001) Studies on bioemulsifier production by *Acinetobacter* strains isolated from healthy human skin. J Appl Microbiol 91(2):290–298

Peleg AY, Seifert H et al (2008) *Acinetobacter baumannii*: emergence of a successful pathogen. Clin Microbiol Rev 21(3):538–582

Pirog TP, Kovalenko MA et al (2004) Physicochemical properties of the microbial exopolysaccharide ethapolan synthesized on a mixture of growth substrates. Microbiology 73(1):14–18

Singh RS, Kaur P (2009) Evaluation of litchi juice concentrate for the production of wine. Nat Prod Rad 8(4):386–391

Sorokulova IB, Pinchuk IV et al (2008) The safety of two *Bacillus* probiotic strains for human use. Dig Dis Sci 53(4):954–963

Szabo NJ, Dolan LC et al (2011) Safety evaluation of *Lactobacillus pentosus* strain b240. Food Chem Toxicol 49(1):251–258

Zhai Y, . Han D et al (2015) Enhancing GDP-fucose production in recombinant *Escherichia coli* by metabolic pathway engineering. Enzym Microb Technol 69(0):38–45

Litchi chinensis: Taxonomy, Botany and Its Cultivars

12

Renu Soni and Shachi Agrawal

Abstract

Litchi chinensis (common name, litchi or lychee) is a tropical to subtropical fruit that was originated in China and Vietnam but now grows in more than 20 countries at the commercial level. China is the largest producer of litchi followed by India. Because of its delicious taste and high nutritional value, it is also called as "King of the Fruits" in China. Litchi belongs to family Sapindaceae, also known as the soapberry family. *Litchi chinensis* subsp. *chinensis* is the only commercially grown subspecies. Litchi is a highly cross-pollinated entomophilous crop. Honey bees are its principal pollinators. The edible part of the fruit is the aril, which is fleshy, white, translucent and juicy. The bud, leaf and shoot development, inflorescence emergence, flowering, fruit development and fruit maturity are the seven principal growth stages of its complete growth cycle. Many litchi cultivars are known in various parts of the world. Some of these, with special or superior characteristics, are grown commercially. Various workers have classified these litchi cultivars differently on the basis of various parameters, viz. morphological traits, isozyme analysis, DNA markers, single nucleotide polymorphism (SNP) markers, etc. Every part of the litchi (leaf, fruit, flower, seed, pericarp) has some medicinal value. Litchi fruits are a good source of vitamin C, minerals (Cu, P, K), B-complex vitamins and antioxidants such as oligonol. This chapter gives an overview of its taxonomy, botany and various cultivars. A brief description of its active constituents and the pharmacological activities is also given here in this chapter.

Keywords

Litchi cultivars • Identification keys • Medicinal uses • Phenology • Sapindaceae • Systematics

R. Soni • S. Agrawal (✉)
Department of Botany, Gargi College, Sirifort Road, New Delhi 110049, India
e-mail: gupta_shachi@yahoo.co.in

M. Kumar et al. (eds.), *Lychee Disease Management*,
DOI 10.1007/978-981-10-4247-8_12

12.1 Introduction

Litchi chinensis Sonn. (litchi) is a long-lived, evergreen, aromatic, delicious tropical and subtropical tree belonging to the family Sapindaceae (Menzel 1985). The genus name "*Litchi*" comes from the Chinese name (li chih) for fruits of the handsome tree, and the specific epithet, *chinensis*, comes from China where it was originated (Stearn 1972). Litchi is one of the dominant trees growing in China and is known as the "King of the Fruits" in China for its delicious taste and nutritional value (Zhang et al. 1997; Morio 2006). The fruit is a fleshy drupe, bright red in colour with a single seed surrounded completely by the fleshy edible aril. The trees look beautiful when laden with fruits. Litchi is an important fruit tree crop in Southeast Asia and is also popular worldwide for its fruits which have an excellent taste, exotic aroma and high nutritional value (Li et al. 2013). It is a highly prized fruit when consumed in both fresh and preserved forms. All parts of the litchi—leaves, flowers, fruits and seeds—are good source of phytochemicals such as cyanidin glycoside, malvidin glycoside, epicatechin, rutin, procyanidin A2, procyanidin B2, saponins, kaempferol, leucocyanidin, and stigmasterol having potential pharmacological activities (Kilari and Putta 2016). As this fruit crop contributes significantly to the economy of several million people in Southeast Asia (Huang et al. 2005), research should be directed at improving agronomic characteristics and fruit quality, especially at the molecular level. This chapter focuses on the taxonomy of the litchi at family, genus and subspecies level, botanical description, pollination and reproduction, and the various cultivars of litchi. A special emphasis has been given to family Sapindaceae, which includes systematics and phylogeny, classification according to various classification systems and its characteristics. Various commercial important cultivars have been described with various classifications based on various parameters. Medicinal uses with the active constituents of the plant parts of litchi are also described in brief.

Scientific name: ***Litchi chinensis* Sonn.**
Synonyms:
***Nephelium litchi* Cambess.**
***Sapindus edulis* Ait.**
***Euphoria litchi* Desf.**
Common name: **Lychee or litchi**
Part used: **Fruit (aril)**
Family: **Sapindaceae**
Subfamily: **Sapindoideae**
Tribe: **Nepheliaea**

12.2 Taxonomy of the Family Sapindaceae

12.2.1 Introduction of Sapindaceae

Litchi chinensis belongs to the family Sapindaceae. Sapindaceae is also known as the soapberry family due to the presence of soap-like triterpenoid saponins in secretory cells of many members of this family (Judd et al. 2007). The important subtropical and tropical members of the subfamily Sapindoideae under the family Sapindaceae are litchi (*Litchi chinensis* Sonn.), longan (*Dimocarpus longan* Lour.), pulasan (*Nephelium mutabile* Blume) and rambutan (*Nephelium lappaceum* L.). All these differ in ecology and fruit morphology. Litchi is the most economically important member of this family. Both litchi and longan, native to South China, grow in the warm subtropics or the tropics at an elevation of 500 m. Fruits come only once in a year. Litchi and longan resemble each other but differ in fruit size and flavour. Fruits of longan are smaller, smoother, yellow to brown and have less acid with a mild flavour. Pulasan and rambutan are strictly tropical trees commonly found in Southeast Asia. Both are similar to litchi, with skin yellow or red; however, protuberances are replaced by spinterns or long hairs. Rambutans may fruit twice a year (Nakasone and Paull 1998; Tongdee 1997; FAO).

12.2.2 Systematics and Phylogeny of Sapindaceae

Cambessèdes, in 1828, originally described the family Sapindaceae (Subhadrabandhu and Stern 2005). However, the first comprehensive systematic treatment was published by Radlkofer (1890, 1933) on the basis of a wide range of evidence such as the absence or presence of a terminal leaflet, pollen morphology, ovules per carpel, fruit structure, and an aril. On the basis of morphology and biogeography, Sapindaceae was treated as distinct from the closely related families Hippocastanaceae and Aceraceae until the late 1980s (Takhtajan 1987; Cronquist 1988; Dahlgren 1989). According to plant characteristics, pollen morphology (Müller and Leenhouts 1976) and geography, the Sapindaceae earlier split into two subfamilies: Dodonaeoideae (austral distribution) and Sapindoideae. On the basis of phylogenetic analysis using molecular sequence data, however (Gadek et al. 1996; Savolainen et al. 2000; APG II 2003; Harrington et al. 2005; Thorne and Reveal 2007; APG III 2009; Buerki et al. 2009, 2010a), Sapindaceae was further subdivided into four subfamilies, namely, Dodonaeoideae Burnett, Hippocastanoideae Burnett, Sapindoideae Burnett, and Xanthoceroideae Thorne and Reveal, which is monogeneric (*Xanthoceras*), thus adopting a monophyletic concept comprising 142 genera having 1900 species distributed among four subfamilies. Recent studies based on phylogenetic analyses using a dataset that includes nuclear (ITS) and plastid (*matK*, *rpoB*, *trnD-trnT*, *trnK-matK*, *trnL-trnF*, *trnS-trnG*) markers provides a broad classification of Sapindaceae which includes traditionally placed taxa in Aceraceae and Hippocastanaceae, achieving monophyly. Using this information from biogeography and morphology, particularly

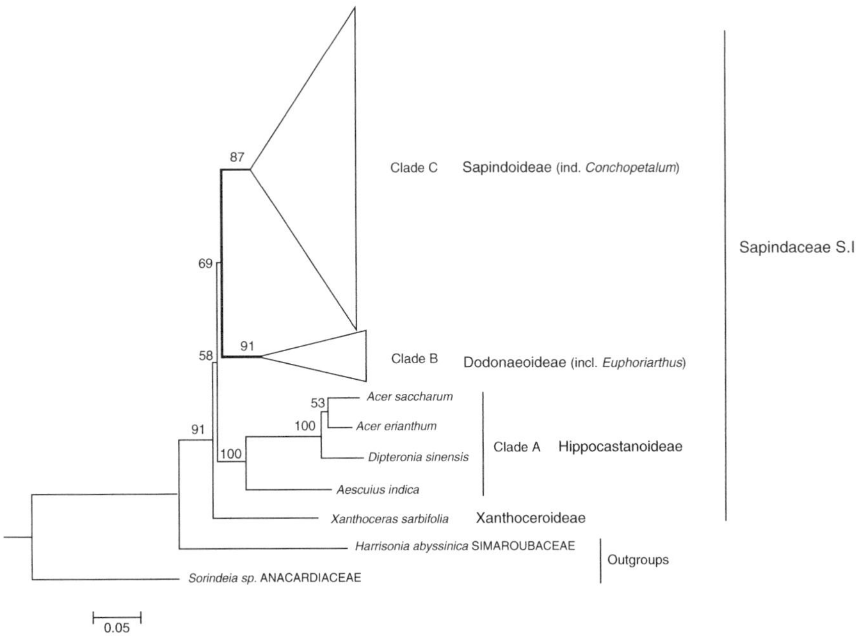

Fig. 12.1 Best maximum-likelihood phylogenetic tree for Sapindaceae inferred from eight nuclear and plastid nucleotide sequences (From Buerki et al. 2009)

with respect to the other taxa placed in Harpullieae, it was found out that the monotypic, temperate Asian genus *Xanthoceras*, historically placed in Sapindaceae tribe Harpullieae, is basal within the group (Fig. 12.1) (Buerki et al. 2009, 2010b), and a new monotypic family, Xanthoceraceae, was described to accommodate this genus. According to Buerki et al. (2010b), there are two different classification approaches for the family Sapindaceae: (1) it includes four subfamilies: Dodonaeoideae, Sapindoideae, Hippocastanoideae and Xanthoceroideae, as proposed by the Angiosperm Phylogeny Group (APG II 2003; APG III 2009) or (2) retaining taxa from the traditionally related families Hippocastanaceae and Aceraceae but excluding *Xanthoceras*. Thus, it describes a new monotypic family, Xanthoceraceae, to accommodate the monotypic genus *Xanthoceras*.

12.2.3 Classification System

Classification of family Sapindaceae according to different classification systems as mentioned in Table 12.1.

Table 12.1 Classification systems of Sapindaceae

Bentham and Hooker's system (1862–1883)	Hutchinson's system (1973)	Takhtajan's system (1980)	Cronquist's system (1981)	APG III (2009)
Dicotyledonae	Dicotyledon	Embryophyta	Magnoliophyta	Angiosperms
Polypetalae	Lignosae	Magnoliopsida	Magnoliopsida	Eudicot
Disciflorae	Sapindales	Sapindales	Rosidae	Rosids
Sapindales	Sapindaceae	Sapindaceae	Sapindales	Sapindales
Sapindaceae			Sapindaceae	Sapindaceae

12.2.4 Characteristics of Family Sapindaceae

Most of the genera of family Sapindaceae are woody trees, shrubs and tendril-bearing lianas found mostly in tropical or subtropical climates. Leaves are alternate, spiral, usually pinnately or palmately compound, unifoliate or trifoliate. Leaflets are entire or serrate with palmate or pinnate venation and are rarely opposite. In some species, hairs or hollow glands are present in the axil of the leaflets. Flowers usually bracteolate; small; mostly white, cream or orange in colour; tetramerous or pentamerous; unisexual or bisexual; actinomorphic or zygomorphic; hypogynous and sometimes hypanthium are also present. Flowers are borne in panicles, racemes or corymb. Flowers consist of four or five distinct sepals or connate sepals and four or five distinct petals often with scale-like appendage on the inner side or sometimes may be apetalous. Stamens are equal or twice the number of calyx lobes, filament often hairy. An extra-staminal or intra-staminal disc is present. Gynoecium consists of two or three superior, syncarpous ovaries. Fruits are fleshy or dried capsules, nuts and berries. Seeds are dark brown to black in colour, commonly enclosed or partially enclosed in fleshy aril. Seeds are usually non-endospermic with folded or curved embryo (Heywood 1978; Yeap 1987; Sharma 2009; Acevedo-Rodríguez et al. 2010; Tindall 2010).

Floral Formula: Br or Ebr + or % ⚥, or ♂, or ♀ $K_{4\text{-}5}$ $C_{4\text{-}5}$ $A_{8\text{-}10}$ $G_{\underline{(3)}}$

12.3 Taxonomy of Genus *Litchi*

Its attractive appearance and nutritious fruits ensure litchi is considered one of the most important genera of the family Sapindaceae and is cultivated all around the world. The French naturalist Pierre Sonnerat was the first person to describe the Chinese fruit tree litchi, and he named it *Litchi chinensis* in his *Voyage Aux Indes Orientales* (1774). In the literature, earlier it was described as *Dimocarpus litchi* Wild and *Sapindus litchi* Roxb. Now these all are regarded as synonyms. The genus *Litchi* includes only one species, *Litchi chinensis*, which has three subspecies based mainly on thickness of twig, flower arrangement, number of stamens and fruit characters (Leenhouts 1978). These subspecies are *Litchi chinensis* subsp. *chinensis* Forest and Kim Starr (synonyms: *Dimocarpus litchi* Lour., *Nephelium litchi* Cambess) which is the only commercialized litchi widely grown in China, North Vietnam and Cambodia where it grows in the wild (Groff 1921; Saxena et al. 2011). The tree has thin twigs; flowers have six stamens and arillate fruits with sweet, juicy and translucent flesh with or without protuberances up to 2 mm (Huang et al. 2005).

Litchi chinensis subsp. *javensis* Leenh (synonym *Litchi chinensis* Sonn. f. *glomeriflora* Radlk) is cultivated only in the Southern Indochina and West Java. It is suited to the humid tropics. The tree has thick twigs; flowers are in sessile clusters with 7–11 stamens. The fruit is smooth or with protuberances up to 1 mm high, similar to cultivated litchi, but with a thinner aril.

Litchi chinensis subsp. *philippinensis* (Radlk.) Leenh. (synonyms: *Euphoria didyma* Blanco, *Litchi philippinensis* Radlk.) is found only in the wild condition in the Philippines and Papua New Guinea. The tree has slender twigs and flowers with six or seven stamens. The fruit is long, oval shaped with thorny protuberances up to 3 mm high. The fruit has an inedible aril which partly covers the seed. From the presence of the smallest pollen grains, this subspecies is considered to be most primitive (Berg and Den 1978).

Neither of these two subspecies, *javensis* or *philippinensis*, is commercially grown (Menzel et al. 1993; Huang et al. 2005).

Key to the Subspecies (Leenhouts 1978)

la. Twigs thick up to 7 mm, few spike-like branching inflorescences, flowers in sessile clusters.. ... ...subsp. **javensis**

1b. Twigs slender up to 3.5 mm, widely branched inflorescences, flowers in lax cymules.. 2

2a. Leaves 1- or 2- or (3) jugate, stamens mostly 7, rarely 6. Fruits thorny, pyramidal, acute warts up to 3 mm high.. ... ...subsp. **philippinensis**

2b. Leaves 2–4-jugate. Stamens mostly 6, exceptionally up to 10. Fruits smooth or rarely with pyramidal, acute warts up to 1 mm high. ... ... subsp. **chinensis**

12.4 Litchi: Origin and Distribution

Litchi is a native of Southern China (particularly the Provinces of Kwangtung, Fukien), Northern Vietnam and Malaysia originated 2300 years ago (Mitra and Pathak 2010). It came to India through Burma and the Northeast region during the eighteenth century, from where it spread to the tropical and subtropical world (Goto 1960; Liang1981). Because of its attractive appearance and delicious fruits, it has become popular all over the world. The top five litchi-producing countries of the world are China, India, Taiwan, Thailand and Vietnam (Ray 1996). Aside from

India, Thailand and Vietnam, China is the largest producer of litchi. India is the second largest producer (Menzel 2005). Guangdong in China is called "the Kingdom of Lychee" as it has the highest and best litchi production. In India major litchi-growing states are Northern Bihar, West Bengal, Uttar Pradesh, Tripura, Haryana and Punjab. In India Bihar has the highest production (74%), then West Bengal, followed by Tripura. In India, because of the different time of maturity of litchi fruits, there is an interesting distribution pattern; the first comes from Tripura and then West Bengal followed by Bihar (Ray 2002; FAO).

12.5 Litchi: Botanical Description

Litchi is an evergreen, polygamous, monoecious, mycorrhizal tree with a dense rounded crown of glossy foliage with grey-black bark and brownish red branches. Leaves are alternate, pinnately compound with 2–9 glossy, leathery oblong-lanceolate pointed leaflets. Hundreds of small greenish white to yellow flowers, which are distinctively fragrant, are borne in terminal branched panicles (Fig. 12.2a) (Sharma 2009; Menzel 2005). The floral axis is of "compound racemose" type but the flowers occur in cymes. On the basis of sex, groups of flowers in a cyme vary, and six types of cymes have been reported in litchi, viz. (1) staminate flowers only, (2) pistillate flowers only, (3) terminal pistillate and lateral staminate flower, (4) terminal staminate and lateral pistillate flowers, (5) terminal staminate with lateral flowers of different sexes, and (vi) terminal pistillate with lateral flowers of different sexes (Sarin et al. 2009). Flowers are polygamous and apetalous with small valvate sepals. Within the same inflorescence, three different types of flowers are found. (1) The true male flower usually has 6–10 viable stamens only, their filaments connecting to the base. (2) The hermaphrodite female flower has a

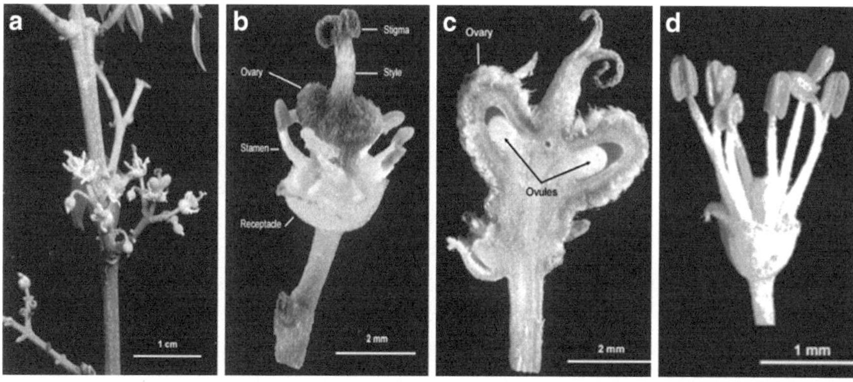

Fig. 12.2 *Litchi chinensis*: showing branches with blossom in panicles (**a**); showing female hermaphrodite flower (**b**); showing LS of female hermaphrodite flower having ovary with two ovules (**c**); showing male hermaphrodite flower (**d**) (From Nacifa et al. 2001)

fully developed bicarpellate pistil which is functional with 6–10 staminodes (Fig. 12.2b). Their sticky stigma is receptive as soon as the flower opens. These flowers when fertilized form fruits. (3) The hermaphrodite male flower has a nonfunctional pistil with undeveloped ovaries and a style without stigma, with well-developed stamens (Fig. 12.2d); these flowers cannot form fruits. The amount of fruit setting in different cultivars depends on the ratio of these different flowers. These three types of flowers mature at different times, over a period of 2–6 weeks in most cultivars.

The anthesis usually occurs during a period of 20–45 days (Chadha and Rajpoot 1969; Pivovaro 1974). A day after anthesis starts, anther dehiscence begins and continues up to 3 days, but not all the anthers dehisce simultaneously (Chaturvedi and Saxena 1965). In dry conditions, pollen grains are barrel shaped and are binucleate at the time of shedding. Pollen grains of male flowers are usually less viable compared to anthers of hermaphrodite flowers (Mustard et al. 1953). The flowers are bicarpellary syncarpous with a bilobed superior ovary. Its one lobe is fertile, which rapidly increases in size and turns erect, and another lobe is usually abortive. An erect style is present between the ovary lobes and has a terminal bifid stigma with revolute branches. In each locule of the ovary, one anatropous ovule is present. Functionally pistillate flowers bloom. Male flowers open first, followed by hermaphrodite female flowers, and then hermaphrodite male flowers open. The developmental stage of the two male flowers overlaps with the female cycle. True male flowers shed pollen first and then the hermaphrodite male flower, which shed most of the viable pollen grains and is responsible for the pollination and fertilization of most of the hermaphrodite female flowers having a viable pistil.

12.6 Litchi: Pollination and Reproduction

Litchi chinensis is a highly cross-pollinated crop as its flowers are hermaphrodite and highly self-sterile but possess nectar. In litchi, pollination and proper fruit setting require insects (Pandey and Yadava 1970), and the insects that visit litchi flowers are mainly bee species (*Apis mellifera*, *A. dorsata*, *A. cerana indica*, *A. florea*), flies, wasps, and beetles (Kumar and Kumar 2014). Thus, the litchi fruit crop is entomophilous (Dhaliwal et al. 1977), and one of the most outstanding beneficial insects on this crop is the honey bee (Groff 1943). Pollination by honeybees is one of the effective and inexpensive methods for improving the yield and quality of litchi fruit. Honey bees may be considered as principal pollinators because they have a significant role in enhancing fruit set, fruit yield and quality of litchi (Butcher 1956, 1957; King et al. 1989). Honey bees forage during daylight and are unlikely to carry pollen grains viable to effect fertilization beyond 1200 (Kraai 1962). The number of flowers visited per minute by any bee species depends on various factors, for example, instinctive foraging behaviour, length of proboscis and floral structure (Free 1993), particularly the corolla depth, type and quantity of floral rewards, density of flowers of particular cultivar of the crop grown, and the time of the day.

Earlier, the Indian honey bee (*Apis cerana indica* Fabricius) was the dominant pollinator for litchi, but recently the European honey bee *Apis mellifera* is being widely used commercially as an efficient pollinator of litchi (Kumar and Kumar 2014).

At the time of lobe initiation in the ovary, the stigma becomes receptive (exhibits up to 75% receptivity), usually a day after anthesis, which continues up to 2 additional days (Chaturvedi and Saxena 1965). Sometimes when the pistil is pollinated by nonviable or undeveloped pollen grains, the result is small and shrivelled abortive seeds, often called "chicken-tongue seed" (Menzel 2005). Usually, of the two ovules present in the ovary, one ovule remains functional, and the other is absent or the embryo sac degenerates (Fig. 12.2c) (Stern et al. 1996), resulting frequently in ripening of only one fruit from the functional ovule (Joubert et al. 1967). Occasionally both ovaries containing locules may develop into fruits, as in the litchi cv. "Deshi" and "Kasba" as reported by Prasad (1977). Depending on the location, cultivar and climate, the fruits mature in 80–112 days and differ in shape being either round, ovoid or heart shape. Fruits are globose with red or yellow tubercled, thin, tough inedible rind (skin) that is green when immature but changes to red on ripening. When left out after harvesting, the skin becomes dry and brown in colour. Within the fruit, a single seed develops from an anatropous ovule after fertilization. In this early stage, the small and light green seed becomes large chocolate in colour at maturity. During the process of maturation, an aril emerges from the base of the seed (Huang et al. 1983). The edible aril is fleshy, white and translucent and develops from the funicle, surrounding a single, large, brown, elliptical inedible seed (Figs. 12.3 and 12.4), which varies in size and shape between cultivars (Singh and Srivastava 1987; Costes 1987). During fruit ripening the aril becomes juicy (Prasad 2000). The flavour and aroma of fruit vary in different cultivars.

Fig. 12.3 Litchi berries in a bunch

Fig. 12.4 Litchi berry showing tubercled rind and white translucent aril enclosing brown seed

Litchi seeds are recalcitrant in nature and are sensitive to moisture stress (Chin et al. 1984; Ray and Sharma 1985; Fu et al. 1990; Singh and Prasad 1991). Within the fruit, seeds remain viable, but after separation from the whole fruit, they become nonviable very rapidly, within 1 or 2 days (Xia et al. 1992a, b). Seeds sown soon after separation from the whole fruit germinate readily. The optimum favourable conditions required for seed germination are a sand bed under shade with regular irrigation. Seed germination percentage can be enhanced by seed sowing in a sand bed under shade as reported by Prasad et al. (1996) in the "Early Bedana" cultivar which is traditionally considered nonviable (Pandey and Sharma 1989) but shows up to 60% germination in such conditions.

12.7 Litchi: Phenology

The evergreen litchi tree grows to 15 m tall, if left unpruned (Cronje 2010); however, under favourable conditions, it may grow up to 30 m. Usually, litchi trees have a thick short trunk and brown-grey bark. Litchi leaves vary in appearance at various points in the growing season or among different cultivars. They are usually 5–15 cm long and 3–5 cm wide, and their colours range from copper red or amaranth to copper yellow or light green when young, which gradually become dark green with time. Under normal conditions, litchi trees undergo two to three batches of vegetative flushes each year. The litchi tree produces a number of shoots each year either as a vegetative flush or as an inflorescence (Subhadrabandhu and Stern 2005). During the period from June to November, lychee leaves produce large amounts of carbohydrates that becomes accumulated in branches and trunks, with the highest starch concentration in small branches (Chen et al. 2004), which provides the necessary energy for floral formation and flowering in early spring. Terminal buds of the stem or, sometimes, after a killing flush or the failure of initial flowering, the axillary buds produce compound raceme panicles. Raceme panicles are 10–40 cm long and contain 200–4000 functionally male and functionally female flowers as well as a few hermaphrodite flowers. The number of flowers and their sex ratio are dependent on cultivar and climatic conditions. Additionally, litchi exhibits dichogamy, in which the peak time of functionally male flowers does not synchronize with that of the functionally female flowers. The litchi fruit is a drupe whose size, shape, colour and weight vary. It derives from a superior ovary and consists of an aril, pericarp, carpopodium, pedicel and five seed parts.

It is very important to understand the various growth stages for the correct timing of general orchard management, particularly for disease and pest management, irrigation, fruit retention, flower thinning, flush control and fertilizer effectiveness. Phenological growth stages of lychee have been described by Wei et al. (2013) using the Biologische Bundesanstalt, Bundessortenamt and Chemische Industrie (BBCH) scale. In this study, the entire growth cycle of lychee was divided into seven principal growth stages: bud, leaf and shoot development, inflorescence emergence, flowering, fruit development and fruit maturity.

Vegetative Growth Vegetative growth occurs after harvesting or during the season when there is no flowering or fruit setting because the climatic conditions are unfavourable. In a typical growth cycle of the litchi tree, bud, leaf and shoot developments are three stages of vegetative growth. In higher altitudes or cooler areas, only one or two flushes occur after harvesting. Successive shoot growth is separated by dormant periods when the expanded leaves are darkened and thickened. Under warmer and sunny conditions, flush growth can be enhanced by regular watering through irrigation or by fertilization (Huang and Hu 2010). The sturdiness of a flush directly impacts the subsequent floral induction and yield. Thus, protecting the terminal bud and young leaves against insects is very important.

Flowering Panicle Development and Flowering Flower induction takes place in the dormant bud, whereas flower initiation takes place in the early part of the growing bud stage (Huang and Chen 2003). For good flowering, there must be an effective cold temperature (<15 °C) for a prolonged period during flower induction. If the temperature is high (>20 °C), then leaf primordia grow but the floral organs atrophy. Excessively high temperature (>25 °C) can even cause all floral primordia to atrophy, leaving behind only leaves (Cronje 2010). After flower initiation, flower panicle and flower develop continuously uninterrupted and lead to anthesis, which lasts about 4–6 weeks depending on the temperature and the cultivar (Menzel 2001). Litchi flowering follows the "male-female-male" pattern. The ratio of male to female flowers depends on the environment and the various cultivars. If temperature is high during flower initiation, then the proportion of female flowers is reduced (Cronje 2010). In southern China, flowering and fruit set coincide with the rainy season. Wet conditions promote fruit setting but sometimes may affect pollinator activity, pollen tube growth and stigma receptivity (Huang and Hu 2010). In contrast, in South Africa, the favourable period of fruit set is characterized by low humidity and low rainfall. Under such circumstances, irrigation can help to ensure good flowering and fruit set (Cronje 2010).

Fruit Development Litchi fruit growth may be divided into three growth stages (Joubert 1986). During the first growth stage, the pericarp, embryo and seed coat develop. Initially, the endosperm is liquid and later on becomes absorbed by the developing embryo. In the second stage of fruit growth, the embryo grows rapidly and aril development begins. During the third growth stage, the aril grows rapidly and maturation processes take place. Litchi fruit maturation occurs in the early part of June. The most obvious aspect of fruit maturation is the change in pericarp colour which is caused by chlorophyll degradation and synthesis of anthocyanin (Wang et al. 2007).

12.8 Litchi Cultivars and Litchi Germplasm

After years of collection, classification, and documentation of Chinese litchi germplasm, more than 500 accessions have been reported (Liu et al. 2015). Of these, some represent commercial litchi cultivars and litchi germplasm which are characterized by superior or special characteristics. Currently, in situ conservation is the most important approach for the germplasm conservation. Rich litchi germplasm resources and abundant cultivars and strains have become established because of the long history of litchi cultivation in China. At present, more than 500 accessions are preserved in the National Litchi Germplasm Gene Bank located at the Institute of Fruit Tree Research, Guangdong Academy of Agricultural Science, Guangzhou, China, which is the largest litchi germplasm gene bank in the world.

In 2010, researchers from the China Litchi and Longan Industry Technology Research System have initiated the litchi genome sequencing and resequencing project (Project No. CARS-33), and many genomic resources have been developed for the litchi that include validated single-nucleotide polymorphism (SNP) sequences and high-density linkage maps (http://litchidb.genomics.cn/page/species/index.jsp). These available genomic resources and high-throughput tools are of great help in cultivar identification standardization and the genetic relationship assessment of the litchi collection.

12.8.1 Classification of Litchi Cultivars

Litchi cultivar nomenclature is too confusing. The same cultivars may have different names in different localities (synonyms), and different cultivars may have the same name (homonyms) (Li 2008). In China, litchi cultivar characterization is basically based on morphological traits, such as floral and fruit characteristics and the harvest season (Wu 1998). However, these morphological traits may interact with environmental conditions, and thus the usefulness of this approach becomes limited (Nielsen 1985) and, using only these morphological characters, it is very difficult to distinguish between closely related germplasms. Therefore, there must be a standardized way for the naming of litchi cultivars. Till now, various workers have classified litchi cultivars differently on the basis of various parameters, such as morphological traits, isozyme analysis, DNA markers, and SNP markers which are described here briefly.

12.8.1.1 Classification of Litchi Cultivars Based on Morphological Traits

On the basis of fruit shape, that is, long oval, heart shaped, short heart shaped and round, Li and Fang (1956) separated Fujian litchi cultivars into four types. In *The Flora of Guangdong Litchi* (Guangdong Academy of Agricultural Sciences 1978), Guangdong litchi cultivars were separated into seven types on the basis of the shape of tortoise shell-like cracking segments and pericarp with sharp protuberances or not as well as the shape of leaf, inflorescence and fruit, maturity time and the quality of fruit: (1) Guiwei, (2) Xiaozhi, (3) Jinfeng, (4) Sanyuehong, (5) Heiye, (6) Nuomici, and (7) Huaizhi. Later, in *The Flora of Guangxi Litchi* (Guangxi Academy of Agricultural Sciences and Guangxi Agricultural School 1986), litchi cultivars were classified into three types and seven groups according to a four-rank classification criterion: first, the characteristics of pericarp; and second, the shape of the fruit, followed by other morphological traits. Then, these classification principles were adopted in *The Flora of Chinese Fruit Tree, the Volume of Litchi* (Wu 1998). On the basis of the pericarp, Chinese litchi cultivars were separated into three types: (1) smooth pericarp, (2) somewhat warty pericarp, and (3) pericarp with sharp protuberances.

12.8.1.2 Classification of Litchi Cultivars by Isozyme Analysis

Isozymes are convenient and reliable genetic markers as they exhibit codominant expression and do not show environmental effects (Torres and Bergh 1980); they have been used to identify cultivars of various fruit trees, such as apple (Weeden and Lamb 1985), avocado (Goldring et al. 1985; Torres and Bergh 1980), grape (Parfitt and Arulsekar 1989), kiwi fruit (Messina et al. 1991), loquat (Degani and Blumenfeld 1986) and mango (Degani et al. 1990).

The PER isozyme of 24 Guangxi litchi cultivars and 35 Guangdong litchi cultivars has been analysed by Liu et al. (1989) and Zhou et al. (2001), respectively. Albeit all 9 litchi cultivars were the same in both these reports, their classification results are different, except for "Dazao." Liu et al. (1989) also noted that litchi's PER isozyme pattern was not affected by the age of leaves.

12.8.1.3 Classification and Identification of Litchi Cultivars by DNA Markers

Cultivar identification and genetic relationship analysis can be done directly using molecular genetic marker technology. Until now a number of systems, for example, random amplified polymorphic DNA (RAPD) (Ding et al. 2000; Chen et al. 2004; Liu and Mei 2005; Wang et al. 2006), amplified fragment length polymorphism (AFLP) (Yi et al. 2003; Peng et al. 2006), sequence-related amplified polymorphism (SRAP) (Zan et al. 2009), inter simple sequence repeat (ISSR) (Wei et al. 2006), and simple sequence repeat polymorphism (SSR) (Yao et al. 2009; Fu 2010; Xiang et al. 2010), have been used for litchi germplasm, especially in China.

RAPD is the most commonly used DNA marker for the classification of litchi cultivars (Ding et al. 2000; Chen et al. 2004, 2005; Liu and Mei 2005; Wang et al. 2006), followed by AFLP (Yi et al. 2003; Peng et al. 2006) and SSR (Li and Zheng 2004). To some extent, the classification results obtained by some authors (Chen et al. 2004; Wang et al. 2006) were similar to those based on morphological traits, whereas those obtained by other authors (Ding et al. 2000; Yi et al. 2003a) were not. The clustering of litchi cultivars reported by Liu and Mei (2005) was consistent with their resemblance of maturity time. However, for most of the litchi cultivars tested, classification results obtained using different DNA markers may be different from each other; even those obtained from the same DNA marker may also not be totally identical. But the results of RAPD and AFLP analysis both indicated that, at molecular level, the genetic diversity within litchi collections is limited (Ding et al. 2000; Yi et al. 2003; Chen et al. 2004). As to the identification of litchi cultivars, six pairs of accessions have been reported to be synonyms: "Nongmei No. 9" and "Qiongshan No. 27" (Chen et al. 2004), "Dazao" and "Zaohong," "Baiye" and "Guahong," "Feizixiao" and "Zhimali," "Ziniangxi" and "Zengchengdaguoli" (Liu and Mei 2005), and "Fengshuang" and "Tunchangfengshuang" (Wang et al. 2006). According to Ding et al. (2000), however, "Dazao" and "Zaohong" are two different cultivars.

12.8.1.4 Classification of Litchi Cultivars Using Single-Nucleotide Polymorphism (SNP) Markers

To overcome the widespread confusion regarding litchi cultivar nomenclature and to collect detailed information of genetic relationships among litchi germplasm, single-nucleotide polymorphism (SNP) markers can be used. In a study by Liu et al. (2015), the potential of single-nucleotide polymorphisms (SNP) for the identification of 96 representative litchi accessions and their genetic relationships in China was evaluated using 155 SNPs that were evenly spaced across the litchi genome. They reported a relatively high level of genetic variation among litchi accessions. The SNP-based multilocus matching identified two synonymous groups, "Heiye" and "Wuye" and "Chengtuo" and "Baitangli 1."

Liu et al. (2015) identified two groups of cultivars that were identical in all 90 SNP loci tested. "Heiye" from Guangdong Province and "Wuye" from Fujian Province were in one group. "Wuye" is an alias of "Heiye," and "Wuye" was introduced to Fujian Province from Guangdong Province about 500 years ago (Wu 1998). "Heiye" and "Wuye" cultivars have been suspected as the same cultivar for many years because of their similar morphological and biological characteristics (Wu 1998). After SNP marker analysis, it has been confirmed that these two cultivars are synonyms. In this analysis, another group was formed by "Chengtuo" from Guangdong Province and "Baitangli 1" from Guangxi Province. This finding was unexpected as the origin of both cultivars was different. However, they showed much similarity in tree and fruit characteristics and ripping period (Wu 1998), so they could be synonymous cultivars or sports (spontaneous somatic mutants).

On the basis of the unweighted pair-group method of arithmetic average (UPGMA) cluster analysis, litchi accessions analysed were divided into four main groups: extremely early maturing, early maturing, middle maturing and late maturing (Liu et al. 2015). Thus, fruit maturation period can also be considered as the primary criterion for litchi taxonomy (Liu et al. 2015).

12.8.2 Major Litchi Cultivars Grown in Different Countries

Many litchi cultivars are known in various parts of the world. In Guangdong, China, 26 major and 40 minor cultivars have been identified. Thirty-three cultivars in India and numerous local selections in Australia, Florida, Taiwan, Thailand and Hawaii have been reported (Singh et al. 2012). Some major cultivars of litchi are growing in different countries as instanced in Table 12.2.

12.8.3 Litchi Cultivars of India

Many cultivars are grown in India. In Bihar, the important varieties are Bedana, China, Dehra, Deshi, McLean, Muzaffarpur, Purbi and Rose. In Uttar Pradesh/ Uttarakhand, Punjab and Haryana, the varieties are Calcutta, Early Large Red, Early Seedless, Gulabi, Khatti, Late Seedless and Rose Scented, whereas the varieties recommended for growing in Punjab and Haryana are Calcutta, Dehra

Table 12.2 Litchi cultivar distribution at the international level

Country	Major cultivars
Bangladesh	Bedana, Bombai, China 3, Muzaffarpur
Brazil	Bengal
China	Bah Lup, Baitangying, Fay Zee Siu, Kwai May Red, No Mai Chee, Souey Tung, Sum Yee Hong
Florida, USA	Brewster, Mauritius
India	Bedana, Bombai, Calcuttia, China, Longia, Rose Scented, Shahi
Indonesia	Local selections
Israel	Floridian, Mauritius
Madagascar, Mauritius and Reunion	Mauritius
Nepal	Calcuttia, China, Dehradun, Mujafpuri, Raja Saheb
Philippines	Sinco, Tai So, UPLB Red
South Africa	Mauritius, McLean's Red
Thailand	Chacapat (Chakrapad), Haak Yip (O-Hia), Kom, Tai So (Hong Huay), Wai Chee (Kim Cheng)
Vietnam	Vaithieu

Table 12.3 Keys for litchi cultivars

	Key	Cultivars
1.	Flush pink, leaf boat shaped, dark green, panicle long, fruits oblong with round apex	
	Colour of fruit deep pink	Shahi/Trikolia
	Rose flavour	Rose Scented
	Colour of fruit light and greenish	Green
	High cracking and big seed	Ajhauli
	Late in maturity	Dehra Dun
2.	Deep pink flush, leaf with twist along the length, curved upward from the midrib and down along their length, panicle long, fruit oblong with pointed apex	
	Colour of fruit pink	China
	Fruits deep pink	Purbi/Mandraji
	Fruits in bunches	Bombaia/Calcuttia
	Early maturity	CHES-2
3.	Dark pink flush, oval-shaped leaves, compact and small panicles. Fruit round, smooth, chicken-tongue seed (aborted seed)	
	Early maturing	Bedana/Early Seedless
	Late maturing	Late Bedana/ Late Seedless
	Deep pink colour and mid-season maturity	Swarna Roopa
4.	Deep pink flush, boat-shaped and dark green long leaves, panicle long, largest fruit, deep in colour	Kasba
5.	Small elongated leaves, light green in colour, panicle compact, fruit medium in size, very late maturity	
	Pulp sweet and excellent flavour	Longia
	Pulp sour	Kaselia/Khatti/ Piyazi

Dun, Muzaffarpur, Rose Scented, Saharanpur and Seedless Late. Bombai, China, Elachi Early and Elachi Late are considered important for its quality as well as for yield and are grown in West Bengal. Singh et al. (2012) have given the basic keys to classify the Indian cultivars of litchi (Table 12.3).

12.8.4 We Represent Some Peculiar Characteristics of Commercially Important Litchi Cultivars of China and India as Mentioned in Table 12.4

Table 12.4 Characteristics of commercially important litchi cultivars

Cultivar name	Producer area	Maturity time	Fruit weight (g)	Seed size/weight (g)	Remarks	References
Baitangying	Guangdong	Late May to early June	34.8	Small	One of early-bearing cultivars characterized by the best fruit quality, high and stable yield	Zhang (1997)
Feizixiao	Guangdong	Early to middle June	30	1.4	A high-quality cultivar matures a little later than early-bearing cultivars but with short storage life	Zhang (1997)
Guiwei	Guangdong	Late June to early July	17	Most seeds are aborted (0.5)	One of the most popularized cultivars	Zhang (1997)
					Characterized by excellent flavour	
Heiye	Guangdong	Late June to early July	19.0	2.2	One of the most important commercial cultivars in China, also suitable for processing	Zhang (1997)
Huaizhi	Guangdong	Middle to late July	20.6	Seeds are separated into two types: medium (2.4) and aborted (0.4)	One of the most important commercial cultivars in China, also suitable for processing	Wu (1998)
Linshanxiangli	Guangxi	Middle July	21.0	Seeds are separated into two types: big (2.8) and small (0.6)	Guangxi's main cultivar with high quality, high yield but not stable, also suitable for processing	Wu (1998)
Nuomici	Guangdong	Late June to early July	25	Most seeds are aborted (0.6)	One of Guangdong's main cultivars characterized by the best fruit quality, but the	Wu (1998)

(continued)

Table 12.4 (continued)

Cultivar name	Producer area	Maturity time	Fruit weight (g)	Seed size/weight (g)	Remarks	References
					pericarp of one strain is liable to crack	
Sanyuehong	Guangdong	Middle to late May	37–42	2.0 g	A noted early-bearing cultivar with big fruit and medium quality, high and stable yield	Wu (1998)
Shahi	North Bihar, Jharkhand, Uttaranchal and Uttar Pradesh regions of India	Second week of May to the first week of June	20.98	3.88 g	The fruits are known for excellent aroma and quality. This cultivar occupies a major area under lychee in India	Singh et al. (2012)
China	Bihar, Jharkhand, West Bengal, Assam in India	End of May to third week of June	20.30	3.83 g	Important cultivar in India that ripens when most of the other cultivars have been harvested	Singh et al. (2012)
Bombai	West Bengal in India	Second week of May	18.93	3.83 g	Fruits of this cultivar are good for canning (Bose, et al. 2001)	Singh et al. (2012)
Bedana	Bihar, Jharkhand, Uttar Pradesh in India	Mid-season	16.33	Small seed	Small seed, large and good quality fruit	Singh et al. (2012)
Rose Scented	North Bihar, Jharkhand, Uttarakhand and Uttar Pradesh in India	Most popular mid-season cultivar, which ripens during last week of May to first week of June	18.44	3.42 g	The fruits have distinct aroma and hence called as Rose Scented	Singh et al. (2012)

12.9 Medicinal Uses of Litchi

Litchi fruits are a good source of vitamin C, dietary fibres, minerals (copper, phosphorus, potassium), B-complex vitamins such as folates, niacin and thiamin, and antioxidants such as oligonol. Fruit is low in sodium and saturated fat (Sakurai 2008). Thus, every part of the litchi tree has medicinal properties (Table 12.5).

12.10 Conclusion and Future Prospects

Various studies in litchi, including taxonomic and morphological studies of vegetative and reproductive characters, have been carried out by various workers. Different reports reviewed in this chapter have revealed that the studies to identify litchi cultivars using different molecular markers have been successful in distinguishing among accessions, in clarifying synonyms and in identifying mislabelled cultivars. The existing confusion over litchi cultivar names has been reduced, but still there is no standard method for litchi nomenclature. Thus, there is a need of a standard and systematic approach for the description of litchi cultivars. A comparison should be made among accessions conserved in different countries to

Table 12.5 Active constituents with pharmacological activities

Source	Active constituents	Pharmacological activities
Leaves	Epicatechin, procyanidin A2, procyanidin B2 (Castellain et al. 2014)	Antioxidant activity (Castellain et al. 2014), hepatoprotective activity (Basu et al. 2012), antiinflammatory and analgesic activity (Chauhan et al. 2014)
Fruits	Benzyl alcohol, (+) – catechin and 5-hydroxymethyl-2-furfurolaldehyde (5-HMF) and hydrobenzoin (Zhou et al. 2012)	Antiinflammatory effect (Yamanish et al. 2014), nootropic activity (Irene et al. 2012), antiviral activity (Ichinose et al. 2013)
Flowers	Contains phenols, flavonoids and tannins	Cardiovascular activity (Yang et al. 2010), antilipase activity (Wu et al. 2013), cytotoxicity (Hwang et al. 2013), antioxidant activity (Yang et al. 2012)
Seeds	Cyanidin glycoside, leucocyanidin, malvidin glycoside and saponins	Antimicrobial and antioxidant activities (Singh et al. 2013)
Pericarp	Bis-(8-epicatochinyl) methane, 5–2-(2-hydroxy-5-(methoxycarbonyl) phenoxy) benzoic acid, butylated hydroxytoluene, epicatechin, dehydrodiepicatechin A, methyl and ethyl shikimate, kaempferol, isolariciresinol, methyl 3,4-dihydroxy benzoate, proanthocyanidin A1, A2, stigmasterol and rutin (Ma et al. 2014)	Antimicrobial activity (Putta and Sastry 2014), anticancer activity (Wang et al. 2006)

provide the whole picture of litchi germplasm diversity and also information on the various germplasms and accessions conserved in ex situ conditions in different collections. This comparison will contribute to a much greater extent to the breeders in parent selection and creating new combinations of superior characters in the coming progenies.

References

Acevedo-Rodríguez P , Welzen PCV, Adema F, Van der Ham RWJM (2010) The families and genera of vascular plants. Flowering plants. Eudicots, vol 10. Springer, Berlin, pp 357–407

Angiosperm Phylogeny Group (APG II) (2003) An update of the angiosperm phylogeny group classification for the orders and families of flowering plants: APG II. Bot J Linn Soc 141:399–436

Angiosperm Phylogeny Group (APG III) (2009) An update of the angiosperm phylogeny group classification for the orders and families of flowering plants: APG III. Bot J Linn Soc 161:105–121

Basu S, Haldar N, Bhattacharya S, Biswas S, Biswas M (2012) Hepatoprotective activity of *Litchi chinensis* leaves against paracetamol-induced liver damage in rats. Am-Euras J Sci Res 7:77–81

Bentham G, Hooker JD (1862–1883) Genera plantarum ad exemplaria imprimis in herbariis kewensibus servata definite, vol 1–3. L. Reeve & Co., London

Berg RG, Den V (1978) Pollen morphology of the genera *Pometia*, *Cubilia*, *Otonephelium*, and *Litchi* (Sapindaceae-Nephelieae). Blumea 24:369–494

Buerki S, Forest F, Acevedo-Rodríguez P, Callmander MW, Nylander JA, Harrington M, Sanmartín I, Küpfer P, Alvarez N (2009) Plastid and nuclear DNA markers reveal intricate relationships at subfamilial and tribal levels in the soapberry family (Sapindaceae). Mol Phylogenet Evol 51:238–258

Buerki S, Phillipson PB, Lowry PP, Callmander MW (2010a) Molecular phylogenetics and morphological evidence support recognition of *Gereaua*, a new endemic genus of Sapindaceae from Madagascar. Syst Bot 35:172–180

Buerki S, Porter P, Lowry, Alvarez N, Razafimandimbison SG, Küpfer P, Callmander MF (2010b) Phylogeny and circumscription of Sapindaceae revisited: molecular sequence data, morphology and biogeography support recognition of a new family, Xanthoceraceae Plant Ecol Evol FastTrack 1–12

Butcher FG (1956) Bees pollinate lychee blooms. Proc Fla Lychee Growers Assoc 3:60

Butcher FG (1957) Pollination insects on lychee blossoms. Proc Fla State Hortic Soc 70(326):328

Castellain RC, Gesser M, Tonini F, Schulte RV, Demessiano KZ, Wolff FR (2014) Chemical composition, antioxidant and antinociceptive properties of *Litchi chinensis* leaves. J Pharm Pharmacol 66:1796–1807

Chadha KL, Rajpoot MS (1969) Studies on floral biology, fruit set and its retention and quality of some litchi varieties. Indian J Hortic 26:124–129

Chaturvedi RB, Saxena GK (1965) Studies on the blossom biology of litchi (*Litchi chinensis* Sonn.) Allahabad Farmer 39:10–13

Chauhan S, Kaur N, Kishore L, Singh R (2014) Pharmacological evaluation of anti-inflammatory and analgesic potential of *Litchi chinensis*. Int J Pharm Pharm Sci 6:116–119

Chen YY, Deng SS, Zhang X, Wei SX, Gao AP, Wang JB, et al (2004) RAPD analysis of genetic relationship among partial litchi germplasms in Hainan Island. Acta Hortic Sin 31:224–22226

Chen YT, Lai ZX, Chen JY, Guo YQ, Cai YQ, Wu JS (2005) RAPD analysis of genetic relationship among some ancient litchi genetic resources. J Fujian Agric For Univ 34 (4):458–463

Chin HF, Hor YL, Lassim MBM (1984) Identification of recalcitrant seeds. Seed Sci Technol 12:429–436

Costes E (1987) Le litchi à la Réunion. Étude des caractéristiques des fruits de la récolte 1985. Fruits 42(7–8):443–453

Cronje RB (2010) Litchi phenology and botanical aspects. In: de Villiers EA, Joubert PH (eds) The cultivation of litchi. ARC Institute for Tropical and Subtropical Crops, Nelspruit, pp 8–22

Cronquist A (ed) (1981) An integrated system of classification of flowering plants. Columbia University Press, New York

Cronquist A (1988) The evolution and classification of flowering plants, 2nd edn. New York Botanical Garden Press, New York

Dahlgren G (1989) An updated system of classification. Bot J Linn Soc 100:197–203

Degani C, Blumenfeld A (1986) The use of isozyme analysis for differentiation between loquat cultivars. HortScience 21:1457–1458

Degani C, El-Batsri R, Gazit S (1990) Enzyme polymorphism in mango. J Am Soc Hortic Sci 115:844–847

Dhaliwal HS, Srivastava S, Adlakha RL (1977) Insect pollination of lychee, *Litchi chinensis* Sonn. in the valley areas of the Indian Himalayas. In: Proceedings of the twenty-sixth international apiculture congress. Apimondia, Bucharest, p 396

Ding XD, Lv LX, Chen XJ, Guan X (2000) Identifying litchi cultivars and evaluating their genetic relationships by RAPD markers. J Trop Subtrop Bot 8:49–54

FAO (1982) Fruit-bearing forest trees: technical notes. FAO-Forestry-Paper No 34:177

Free JB (1993) Insect pollination of crops, 2nd edn. Academic, London, p 684

Fu JX (2010) Development and application of microsatellite markers in litchi, longan and selection of excellent individuals in litchi populations. D.Phil. thesis, South China Agricultural University

Fu JR, Zhang BZ, Wang XP, Qiao YZ, Huang XL (1990) Physiological studies on desiccation, wet storage and cryo-preservation of recalcitrant seeds of three fruit species and their excised embryonic axes. Seed Sci Technol 18:743–754

Gadek PA, Fernando ES, Quinn CJ, Hoot SB, Terrazas T, Sheahan MC, Chase MW (1996) Sapindales: molecular delimitation and infraordinal groups. Am J Bot 83:802–811

Goldring A, Zamir D, Degani C (1985) Duplicated phosphoglucose isomerase genes in avocado. Theor Appl Genet 71:451–454

Goto (1960) Technical papers of the first Pacific Rim food conference, Honolulu, HI, pp 15–23

Groff GW (1921) The lychee and lungan. Orange Judd, New York

Groff GW (1943) Some ecological factors involved in successful lychee culture. Proc Fla State Hortic Soc 56(34):155

Harrington MG, Edwards KJ, Johnson SA, Chase MW, Gadek PA (2005) Phylogenetic inference in Sapindaceae sensu lato using plastid matK and rbcL DNA sequences. Syst Bot 30:366–382

Heywood VH (ed) (1978) Flowering plants of the world. Oxford University Press, London

Huang HB, Chen HB (2003) A phasic approach towards the floral formation in *Litchi chinensis* Sonn. J Fruit Sci 20(6):487–492

Huang XM, Hu GB (2010) Management for early bearing, high quality and high yield in Feizixiao litchi. In: Chen HB (ed) Practical techniques in litchi production. Guangdong Science and Technology Press, Guangzhou, pp 60–81

Huang HB, Jiang SY, Xie C (1983) The initiation of aril and ontogeny of fruit in *Litchi chinensis* Sonn. J South China Agric Coll 4:78–83

Huang X, Subhadrabandhu S, Mitra SK, Ben-Arie R, Stern RA (2005) Origin, history, production and processing. In: Menzel CM, Waite GK (eds) Litchi and longan: botany, production and uses. CABI Publishing, Nambour, pp 1–24

Hutchinson J (ed) (1973) The families of flowering plants, arranged according to a new system based on their probable phylogeny, 2 vols (3rd edn). Oxford University PressOxford

Hwang JY, Lin JT, Liu SC, Hu CC, Shyu YS, Yan DJ (2013) Protective role of litchi (*Litchi chinensis* Sonn.) flower extract against cadmium- and lead-induced cytotoxicity and transforming growth factor β1-stimulated expression of smooth muscle α-actin estimated with rat liver cell lines. J Funct Foods 5:698–705

Ichinose T, Musyoka TM, Watanabe K, Kobayashi N (2013) Evaluation of antiviral activity of Oligonol, an extract of *Litchi chinensis*, against betanodavirus. Drug Discov Ther 7:254–260

Irene PR, Babu DJ, Rao NV, Sheikh RA (2012) Nootropic activity of fruit extracts of *Litchi chinensis* Sonn (Sapindaceae). Int J Pharmacol Technol 4:4795–4804

Joubert AJ (1967) Die bloei, embriosak-, embrioem verugontwikkeling von. *Litchi chinensis* Sonn. cultivar HLH Mauritus. Thesis, University of Witwatersrand, Johannesburg, South Africa

Joubert AJ (1986) Litchi. In: Monselise SP (ed) Handbook of fruit set and development. CRC Press, Boca Raton, pp 233–246

Judd WS, Campbell CS, Kellogg EA, Stevens PF, Donogue MJ (2007) Plant systematics: a phylogenetic approach. Sinauer, Sunderland

Kilari ES, Putta S (2016) Biological and phytopharmacological descriptions of *Litchi chinensis*. Pharmcognosy Rev 10(19):60–65

King J, Exley EM, Vithanage V (1989) Insect pollination for yield increases in lychee. In: Batten DJ (ed) Proceedings of the fourth Australian conference on trees and nut crops. Exotic Fruit Growers Association, Lismore, pp 142–145

Kraai A (1962) How long do honey-bees carry germinable pollen on them? Euphytica 11:53–556

Kumar R, Kumar V (2014) Impact of pollination by European honey bee, *Apis mellifera* L. on the yield and quality of litchi (*Litchi chinensis* Sonn.) fruits in India. Pest Manag Hortic Ecosyst 20 (2):127–132

Leenhouts PW (1978) Systematic notes on the Sapindaceae-Nephelieae. Blumea 24:395–403

Li JG (2008) The litchi. China Agriculture Press, Beijing

Li LR, Fang Y (1956) The research of litchi and longan. Scientific Press, Beijing, pp 1–10

Li MF, Zheng XQ (2004) Development of SSR markers in litchi (*Litchi chinensis*). Hereditas 26 (6):911–916

Li CQ, Wang Y, Huang XM, Li J, Wang HC, Li JG (2013) De novo assembly and characterization of fruit transcriptome in *Litchi chinensis* Sonn. and analysis of differentially regulated genes in fruit in response to shading. BMC Genomics 14:552

Liang J (1981) The litchi: origin, utilization and development of its culture. J Agric Trinidad Bot Appl 28:259–270

Liu CM, Mei MT (2005) Classification of lychee cultivars with RAPD analysis. Acta Hortic 665:149–1159

Liu HB, Liu YQ, Liu RG (1989) The peroxidase isozymes analysis of litchi cultivar. J Guangxi Agric Sci 5:27–29

Liu W, Xiao Z, Bao X, Yang X, Fang J, Xiang X (2015) Identifying litchi (*Litchi chinensis* Sonn.) cultivars and their genetic relationships using single nucleotide polymorphism (SNP) markers. PLoS One 10(8):e0135390

Ma Q, Xie H, Li S, Zhang R, Zhang M, Wei X (2014) Flavonoids from the pericarps of *Litchi chinensis*. J Agric Food Chem 62:1073–1071078

Menzel CM (1985) Propagation of lychee: a review. Sci Hort ic 25:31–48

Menzel CM (2001) The physiology of growth and cropping in lychee. South Afr Litchi Growers Assoc Yearb 12:9–14

Menzel C (ed) (2005) Litchi and longan: botany, production and uses. CABI, Wallingford, p 26

Menzel CM, Simpson DR, Watson BJ (1993) Fruits of tropical climates: fruits of the Sapindaceae. In: Caballero B, Trugo L, Finglas P (eds) Encyclopedia of food science. Food technology and nutrition. Academic Press, London, pp 2114–2120

Messina R, Testolin R, Morgante M (1991) Isozymes for cultivar identification in kiwifruit. HortScience 26:899–902

Mitra SK, Pathak PK (2010) Litchi production in the Asia-Pacific region. Acta Hortic 863:29–36

Morio M (ed) (2006) A kaleidoscope of China. China International Press, Beijing

Müller J, Leenhouts PW (1976) A general survey of pollen types in Sapindaceae in relation to taxonomy. In: Ferguson IK, Müller J (eds) The evolutionary significance of the exine. Academic, London, pp 407–445

Mustard MJ, Liu SY, Nelson RO (1953) Observations of floral biology and fruit setting in lychee varieties. Proc Fla State Hortic Soc 66:212–220

Nacifa ST, Paolib AAS, Salomão LCC (2001) Morphological and anatomical development of the litchi fruit (*Litchi chinensis* Sonn. cv. Brewster). Fruits 56:225–233

Nakasone HY, Paull RE (1998) Litchi, longan and rambutan. In: Tropical fruits. CAB International, Wallingford, pp 173–207

Nielsen G (1985) The use of isozymes as probes to identify and label plant varieties and cultivars. In: Rattazzi MC, Scandalios JG, Whitt GS (eds) Isozymes: current topics in biological and medical research. Liss, New York, pp 569–579

Pandey RM, Sharma HC (1989) The litchi. ICAR, New Delhi

Pandey RS, Yadava RPS (1970) Pollination of litchi (*Litchi chinensis*) by insects with special reference to honeybees. J Apic Res 9(2):103–105

Parfitt DE, Arulsekar S (1989) Inheritance and isozyme diversity for GPI and PGM among grape cultivars. J Am Soc Hortic Sci 114:486–491

Peng HX, Cao HQ, Zhu JH, Liao HH, Huang FZ, Li JZ et al (2006) Studies on the application of AFLP molecular markers on genetic diversity and classification of good and rare litchi resources in Guangxi. Southwest China J Agric Sci 19:108–111

Pivovaro SZ (1974) Studies on the floral biology and the influence of growth regulators on fruit set, size and drop of *Litchi chinensis* Sonn. M.Sc. thesis, Hebrew University, Rehovat

Prasad US (1977) Physiological studies on ripening of litchi of Bhagalpur. Ph.D. thesis, Bhagalpur University, Bhagalpur

Prasad US (2000) Growth and ripening of litchi fruits. In: Pandey AK, Dhakal T (eds) Advances in plant reproductive biology. Narendra, Delhi

Prasad JS, Kumar R, Mishra M, Singh AK, Prasad US (1996) Characteristics of litchi seed germination. Hortic Sci 31:1187–1189

Putta EK, Sastry N (2014) Screening of methanolic fruit pericarp extract of *Litchi chinensis* against various disease conditions. AJPHR 2:300–306

Radlkofer L (1890) Ueber die Gliederung der Familie der Sapindaceen. Sitzungsberichte der Königl Bayerischen Akademie der Wissenschaften zu München 20:105–379

Radlkofer L (1933) Sapindaceae. In: Engler A (ed) Das Pflanzenreich: Regni Vegetabilis Conspectus (IV) 165 (Heft 98ah). Verlag von Wilhelm Engelmann, Leipzig

Ray PK (ed) (1996) Litchi. Breeding of tropical and subtropical fruits. Narosa, Delhi, pp 129–142

Ray PK (ed) (2002) Breeding tropical and subtropical fruits. Springer, Berlin, pp 129–142

Ray PK, Sharma SB (1985) Viability of *Litchi chinensis* seeds when stored in air and in water. J Agric Sci Camb 104:247–248

Sakurai T, Nishioka H, Fujii H, Nakano N, Kizaki T, Radak Z, Izawa T, Haga S, Ohno H (2008) Antioxidative effects of a new lychee fruit-derived polyphenol mixture, oligonol, converted into a low-molecular form in adipocytes. Biosci Biotechnol Biochem 72(2):463–476

Sarin NB, Prasad US, Kumar M, Jain SM (2009) Litchi breeding for genetic improvement. In: Jain SM, Priyadarshan PM (eds) Breeding plantation tree crops: tropical species. Springer, New York, pp 217–245

Savolainen V, Fay MF, Albach DC, Backlund A, Van der Bank M, Cameron KM, Johnson SA, Lledó MD, Pintaud JC, Powell M, Sheahan MC, Soltis DE, Soltis PS, Weston P, Whitten WM, Wurdack KJ, Chase MW (2000) Phylogeny of the eudicots: a newly complete familial analysis based on rbcL gene sequences. Kew Bull 55:257–309

Saxena S, Hajare SN, More V, Kumar S, Wadhawan S, Mishra BB, Parte MN, Gautam S, Sharma A (2011) Antioxidant and radioprotective properties of commercially grown litchi (*Litchi chinensis*) from India. Food Chem 126:39–45

Sharma G (ed) (2009) Systematics of fruit crop. New India Publishing, New Delhi, pp 193–198

Singh AK, Prasad US (1991) Dehydration pattern and viability loss in seeds of two cultivars of litchi (*Litchi chinensis* Sonn.) Seed Res 19:41–43
Singh ABA, Srivastava S (1987) Variation in quality traits of litchi (*Litchi chinensis* Sonn.) cultivars. J Food Sci Technol 4(5):264–266
Singh A, Pandey SD, Nath V (2012) The world litchi cultivars. Technical bulletin 007 NRC for litchi. Mushahari, Muzaffarpur
Singh JP, Chandel R, Mishra B, Suneetha V (2013) Evaluation of antimicrobial and antioxidant property of lychee's seed for therapeutic purpose. Int J Pharm Sci Rev Res 19:72–76
Stearn WT (1972) A.W. Smith: a gardener's dictionary of plant names: a handbook on the origin and meaning of some plant names. Littlehampton Book Services, New York
Stern RA, Eisenstein H, Voet H, Gazit S (1996) Anatomical structure of two day old litchi ovules in relation to fruit set and yield. J Hortic Sci 71(4):661–671
Subhadrabandhu S, Stern RA (2005) Taxonomy, botany and plant development. In: Menzel CM, Waite GK (eds) Litchi and longan. Comwell Press, Trowbridge, pp 25–34
Takhtajan A (1980) Outline of the classification of flowering plants (Magnoliophyta). Bot Rev 46 (3):225–359
Takhtajan A (1987) Systema Magnoliophytorum. Soviet Sciences Press, Leningrad
Thorne RF, Reveal JL (2007) An update classification of the class Magnoliopsida ("Angiospermae"). Bot Rev 73:67–182
Tindall HD (2010) Sapindaceous fruits: botany and horticulture. In: Janick J (ed) Hortic Rev 16:143–189
Tongdee SC (1997) Longan. In: Mitra S (ed) Postharvest physiology and storage of tropical and subtropical fruits. CAB International, New York, pp 335–345
Wang JB, Deng SS, Liu ZY, Liu LZ, Du ZJ, Xu BY, et al (2006a) RAPD analysis on main cultivars of litchi (*Litchi chinensis* Sonn.) in Hainan. J Agric Biotechnol 4:391–39396
Wang X, Yuan S, Wang J, Lin P, Liu G, Lu Y (2006b) Anticancer activity of litchi fruit pericarp extract against human breast cancer in vitro and in vivo. Toxicol Appl Pharmacol 215:168–178
Wang HC, Huang HB, Huang XM (2007) Different effects of abscisic acid and ethylene on the fruit maturation of *Litchi chinensis* Sonn. Plant Growth Regul 52:189–198
Wei SX, Chen YY, Xie ZS, Zhu M, Luo SR (2006) ISSR analysis of parts of lychee germplasm in Hainan province. Chin J Trop Crops 27:51–555
Wei YZ, Zhang HN, Li WC, Xie JH et al (2013) Phenological growth stages of lychee (*Litchi chinensis* Sonn.) using the extended BBCH-scale. Sci Hortic 161:273–277
Wu SX (1998) Encyclopedia of China fruits: litchi. China Forestry Press, Beijing
Wu YH, Chiu CH, Yang DJ, Lin YL, Tseng JK, Chen YC (2013) Inhibitory effects of litchi (*Litchi chinensis* Sonn.) flower-water extracts on lipase activity and diet-induced obesity. J Funct Foods 5:923–929
Xia Q, Chen HRZ, Fu JR (1992a) Effect of desiccation, temperature and other factors on the germination of lychee (*Litchi chinensis* Sonn.) and longan (*Euphoria longan* Steud.) seeds. Seed Sci Technol 20:119–127
Xia Q, Chen HRZ, Fu ZJR (1992b) Moist storage of lychee (*Litchi chinensis* Sonn.) and longan (*Euphoria longan* Steud.) seeds. Seed Sci Technol 20:269–2279
Xiang X, Ou LX, Chen HB, Sun QM, Chen JZ, Cai CH, et al (2010) ESTSSR analysis of genetic diversity in 96 litchi (*Litchi chinensis* Sonn.) germplasm resources in China. Genomics Appl Biol 29:1082–101092
Yamanishi R, Yoshigai E, Okuyama T, Mori M, Murase H, Machida T (2014) The anti-inflammatory effects of flavanol-rich lychee fruit extract in rat hepatocytes. PLoS One 9: e93818
Yang DJ, Chang YY, Hsu CL, Liu CW, Wang Y, Chen YC (2010) Protective effect of a litchi (*Litchi chinensis* Sonn.)-flower-water-extract on cardiovascular health in a high-fat/cholesterol-dietary hamsters. Food Chem 119:1457–1464
Yang DJ, Chang YZ, Chen YC, Liu SC, Hsu CH, Lin JT (2012) Antioxidant effect and active components of litchi (*Litchi chinensis* Sonn.) flower. Food Chem Toxicol 50:3056–3061

Yao QR, Zhao CZ, Wang WQ (2009) Analysis of genetic relationship of Hainan litchi germplasm resources by SSR marker. B Bot Res 29:628–632
Yeap CK (1987) The sapindaceous fruits and nuts. Yearbook of the West Australian Nut and Tree Crops Association 12:16–33
Yi GJ, Huo HQ, Chen DC, Huang ZR, Cai CH, Qiu YP (2003) Studies on genetic relationship among litchi varieties by using AFLP. Acta Hortic Sin 30:399–403
Zan FG, Wu ZD, Zeng Q, Zhang HY, Li MF, Zheng XQ (2009) Genetic diversity analysis of litchi germplasm by SRAP markers. Mol Plant Breed 7:562–56568
Zhang Z, Yuan P, Wang B, Qiu Y, Li J (1997) Litchi: pictorial narration of cultivation. In: Zhang Z (ed) Pomol Res Inst Guangdong Acad Agric Sci China
Zhou Y, Wang H, Yang R, Huang H, Sun Y, Shen Y (2012) Effects of *Litchi chinensis* fruit isolates on prostaglandin E (2) and nitric oxide production in J774 murine macrophage cells. BMC Compl Altern Med 12:12

Procedural Insights on In Vitro Propagation of *Litchi chinensis* (Sonn.) 13

Vivek Kumar, Manoj Kumar, Shivesh Sharma, Ajit Varma, and Neera Bhalla-Sarin

Abstract

Systemic problems in litchi have been persistent while propagating litchi plants through in vitro techniques at commercial scale. It has been reported that fungal attacks start on the tip of nodal segments and spread throughout the culture vessels, which revealed the symptoms of an endogenous pathogen because fungal contamination goes beyond the control when they receive highly rich nutrient medium. Here the authors suggest that it is worthwhile to screen for specific antifungal agents; therefore, it is essential to evaluate them before applying several antagonists for their elimination from the culture. We address taxonomic information for in vitro-raised fungi and suggests ways to eliminate such bottlenecks of contamination in litchi tissue cultures. This chapter reports a methodical approach and explains a specific protocol for indexing the endogenous contamination and browning effect of phenolic leaching, henceforth suggesting control measures.

Keywords

Endogenous pathogen • Fungal contamination • In vitro • Browning • Litchi

V. Kumar • M. Kumar (✉) • A. Varma
Amity Institute of Microbial Technology, Amity University, Noida, Uttar Pradesh-201311, India
e-mail: manojjnu@gmail.com

S. Sharma
Department of Biotechnology, MLN National Institute of Technology, Allahabad, India

N. Bhalla-Sarin
School of Life Sciences, Jawaharlal Nehru University, New Delhi-67, India

M. Kumar et al. (eds.), *Lychee Disease Management*,
DOI 10.1007/978-981-10-4247-8_13

13.1 Introduction

An attractive and rare tropical fruit tree, *Litchi chinensis* Sonn. (belonging to the family Sapindaceae) that bears luscious red fruits originated more than 2000 years ago in low elevation of Southern China in the provinces of Kwangtung and Fukien. One of the most environmentally sensitive tropical tree fruit crops, litchi is adapted to areas of the world characterized by warm subtropics and elevated tropics having cool dry winters and warm wet summers (Menzel 1985). China, India, Southeast Asia, and South Africa are among the major producer countries.

Its common names are lychee, litchi, leechee, lichee, or lichi. The litchi, or lychee, tree is native to Southern China and Southern Asia. It was widely cultivated for its prized fruit even before 1766 BC and is one of the most important fruit crops (Menzel 1985). It reached Hawaii in 1873 and Florida in 1883 and was then conveyed to California in 1897. The most popular varieties grown in South Florida are the Brewster, the Hakip, and the Mauritius.

Morphologically the fruit is a dry nut, but the presence of a white, thick, juicy, translucent aril (or pulp, the edible part which is high in vitamin C) surrounding the seed makes it fleshy. It also contains magnesium, and phosphorus, as well as small amounts of iron, zinc, copper, manganese, selenium, thiamin, riboflavin, niacin, calcium, vitamin B_6 and vitamin E. It may be taken fresh, frozen, canned in syrup or dried to produce "litchi nuts." The trees make beautiful landscape specimens with their dark green leaves and bright red fruits.

More than 350,000 tonnes of litchi fruit is being produced annually with the largest share coming from China, India and Taiwan, and the production has considerably increased in South Africa, Australia, Southeast Asia, Vietnam, USA and Israel (Underhill and Critchley 1993).

In China, the most important fruit-producing city is Gaozhao in South China on the banks of Pearl River, having boundless expanses of litchi plantations with beautiful dark green dome-shaped canopies. The area under litchi cultivation is 38,000 ha with an annual production of 84,000 tons. Other important litchi-growing areas are Dongguan and Shenzhen, famous for several elite litchi cultivars such as Smile of the Emperor's Concubine (Fezixiao), Guiwei and Nuomici. In India, Bihar is the most important litchi-growing state, which contributes 77% of total litchi production in the country. The area of cultivation is spread over 22,503 ha with an annual production of 2.7 lakh metric tonnes. The main litchi-growing districts in Bihar belong to Muzaffarpur, Vaishali, Sitamarhi, Gopalganj, Champaran, Saran, Siwan, Samastipur, Katihar and Bhagalpur. According to the State Horticulture Department, at least 1000 ha of cropping areas are being added annually.

Bihar contributes 70% of the total litchi fruit production in India, and some parts of Punjab, Uttaranchal, UP and Jharkhand are also the source of litchi in India. Litchi is mostly harvested during May–June, and yield is around 80–100 kg per tree per year in India. The overall yield was 5900 kg per hectare, whereas area under litchi cultivation was 56,538 ha (Directory of Indian Agriculture).

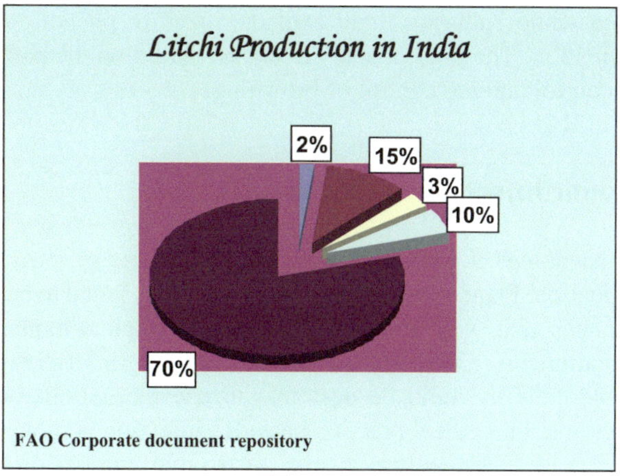

In vitro oxidative browning of cultures, contamination, vitrification, and high mortality during acclimatization are some of the problems associated with micropropagation of woody fruit trees. In vitro oxidative browning can be controlled using different phenol binding agents, modifying the redox potential of the media, quick subculturing, keeping cultures in the dark and through explant waxing (George and Sherrington 1984; Mishra et al. 1998). The conditioning of a stock plant has been very important for the establishment of shoot tips and nodal explants. New vegetative growth was found to be used as an explant in jackfruit, guava (Amin and Jaiswal 1988), persimmon (Mishra 1982) and (Amin 1992) and a number of other fruit crops. Substantial numbers of micropropagated plants do not survive from in vitro condition to greenhouse or field environment. The plantlets develop within the culture vessels under low levels of light and aseptic conditions on a medium containing ample sugar and nutrient to allow for heterotrophic growth and in an atmosphere with high relative humidity.

Micropropagation of litchi has already been reviewed in the recent past (Kumar et al. 2006; Kantharajah et al. 1989; Kantharajah et al. 1992; Das et al. 1999; Chandra and Padaria 1999; Kumar et al. 2004). These reports evidenced that when using explants from adult trees, several problems impede the development of effective protocols of micropropagation, among which are (1) the heavy oxidation of tissue when explants (shoot tips, meristems) were collected from infield or greenhouse plants, (2) the difficulty of getting sterile shoots when nodal explants were used, and (3) the arduous task of establishing shoot cultures are the most common. During the past decade, many trials have been made towards the solution of these problems and the optimization of the various steps involved in litchi micropropagation. The exudation of phenolics from the cut ends of litchi explants greatly hinders their regenerative capacity in any in vitro growth medium.

During the initiation phase, there may be several problems such as high contamination and low sprouting rates, exudate production, and browning in explants as in the case of avocado (Cooper 1987; Zirari and Lionakis 1994). Two major problems are encountered while culturing litchi explants in vitro: phenolic exudation and fungal contamination. The present study is mainly based on the pathogen identification and its control and the control of browning.

13.1.1 Systemic Infection

One problem encountered while culturing litchi explants in vitro was systemic fungal contamination. Plant tissue cultures are sometimes found to be contaminated by microorganisms that were either present in the original explant or arise as laboratory contaminants. Generally contaminated tissue is discarded, but in the case of valuable cells it would be desirable to use the antimicrobial agents to eliminate infections. In medical practice bacterial infection is more common than fungal, but with plants the reverse is true, as a large number of fungicides are available which are effective against plant fungal pathogens. Kumar et al. (2006) and Roy and Fantes (1982) evaluated the benzimidazole fungicides which are in widespread use as systemic agricultural fungicides and appear to act by interfering with fungal microtubules and hence fungal cytokinesis. Shields et al. (1984) tested benomyl, carbendazim (MBC), thiabendazole (TBZ) and fenbendazole (FBZ) on (1) fungi arising as laboratory contaminants of plant tissue culture, (2) plating efficiency of protoplast-derived cells of haploid *Nicotiana plumbaginifolia*, (3) callus culture of *N. tabacum*, (4) root cultures of *N. tabacum*, (5) germination of tobacco seeds and (6) tobacco seedling growth. They compared the effect of fungicides tested on different explants and the phytotoxicity level.

13.1.2 Subsidiary Problem with Contamination: Phenolic Exudation

Leaching of phenolic compounds results in browning of the medium and inhibition of the growth of explants (nodal cuttings and leaf discs). Pretreatment of litchi explants with adsorbents using liquid culture helps in overcoming this problem. Similarly, in-plant regeneration of *Mangifera indica* using liquid culture reduced the phenolic exudation (Kumar et al. 2006; Das et al. 1999). To minimize tissue browning of *Roseomonas mucosa* explants observed in preliminary experiments, segments from hypocotyls (0.5 cm) and leaves (1 cm^2) were immersed in Petri dishes for 15 min in sterile antioxidant solutions containing either citric acid, ascorbic acid, or cysteine (0, 25, 50, or 100 mg/l) (Encina et al. 1994) and in solution containing adsorbent polyvinylpyrrolidone (PVP-40) (0, 100, 500, and 1000 mg/l) (Grattapaglia et al. 1998). Inclusion of 0.1% polyvinylpyrrolidone (PVP-360) in the media reduced phenolic exudation (Thimmappaiah et al. 2002). To prevent browning of callusing explants of *Dendrocalamus latiflorus*, several antioxidants and adsorbing agents were tested. The most effective was the application of activated charcoal (2 g/l) and incubation in the dark (Zamora et al. 1988).

Inhibitors of the enzyme phenylalanine ammonia lyase (PAL) in the phenylpropanoid pathway were used to investigate the role of phenolic metabolism in the browning of lettuce tissue. Browning of the cut ends and uncut surfaces was measured using a visual score and CIE colour values (Peiser et al. 1998). They immersed the methodology to measure the changes in absorbance at 340 nm of aqueous extracts of the explants (midrib of leaves). Browning of the cut and uncut surfaces was determined by visual evaluation and with the colourimeter; absorbance at 340 nm was essentially an average of the browning in all the tissues.

13.1.3 Surface Sterilization

Single-node cuttings from the side branches of the field-grown plants were collected. The explants were washed first under running tap water followed by a detergent, either extran (5% v/v) or savlon (1% v/v), for 15–20 min. After a thorough wash in sterile distilled water, bavistin treatment (1% w/v) was given for 25 min. Again, a repeated thorough wash with sterile distilled water followed by mercuric chloride treatment (0.2% w/v) was given for 5 min. To remove the disinfectants from the surface layer of explants, these were rinsed three or four times with double-distilled sterile water. Nodal segments were cut into appropriate sizes (2 cm) and implanted on sterile pre-culture liquid medium. All subsequent operations were carried out inside a laminar airflow cabinet.

13.1.4 Pre-treatment

Three media, viz. the Murashige and Skoog medium (Murashige and Skoog 1962), Woody Plant Medium, and Gamborg and Phillips (1995), which all gelled with 0.8% agar or 0.24% phytagel, were tested for regeneration from nodal segments, apical shoot meristems, and leaf discs. In these preliminary studies, liquid WPM supplemented with 3% sucrose at pH 5.8 was tested as the pre-culture medium. Disinfested explants were inoculated on media supplemented with various concentrations of BAP, kinetin, IAA and NAA. The cultures were maintained at cold temperature (4°C) for a week initially and then transferred at 25°±°2 C on a 16-h photoperiod (50–70 μmol m^{-2} s^{-1}, cool white fluorescent lamps). The explants were also placed on semi-solid medium gelled with agar or phytagel or liquid WPM medium and subcultured after 12, 24, or 48 h on the same medium. The explants were placed in WPM liquid medium using the filter paper bridge technique for 1 week. The WPM medium containing 0.1–1% PVP and 0.1 1% activated charcoal. A good quality filter paper (Whatman 9) was used for making the bridge for culture medium to maintain its nature.

Another pretreatment involved placing explants in 100 ml liquid medium supplemented with 1–2% PVP in 250-ml conical flasks on an automated shaker at 100 rpm for 24, 48, and 76 h with a replacement of the liquid medium at every 6 h. At the same time explants were also treated with citric acid, ascorbic acid, or cysteine (0, 25, 50, or 100 mg/l). After this treatment, the explants were washed thoroughly with distilled water and placed on agar-gelled WPM medium. As a control, untreated explants (control) were inoculated on WPM medium.

After these pre-treatments, respectively, the explants were transferred to fresh medium supplemented with growth regulators BAP (11 μM), Kn (2.30 μM), GA_3 (0.60 μM), and adsorbent (PVP) (0.2%). One explant was inoculated in each 200 × 25 mm tube containing 20 ml WPM medium and sealed with a cotton plug.

13.1.5 Growth of Fungi

Isolates of various fungi contaminating the litchi cultures in the laboratory were collected and maintained on solidified YEPD oatmeal medium (Shields et al. 1984). They were subcultured either by taking a loop of spores or a piece of mycelium agitating in 1 ml Muller Hinton's broth, dropping 0.1 ml of this suspension onto fresh agar plates and spreading the solution with a glass spreader. Fungi were cultured at 26 C in the dark and subcultured every 2–3 weeks.

13.1.6 Tests of Fungicides on Fungal Growth

Fungicides were dissolved (after heating whenever necessary) in DMSO at 10–50 mg/ml and stored at −20 °C. Appropriate volume (10–50 μg/l) of the concentrated solution of each fungicide was added to molten agar medium which was poured into plates immediately. Fungi were spread on these plates as described. Fungicides were included both in YEPD agar or oatmeal medium and on semi-solidified plant tissue culture medium [WPM + BAP (11 μM) + Kn (2.30 μM) + GA_3 (0.60 μM) + bavistin (30 μg/l) + CW (15%) + CH (400 mg/l) + PVP (0.2%)]. Sucrose (3%) was added to the culture medium. Growth of fungi was assessed every 2 days for 4 weeks.

13.1.6.1 Fungicides Used

1. Benomyl and carbendazim (Bavistin-MBC) from Dupont (U.K.) Ltd.
2. Thiabendazole (TBZ) from Merck
3. Fenbendazole (FBZ) from Hoechst
4. Imidazole from a local supplier (CDH)

13.1.7 Pathogen Culture and Identification

A detailed taxonomic study of pathogen cultures isolated from litchi tissue was carried out at ITCC, Mycology Department, Division of Plant Pathology, IARI, Pusa Institute, N. Delhi, as given below.

13.2 Schematic Representation: Morpho-Taxonomy Studies of Pathogens Under Stereomicroscope

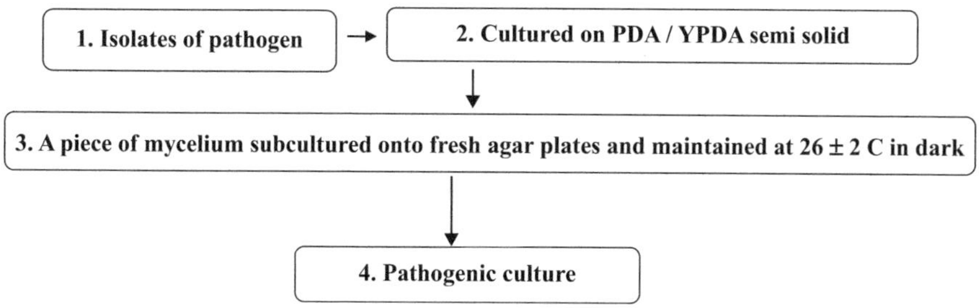

13.3 Overcoming Limitations: Contamination and Browning

Contamination and browning of litchi explants (nodal cuttings, apical shoot meristems and juvenile leaf discs) were the great limitations during initiation of cultures under in vitro conditions. The explants were mainly contaminated by endogenous pathogens. The presence of phenolic compounds and probably a higher polyphenol oxidase activity caused explant browning, affecting vegetative propagation and limiting morphogenic responses (Fig. 13.1a–d). Endogenous contamination was the main factor limiting the culture establishment, contamination starting after the fourth day of culture initiation which persisted until the third subculture (Fig. 13.2a, b). Contamination varied from 30 to 90% depending on the season. Apical shoot tips usually had lower contamination rates compared to the nodal and leaf explants.

13.3.1 Effects of Tree Age, Explant Type, and Source on Shoot Proliferation

Chalupa (2002) investigated the influence of tree age, explant source and genotype on micropropagation of mature trees of Sorbus aucuparia L. Experiments demonstrated the feasibility of using juvenile parts of mature trees for in vitro propagation of selected genotypes. Explants from lower branches and from epicormic shoots of mature trees exhibited high multiplication coefficients of microshoots cultured on modified MS agar nutrient medium supplemented with cytokinin (BA, TDZ) and auxin (IBA). Mature trees are characterized by decreasing capacity for vegetative propagation. However, different parts of mature tree are often in a different degree of maturity. Advances in micropropagation of broad-leaved forest tree species (Chalupa 1979; Chalupa 1983) have opened great opportunities for mass propagation of selected valuable genotypes of forest trees.

Fig. 13.1 (**a–d**) Diseased explants showing abnormal response on woody plant medium (McLloyd Medium-McCown and Sellmer (1987)) without additives (PVP, BAP, etc.) (From Kumar M, 2006 PhD thesis)

Promising results have been mostly obtained with micropropagation of juvenile plant material (Kumar et al. 2006).

Our discussion on litchi is mainly based on the selection of explants, composition of nutrient media, concentration of phytohormones and methods used for micropropagation which have significant effects on shoot multiplication rates, rooting and quality of generated plantlets.

13.3.2 Agar Versus Filter Paper Approach

The nature of the support provided to the explants was found to be a critical factor for the success of multiple shoot induction in litchi (Kumar et al. 2006; Das et al.

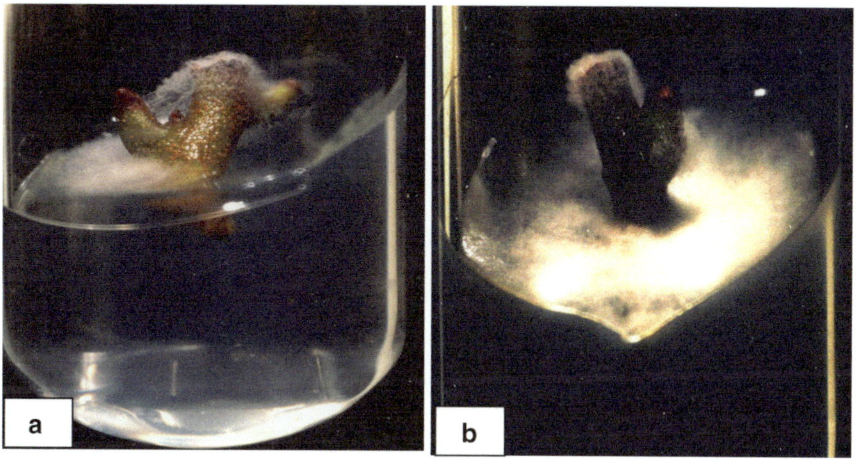

Fig. 13.2 (**a**, **b**) Fungal contamination of explants, after 4 days of inoculation (From Kumar M, 2006 PhD thesis)

1999). Litchi seeds cultured on BAP-supplemented MS semi-solid medium turned black, and their germination was completely inhibited. It is speculated that browning of the cut surface of some fruits is caused by a group of enzymes called polyphenoloxidases. These enzymes are released by the broken cells, and they catalyse the reaction between colourless molecules called polyphenols and molecular oxygen. This reaction creates coloured compounds, and these new compounds can spontaneously cross-react with one another to form black-brown complexes called melanins/quinone (http://www-saps.plantsci.cam.ac.uk/osmoweb/ppo.htm). The excess phenolics released by litchi may not diffuse rapidly on agar-solidified medium, and the resulting local concentration of phenolics around the seeds might be autotoxic.

13.4 Problem Identification

13.4.1 Symptoms

The diseased explants (that is, leaf source) of litchi had dark-brown spots with a blackish-brown border covering an area of 2–4 mm in diameter throughout the surface (Fig. 13.3). On oatmeal medium, growth of the culture reached a diameter of 70 mm in 20 days at 25°C. These spots coalesced, forming bigger spots, covering the entire surface of nodal cuttings and leaf surface. After 20 days, the fruiting structures (pycnidia) were observed which were superficial, spread on the infected area, numerous, solitary or in groups, dark brown and ostiolate. The pycnidial formation starts after 15 days of subculturing. The pycnidia were many, dark coloured, ostiolate, embedded as well as superficial on the surface of the medium. The colony on the YPDA/PDA in the centre was initially cottony white, turning

Fig. 13.3 Diseased leaf of litchi showing *dark brown* spot (From Kumar M, 2006 PhD thesis)

olive brown on maturity. The reverse of the colony was dark brown and released light brownish pigment into the medium.

13.4.1.1 Symptom I

The pycnidial formation was scanty; the pycnidial neck and the slime were prominent 100–150 μm in diameter. The spores come out by oozing. Conidiophores were branched and septate at the base and above. Conidia of two basic types, α-conidia and β-conidia, were present. The young α-spores were oval, hyaline, nonseptate and attained spindle shape with age, measuring 10–12 × 3–5 μm with oil globules prominent at both ends of the spores. β-spores were not much hooked at terminal end, were mostly straight, nonseptate, hyaline and scanty, measuring 40–70 × 3–5 μm. On the basis of these taxonomic characteristics, the pathogen from *Litchi chinensis* was identified as unknown endophytic pathogen (unpublished data).

13.4.1.2 Symptom II

Colonies usually darkly pigmented with white aerial mycelium, consisting of numerous black sclerotia and light brown coloured conidial masses present with dark-brown reverse side were present. Sclerotia were usually abundant, setose, spherical and were often confluent. Conidia were straight, fusiform and attenuated at the ends, 16–22 × 3–4 um. Appressoria were common, clavate and brown, 11–16.5 × 6–9.5 (unpublished data).

This taxonomic study showed a close similarity to symptomatic pathogen (*Fungi imperfecti*) species, and therefore it was observed as unknown pathogen; the genetic aspect is still under study.

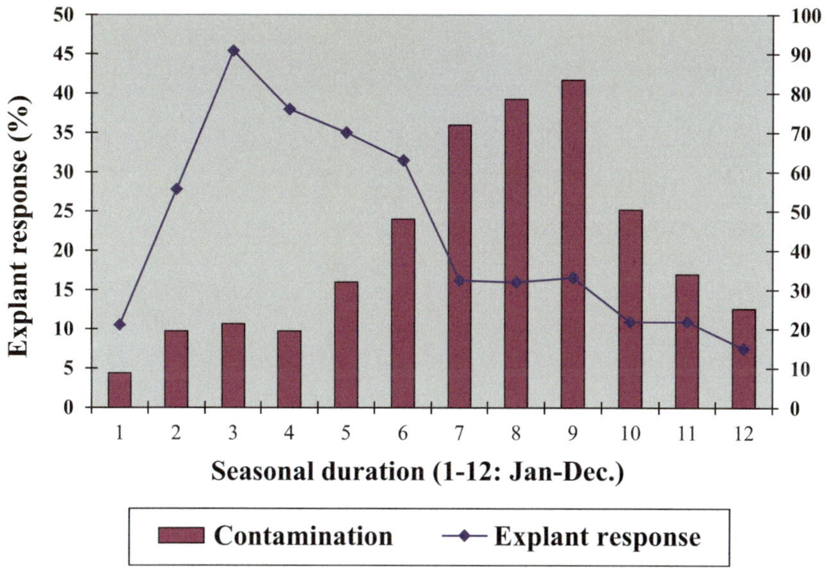

Fig. 13.4 Seasonal effect on litchi micropropagation (From Kumar M, 2006 PhD thesis)

13.5 Establishment of Aseptic Nodal Segment Cultures

13.5.1 Seasonal Effect

The rate of contamination and bud break was highly dependent on the season during which material was collected. By the sterilization procedure described in Section 4.2.1, the cultures initiated in the months February–May showed higher bud break (45%) and less contamination (20%) than those raised in the months June–January. Because June–September is the period coinciding with the rainy season in India, 90% of explants showed contamination. By November–January the shoots become old, and it is difficult to break down the mature dormant state of the buds. Therefore, routinely, the cultures were raised in February–May because of the least contamination and the best shoot growth recorded during the season (Fig. 13.4).

13.5.2 Control of Pathogen

An elaborate procedure was followed as described by Kumar et al. (2006) to eliminate the fungal contamination. Among four fungicides tested, bavistin was found to be the most effective and the least toxic to the plant cells at a concentration of 30 µg/l. Hence a contamination-free regeneration system was established for developing the multiple shoots from the nodal explants of litchi in vitro (Fig. 13.5).

Fig. 13.5 Shoot induction from nodal explants and apical shoot meristems cultured on WPM + BAP (11 μM) + Kn (2.30 μM) + GA3 (0.60 μM) + bavistin (30 μg/l) + PVP (0.2%) (From Kumar M, 2006 Ph D thesis)

13.5.3 Effect of Benzimidazole Fungicides

The fungicides tested were carbendazim (bavistin-MBC), thiabendazole (TBZ), fenbendazole (FBZ) and imidazole. Although FBZ was used primarily as an anthelminthic, carbendazim (bavistin-MBC) had good antifungal activity, but bavistin was found to be the most effective and the least toxic to the plant cells at 30 μg/l (Fig. 13.5). Because the continuous presence of fungicide is not promotive to the plant cell growth, after one subculture, bavistin was excluded from the medium. The pathogen was completely eradicated from the explants, and in vitro cultures were well established.

13.5.4 Control of Phenolics

The protocol standardized for treatment during explant excision allowed tissues to stay green during the two initial subcultures. The liquid pre-culture medium WPM + BAP (11 μM) + Kn (2.30 μM) + GA_3 (0.60 μM) + bavistin (30 μg/l) + PVP (0.2%) using the filter paper bridge technique was found to be very effective. No sucrose was included in this medium.

13.5.5 Use of Filter Paper Bridge Technique and PVP

Nodal segments cultured on WPM semi-solid medium supplemented with BAP, KIN, GA3, PVP and CW turned black. Similarly, leaf discs cultured on MS and B5 semi-solid medium turned dark brown at the initial stage. It was observed that the excess phenolics released by nodal segments and leaf discs of litchi may not diffuse rapidly on agar-solidified medium and that resulting local concentration of

Fig. 13.6 (**a, b**) Nodal explant on filter paper bridge (From Kumar M, 2006 Ph D thesis)

phenolics around the explant might be autotoxic. Therefore, a method using nodal segments placed on filter paper bridges in liquid medium proved to be better at the initial stage because the rapid diffusion of phenolics in liquid medium prevented toxic levels from being reached, and the explants showed a better response (Fig. 13.6a, b). Several antioxidants tested (viz. citric acid, ascorbic acid and cysteine) against browning of explants at initial stage, successfully optimized under *in vitro* conditions and found to be promotive for a healthy propogation of Lychee. Before inoculation, pretreatment of explants with higher concentration of PVP (1–2%) was essential.

Inclusion of PVP (0.2%) into liquid medium followed by transferring into the PVP-free semi-solid culture medium was found to be effective in direct regeneration of shootlets from the nodal explants as well as induction of friable and creamish nodular calli from the leaf discs (Fig. 13.7a–d).

During pre-culture, fresh WPM liquid medium supplemented with BAP (11 μM l) + kinetin (2.30 μM) + GA3 (0.60 μM) was provided to culture at every 3–4 days to compensate for the loss of liquid through transpiration and evaporation. Sucrose was avoided during pre-culture for aseptic culture establishment, after two subcultures of 3% sucrose were added.

13.5.6 Use of Charcoal

Charcoal has been used as a purifying and decolourizing compound for liquids since the eighteenth century. Charcoal used before the present century was produced by pyrolysis only. Activated charcoal adsorbs inhibitory substances

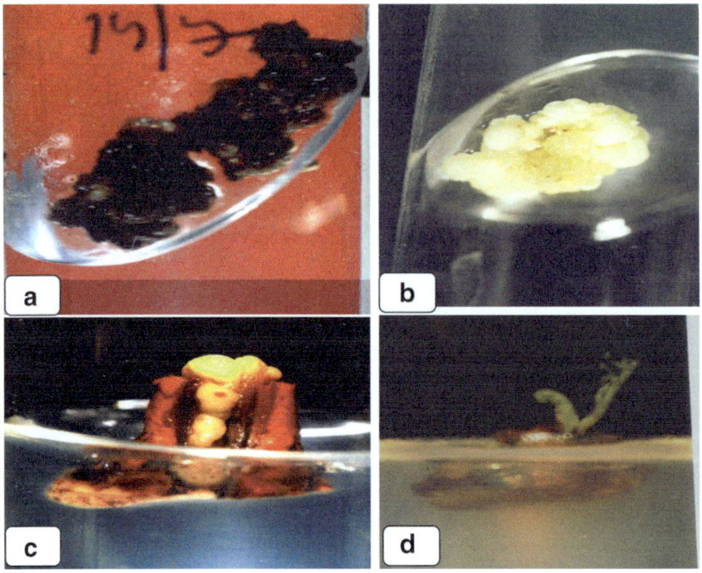

Fig. 13.7 (**a**–**d**) Leaf explant cultured on (**a**) B5 medium without adsorbent (PVP), (**b**) with adsorbent (PVP), (**c**) nodal explants on WPM medium without adsorbent, (**d**) nodal explant on WPM medium with adsorbent (From Kumar M, 2006 Ph D thesis)

accumulating in the culture medium (Fridborg et al. 1975; Fridborg et al. 1978; Weatherhead et al. 1978; Theander and Nelson 1988) and is thus often used to reduce the oxidation of phenolic compounds (Carlberg et al. 1983; Liu 1993; Teixeira et al. 1994) in tissue culture to improve cell growth and development. Pan and Van (1998) reviewed the use of activated charcoal in culture medium and stated that charcoal has a dual role in somatic embryogenesis; sometimes it improves, and at other times it inhibits, the growth of plantlets obtained through somatic embryogenesis. Activated charcoal can provide a dark environment and adsorb substances presumed to be deleterious and/or inhibitory to in vitro cultures, but adsorption of growth regulators (being supplied to the tissue) by activated charcoal could also occur at the same time (Pan and Van 1998), which can inhibit the desired response.

13.5.7 Use of Coconut Water, Polyamines, and Amino Acids

Coconut water has been shown to stimulate shoot proliferation in many species of plants. It contains useful compounds such as nicotinic acid, auxin, gibberellins, pyridoxine and thiamin. Coconut water is an agricultural by-product that can be used not only as a beverage and as a growing medium in tissue culture but also as a growth regulator (Dix and Van 1982).

In plants, polyamines are involved in a wide range of important processes including cell division, protein synthesis and DNA replication and have important functions in various morphogenic responses (Bais and Ravishankar 2002). Mehmet (2004) reported that inclusion of polyamines in the medium improved shoot elongation in hazelnut (*Corylus avellana* L.) micropropagation. Polyamines were found to have a strong effect on both shoot elongation and on number of buds per shoot. Polyamines stimulated mean shoot elongation by 83% and increased the mean number of buds per shoot by 41% as compared to controls. In the presence of polyamines, shoot elongation continued up to 4.0 cm, while in the absence of polyamines shoot elongation reached only 2.0 cm. Results indicated that polyamines in the culture medium could ease the establishment of cultures and enhance the morphogenic capacity of mature explants. Furthermore, it has been reported (Bais et al. 2000) that exogenous application of polyamine (putrescine) may reduce the production of unwanted ethylene and enhance morphogenesis. Therefore, the objective of this study was to explore whether the morphogenic capacity of mature litchi tissues could be increased by the inclusion of polyamines in the culture medium.

Amino acids are commonly included in the organic supplement. The most frequently used is glycine, although arginine, asparagine, aspartic acid, alanine, glutamic acid, glutamine and proline are also used. Amino acids provide a source of reduced nitrogen and, as do ammonium ions; their uptake causes acidification of the medium. Casein hydrolysate has been used as a relatively inexpensive source of a mix of amino acids (Price and Smith 1979; Chaturvedi et al. 2004).

Using these additives enhanced the capacity of immature embryos to produce a high amount of callus with excellent embryogenic nature (friable, easily subcultured, and regenerated almost exclusively via somatic embryogenesis).

13.6 Pathogenic Nature

The explants that were contaminated turned from green to brown and subsequently died. Similar problems in micropropagation of mature hazelnut have been reported by other researchers (Diaz-Sala et al. 1990) are known to be common in most woody plants (Preece and Sutter 1991). Few diseases have been reported from any litchi-growing locality. The glossy leaves are very resistant to fungi. In Florida, litchi trees are occasionally subject to green scurf, or algal leaf spot (*Cephaleuros virescens*); leaf blight (*Gloeosporium* sp.); dieback, caused by pathogen spp. and mushroom root rot (*Clitocybe tabescens*), which was most likely to attack litchi trees planted where oak trees formerly stood. Old oak roots and stumps were found thoroughly infected with the fungus (Morton et al. 1987).

The taxonomy of pathogens has been largely concerned with classical descriptive criteria such as conidial shape and size and presence/absence and morphology of setae. Supplementary criteria were included for type of damage to the host caused by this species and the identity of the host itself. Hardly any report of these taxa has been described for pathogen on litchi, although several hundred taxa have been described in other fruit crops.

13.6.1 Control of Pathogens

Plant pathologists tested a number of agricultural and medically useful fungicides for their effects on numerous fungi as well as on plant cells and tissues. These compounds are in widespread use as systemic fungicides and appear to act by interfering with fungal microtubules and hence fungal cytokinesis (Roy and Fantes 1982). The aim of this study was to see if fungicides could be safely used in litchi tissue culture to rid cultures of fungal contamination. The ideal fungicide would kill all species of contaminating fungi without harming plant tissues. Although no single fungicide combined both these properties, some came close to this ideal. The fungicide which appeared most useful in this study was carbendazim (bavistin-MBC) (30 μg/l) among the tested fungicides. It is possible that the toxicities vary from tissue to tissue and species to species (Brown et al. 1982). In the present study (with the exception of the bavistin), there is good concordance between phytotoxicities observed with nodal segments, apical shoot meristems and juvenile leaf discs of litchi.

13.7 Recommendation

For dominance of seasonal effects on litchi tissue culture (Kumar et al. 2006), researchers also found similar effectivity in various plant species such as neem (Chaturvedi et al. 2004), apple (Hutchinson 1984), sweet gum (Sutter and Barker 1985) and guava (Amin and Jaiswal 1988). Yu and Chen (1998) reported that exudation of phenolics and contamination depended on explant growing conditions of the donor plants during favourable and unfavourable seasonal conditions.

We address the seasonal effects on the litchi mother plants which were very critical for the regeneration of explants taken from them. It has been observed that explants of those that are taken in the months of February–May showed the best response in culture.

A series of changes convert the colourless polyphenols to the brown discolouration during the media-oriented treatments of explants: these include enzyme oxidation, rearrangement of groups, nonenzymatic oxidation and polymerization. In the recent past, many unsuccessful trials were made to regenerate litchi via clonal propagation and somatic embryogenesis (Kumar et al. 2006; Das et al. 1999; Kantharajah et al. 1992). Its micropropagation is hindered by the excision of explants leading to the eventual death of excised cultured tissues. Browning of the cut surface of explant is caused by oxidation of phenolic compounds by the existing polyphenol oxidase. Jiang and Fu (1999) observed that browning in litchi fruit might occur because of both reduction in antioxidant capacities and an increase in oxidation of phenolics catalysed by polyphenol oxidase and peroxidase enzymes. In the present study, pretreatment in liquid medium containing PVP (0.2%) with the filter paper bridge technique was found to be a very convenient and effective method to eliminate the browning problem from the culture. Three methods of phenolic control using rapid subculturing, filter paper bridge technique and liquid shaker culture were reported earlier by Kumar et al. (2004). Researchers reported the use of liquid shaker culture to reduce the phenolic exudation in mango tissue culture, but in the present study this was not found suitable. The nature of the support provided to the nodal segments was found to be critical for the success of this method in multiple shoot induction of litchi (Das et al. 1999). These results are similar to those observed in seedlings of *Vigna mungo* and *Cajanus cajan* (Shiva et al. Shiva et al. 1994) where multiple shoot initiation from the cotyledonary node was possible only through the direct contact of the nodal region with medium.

References

Amin MN (1992) In vitro rooting of jackfruit (*Artocarpus heterophyllus*) microcuttings derived from mature trees. Plant Tissue Cult 2:129–133

Amin MN, Jaiswal VS (1988) Micropropagation as an aid to rapid cloning of a guava cultivar. Sci Hortic 36:89–95

Bais HP, Ravishankar GA (2002) Role of polyamines in the ontogeny of plants and their biotechnological applications. Plant Cell Tissue Organ Cult 69:1–34

Bais HP, Sudha G, Ravishankar GA (2000) Putrescine and silver nitrate influence shoot multiplication, in vitro flowering and endogenous titers of polyamines in *Cichorium intybus* L. cv. Lucknow local. J Plant Growth Regul 19:238–248

Brown DM, Groom CL, Cvitanik M, Brown M, Cooper JL, Arditti J (1982) Plant Cell Tissue Organ Cult 1:165–180

Carlberg I, Glimelius K, Eriksson T (1983) Improved culture ability of potato protoplasts by use of activated charcoal. Plant Cell Rep 2:223–225

Chalupa V (1979) *In vitro* propagation of some broad-leaved forest trees. Commun Inst For Cechosloveniae 11:159–170

Chalupa V (1983) Micropropagation of conifer and broadleaved forest trees. Commun Inst For Cechosloveniae 13:7–39

Chalupa V (2002) *In vitro* propagation of mature trees of *Sorbus aucuparia* L., and field performance of micropropagated trees. J For Sci 48(12):529–535

Chandra R, Padaria JC (1999) Litchi shoot bud culture for micropropagation. J Appl Hortic Lucknow 191:38–40

Chaturvedi R, Razdan MK, Bhojwani SS (2004) *In vitro* clonal propagation of an adult tree of neem (*Azadirachta indica*) by forced axillary branching. Plant Sci 166:501–506

Cooper PA (1987) Advances in micropropagation of avocado (*Persea americana* Mill.) Acta Hortic 212(II):571–575

Das DK, Prakash NS, Sarin NB (1999) Multiple shoot induction and plant regeneration in litchi (*Litchi chinensis* Sonn.) Plant Cell Rep 18:691–695

Diaz-Sala C, Rey M, Rodriguez R (1990) "In vitro" establishment of cycloclonal chain from nodal segments and apical buds of adult hazel (*Corylus avellana* L.) Plant Cell Tissue Organ Cult 23:151–157

Dix L, Van SJ (1982) Auxin and gibberellin-like substances in coconut milk and malt extract. Plant Cell Tissue Organ Cult 1:239–245

Encina CL, Barcelo A, Herrero-Castan A, Pliego-Afaro F (1994) In vitro morphogenesis of juvenile *Annona cherimola* Mill. Bud explants. J Hortic Sci 69:1053–1059

Fridborg G, Gand, Eriksson T (1975) Effects of activated charcoal on morphogenesis in plant tissue cultures. Physiol Plant 34:306–308

Fridborg G, Pedersen M, Landstrom LE, Eriksson T (1978) The effect of activated charcoal on tissue cultures: adsorption of metabolites inhibiting morphogenesis. Physiol Plant 43:104–106

Gamborg OL, Phillips GC (1995) Media preparation and handling. In: Gamborg OL, Phillips GC (eds) Plant cell, tissue and organ culture: fundamental methods. Springer, Berlin, pp 21–34

George EF, Sherrington PD (1984) Plant propagation by tissue culture. Handbook and directory of commercial laboratories. Exegetics, Basingstoke

Grattapaglia D, Machado MA, Micropropagac A (1998) In: Torres AC, Caldas LS, Buso JA (eds) Cultura de tecidos e transformação genetica de plantas. Embrapa-SPI/Embrapa-CNPH, Brasilia, pp 183–260

Hutchinson JE (1984) Factors affecting shoot proliferation and root initiation in organ culture of apple 'Northern Spy'. Sci Hortic 22:347–358. In hazelnut (*Corylus avellana* L.) micropropagation. Turk J Agric 28:189–194

Jiang M, Fu JR (1999) Biochemical and physiological changes in browning of litchi fruit caused by water loss. J Hortic Sci Biotechnol 74:43–46

Kantharajah AS, McConchie CA, Dodd WA (1989) The possible contribution of tissue culture to the lychee industry. Proceedings of the Second National Lychee Seminar, Cairns, pp 59–65

Kantharajah AS, Dodd WA, McConchie A (1992) In vitro embryo culture and induction of multiple shoots in lychee (*Litchi chinensis* Sonn.) Ann Bot 70:153–156

Kumar M, Prakash NS, Muthusamy A, Prasad US, Sarin NB (2004) Problems and prospective of mass scale production of litchi (*Litchi chinensis* Sonn.) using in vitro cultures. Proceedings of National Seminar on Recent Advances in Production and Post-Harvest Technology of Litchi for Export. Faculty of Horticulture, Bidhan Chandra Krishi Viswavidyalaya, Mohanpur, Nadia, West Bengal, pp 12–17

Kumar M, Prakash NS, Prasad US, Sarin NB (2006) A novel approach of regeneration from nodal explants of field grown lychee (*Litchi chinensis* Sonn.) fruit trees. J Plant Sci 5(3):321–327

Liu MSC (1993) Plant regeneration in cell suspension cultures of sugarcane as affected by activated charcoal, medium composition and tissue source. Taiwan Sugar, pp 18–25

Mehmet NN (2004) Inclusion of polyamines in the medium improves shoot elongation. Turk J Agric For 28:1898–194

Menzel CM (1985) Propagation of lychee: a review. Sci Hortic 21:201–215

Mishra HN, Mishra RS (1998) Diallel analysis for variability in bitter gourd (*Momordica charantia*). Indian J Agric Sci 68(1):18–20

Mishra KA (1982) The vegetative propagation of persimmon. Sci Hortic 17:125–127

Morton J, Julia F, Morton M (1987) Lychee. In: Fruits of warm climates. Creative Resource Systems, Winterville, pp 249–259

Murashige T, Skoog F (1962) A revised medium for rapid growth and bioassays with tobacco cultures. Physiol Plant 15:473–497

Pan MJ, Van JS (1998) The use of charcoal in in vitro culture: a review. Plant Growth Regul 26:155–163

Peiser G, López-Gálvez G, Cantwell MI, Saltveit ME (1998) Phenylalanine ammonia lyase inhibitors do not prevent russet spotting lesion development in lettuce midribs. J Am Soc Hortic Sci 123(4):687–691

Preece JE, Sutter EJ (1991) Acclimatization of micropropagated plants to the greenhouse and field. In: Debergh PC, Zimmerman RH (eds) Micropropagation, technology and application. Kluwer Academic, London, pp 71–93

Price HJ, Smith RH (1979) Somatic embryogenesis in suspension cultures of *Gossypium klotzschianum* Anderss. Planta 145:305–307

Roy D, Fantes PA (1982) Benomyl resistants mutants of *Schizosaccharomyces pombe* cold sensitive for mitosis. Curr Genet 6:195–201

Shields R, Robinson JS, Anslow AP (1984) Use of fungicides in plant tissue cultures. Plant Cell Rep 3:33–36

Shiva PN, Pental D, Sarin NB (1994) Efficient regeneration of multiple shoot from cotyledonary node of pigeon pea (*Cajanus cajan*). Plant Cell Rep 13:623–627

Sutter EG, Barker PB (1985) *In vitro* propagation of mature *Liquidambar styraciflua*. Plant Cell Tissue Organ Cult 5:13–21

Teixeira JB, Sondahl MR, Kirby EG (1994) Somatic embryogenesis from immature inflorescences of oil palm. Plant Cell Rep 13:247–250

Theander O, Nelson DA (1988) Aqueous, high-temperature transformation of carbohydrates relative to utilization of biomass. Adv Carbohydr Chem Biochem 46:273–326

Thimmappaiah, Shirly RA, Sadhana PH (2002) In vitro propagation of cashew from young trees. In Vitro Cell Dev Biol 38:152–156

Underhill SJR, Critchley C (1993) Lychee pericarp browning caused by heat injury. HortScience 28(7):721–722

Weatherhead MA, Burdon J, Henshaw GG (1978) Some effects of activated charcoal as an additive to plant tissue culture media. Z Pflanzenphysiol 89:141–147

Yu CH, Chen ZG (1998) Embryogenic suspension culture and protoplast isolation in litchee. Chin J Trop Crops 19:16–20

Zamora AB, Damasco OP, Gruezo SM (1988) Control of browning in cultured tissues of bamboo (*Dendrocalamus latiflorus* cv. Machiku). In: Programme and Abstracts of 1st Natl. Symp. Plant Tissue Culture in Philippines. Agriculture and Forestry, Los Banos, Philippines

Zirari A, Lionakis SM (1994) Effect of cultivar, explant type, etiolation pretreatment and the age of plant material on the in vitro regeneration ability of avocado (*Persea americana*). Acta Hortic 365:69–76

Lychee (*Litchi chinensis*): Biochemistry, Panacea, and Nutritional Value

14

Jasra Anjum, Rafiq Lone, and Khursheed Ahmad Wani

Abstract

Lychee is an evergreen tree of the genus *Litchi* in the soapberry family, mostly grown in China, India, Thailand, Vietnam, and the rest of tropical Southeast Asia, and commercially propagated through air layering by vegetative propagation. Lychee can successfully grow at higher altitudes with sufficient moisture and in an acidic soil environment but is prone to severe frost. Lychee has the ability to spread and produce good foliage growth in the presence of sufficient organic matter including nitrogen, phosphorus, and potassium in the soil.

The lychee contains a good amount of polyphenolic compounds, pigments such as cyanidin 3-rutinoside, cyanidin glucoside, quercetin 3-rutinoside (rutin), and quercetin glucoside, and tannins that include polymeric proanthocyanidins. The consumption of lychee in adequate amounts may help in fighting different types of body ailments as it protects from heart diseases, normalizes blood pressure and heart rate, prevents cancer, improves digestive system health, etc. Lychee is also provided with minerals (potassium and copper) that help in maintaining body fluid balance and heartbeat control and maintaining blood pressure. However, studies have also shown that it may also cause adverse effects such as hypoglycemia and encephalopathy.

J. Anjum
School of Studies in Botany, Jiwaji University, Gwalior, India
e-mail: anjum22jasra@gmail.com

R. Lone
Department of Botany, Sant Baba Bhag Singh University, Khiala, Jalandhar, Punjab 144030, India
e-mail: rafiqlone@gmail.com

K.A. Wani (✉)
Department of Environmental Science, ITM University, Gwalior, India
e-mail: Kuurshidevs@gmail.com

M. Kumar et al. (eds.), *Lychee Disease Management*,
DOI 10.1007/978-981-10-4247-8_14

Keywords
Lychee • Ecology • Nutrients • Health • Minerals

14.1 Introduction

Lychee (*Litchi chinensis* Sonn.) is one of the most popular fruits, belonging to the family Sapindaceae, and is a very delicate fruit commercially valued for its exotic aroma, juicy arils, and nutritional benefits and cultivated in many countries. The major lychee production areas in the world are China, Taiwan, Vietnam, Thailand, India, South Africa, and the Malagasy Republic (Menzel et al. 1988). The juicy fruit is eaten directly, and juice, vinegar, jelly, and wine are also manufactured from the fruit (Salunke and Desai 1984). Lychee flowers are an important nectar source, and bees can turn the nectar into honey (Baltrušaityte et al. 2007) that contains phenolic compounds with antioxidant activity (Aljadi and Kamaruddin 2004).

Lychee has received much attention for its pharmacological activities against various diseases as it displays different biological activities which justify its ethnopharmacological utilization in different cultures. It contains 80% water, 0.4–0.9 g kg^{-1} of vitamin C, 0.2–1.1% acidity, and 11.8–20.6% total soluble solids (Paull et al. 1984). Different varieties of lychee with diverse colors are available in the market; however, in the international market, red cultivars are mostly preferred. The red color is imparted by the presence of anthocyanins, and cyanidin-3-rutinoside and cyanidin-3-glucoside have been reported as the major pigments in the pericarp (Lee and Wicker 1991a, b; Rivera-López et al. 1999). pH is important in imparting color to lychee which is the result of anthocyanins, dependent on pH, which influences their structure. At acid pH, the flavylium form is stable and is colored red but may be converted to basic pH by the carbinol base (colorless) to chalcone (yellow) or to its quinonic base (blue) (Perret 2001). The high dehydration rates and browning of the rind may cause the loss of red color in lychee fruit that is commonly seen during postharvest. Initially the mesocarp cells of the pericarp become brown, and later on, epicarp and endocarp tissues also turn brown in color (Scott et al. 1982; Underhill and Critchley 1994). Enzymatic degradation of the anthocyanins has been reported to be vital in browning as anthocyanase is hydrolyzed to anthocyanidins (Holcroft and Mitcham 1996; Zhang et al. 2000; Jiang et al. 2004a), which are then oxidized by polyphenol oxidases (PPO) and/or peroxidases (POD) to *o*-quinones; this is supported by the research carried by Akamine (1960), Huang and Wang (1990), Jiang (2000), Finger et al. (1997), and Zhang et al. (2005). Chilling, heat injury, senescence, and attack by insects or pathogens are also reported to cause loss of compartmentalization between enzymes and their substrates (Scott et al. 1982; Underhill and Critchley 1994). This plant has been studied for its nutritive value (Kalgaonkar et al. 2010) and antiinflammatory (Besra et al. 1996) and antioxidant (Li and Jiang 2007; Sakurai et al. 2008; Kong et al. 2010) activities. It is reported to have one of the highest reported polyphenol concentrations of any fruit (Brat et al. 2006). It is also rich in minerals such as P, K, Mg, Fe, Zn, Mn, and Cu (Wenkem 1990), although results for P, Ca, K, and Mg would be influenced by fertilizer applications (Wall 2006).

Keeping the commercial and nutritional values in consideration, various efforts have been made to preserve the fruit at a commercial level. Sivakumar and Korsten (2006) demonstrated that 1-MCP pretreatment with CA-1 condition (3% O_2 and 7% CO_2) at 2 °C for 21 days is more effective to prevent pericarp browning. The treatment is helpful in delaying senescence, maintaining anthocyanin content, and limiting oxidation enzyme activity, acceptable taste and flavor, and overall acceptability. Collin et al. (Ducamp-Collin et al. 2008) argued that treatment with citric acid and chitosan will bring changes in polyphenol oxidase (PPO), peroxidase (POD) and anthocyanase, and on the anthocyanin content of the pericarp that is responsible for pericarp browning. The major anthocyanins are cyanidin-3-rutinoside and cyanidin-3-glucoside. The two cultivars (Kwai May and Wai Chee) used in their study differ in anthocyanin and oxidative enzyme composition, and they responded differently to the acid and chitosan treatment. The red color of Kwai May was better preserved during storage than that of Wai Chee, and this treatment could be a replacement for current sulfur treatments used to treat lychees transported by sea. Joas et al. (2005) proposed a new treatment method for preservation of lychee that is based on the use of organic acids (citric or tartaric) in association with chitosan coating. However, treatment with HCl at a high concentration of excessive duration may damage lychee pericarp by the breakdown of enzyme–substrate compartmentalization, leading to an increase in enzyme activity. Jiang et al. (2004a, b) observed that the chitosan citric acid treatment increases the shelf life of the fruit, and the coloration does not change during several months. The differences in enzyme activity are the cause of browning among different cultivars, and the rate of browning may differ because of the difference in enzymatic activity as shown by the experiments of Chen and Huang (2001) on Nuomici, a cultivar that browns easily, and Guiwei, a cultivar which browns more slowly. Browning may also occur because of stress caused by climatic conditions, maturation, diseases, desiccation, and temperature shock (Underhill et al. 1997).

Availability of the fresh fruit is limited because of its short production period and shelf life. The most common process to preserve lychees is canning. As with many other fruits and vegetables, a pink discoloration also occurs in canned lychee (Wu and Chen 1999). This phenomenon is not only of sensory importance but also leads to nutritional losses. Increasing consumer demand for safe, high-quality, fresh-like products that are shelf stable, minimally processed, and additive-free stimulated the interest of the food manufacturing industries (Weemaes et al. 1998). The high-pressure process involved in lychee preservation has good potential for the development of new processes for food preservation or product modifications (Cano et al. 1997). For the development of high-pressure processed fruits and vegetables, it is essential to know the influence of high pressure on the activities of such enzymes as polyphenol oxidase (PPO), peroxidase (POD), and lipoxygenase. Effects of high-pressure treatments on enzymes can be either reversible or irreversible, and inactivation relates to conformational changes in the protein structure. The effects are dependent on the type of enzyme, nature of the substrate, pressure, temperature, and processing time. Enzymatic reactions may be enhanced or inhibited by pressure, depending on whether the volume change associated with the reaction is positive or negative. Pressure-induced changes in the catalytic rate may be caused by changes in the enzyme–substrate interaction, changes in the reaction mechanism, the effect on a particular rate-limiting step, or the overall catalytic rate (Ludikhuyze and Hendrickx 2001).

The present chapter is an attempt to review the ecology, nutritional benefits, and pharmacological activities of lychee with emphasis on its biochemistry. It is projected that the present review can be used to corroborate our knowledge toward medicinal practices and biological activities of *L. chinensis*, so that it can be exploited by modern medicine after safety verification and clinical trials.

14.2 History

Hsiang Li Chih, a native of Fukien, China, was a famous Chinese scholar, a calligrapher, an engineer, and is supposed to be the first explorer of lychee. The work of Hsiang Li Chih was compiled and was carved on stones, wooden blocks, and silk cloth in 1059 AD, and later on, many treatises on lychee were written by different workers. These works were not considered in scientific literature as it was believed they are full of folklore and superstitions, although it is thought that they contained more information on the ecology of lychee than many botanical writings.

The history of lychee cultivation can be traced back to the year 1782 when the Frenchman Pierre Sonnerat, in "Voyage aux orientales et a la Chine," made the first careful and complete description of lychee. Following the suggestions of Osbeck, Sonnerat named the plant *Lychee chinensis* with priority over more than ten synonyms that were published later. The name 'lychee' was Latinized to Lychee as a generic name and *chinensis* as the specific name by Sonnerat. However, few modern taxonomists placed lychee within genus *Nephelium*, but because of the different characteristics of the flowers of *Nephelium*, the name *Lychee chinensis* stood as such for all time. Written with two Chinese characters, it signifies that the fruit originates from the tree with knives, leaving a portion of the branch attached to the fruit cluster. The specific name *chinensis* reflects the country name, although it is strictly found in Southern China. The origin of lychee is China and Northern Vietnam, where it has been cultivated for more than 3000 years (Maity and Mitra 1990). It breeds well in low elevations in Kwangtung and Fukien Provinces in Southern China and along the rivers, very close to the seacoast in Hainan Island in Northern Vietnam, west of Guangdong below 500 m in hilly areas of Leizhou Peninsula and east of Guangxi (Menzel and Simpson 1994). *Litchi chinensis* is usually cultivated as a commercial crop in tropical and subtropical areas (Saxena et al. 2011; Xu et al. 2010). Several lowland rainforest areas in Hainan Island are dominated by wild trees and are also found in American subtropics, Burma (Myanmar), India, Southern Hemisphere (Madagascar, Mauritius, and South Africa), Australia, Brazil, Honduras, Israel, Mexico, New Zealand, Reunion, Taiwan, Thailand, and Zanzibar.

14.3 Ecology

Litchi chinensis is an evergreen, round-topped medium-sized tree with a smooth, gray trunk and limbs that may reach 10–15 m height with yellowish-white flowers. Leaves of lychee are leathery, pinnate and divided into four to eight pairs of elliptical or lanceolate, acuminate, glabrous leaflets. The leaves are 5–7 cm long and red at earlier stages; after development, they turn shiny and bright green. A rough leathery rind or pericarp encloses the fruits that are oval, heart shaped, or nearly round. The edible portion of the fruit is milky white, translucent, firm, and juicy having sweet, fragrant, and delicious flavor with 1- and 2-cm seeds embedded in the aril. The seeds with brown or reddish-brown color are globose or oblong eggs and have a smooth and glossy surface (Menzel 2002; Nacif et al. 2001).

It is produced in different countries of world, but it has not been recognized as a major horticultural crop (Knight 1980; Sampson 1980) because of failure of initiating flowers that has been attributed to several factors, although often there has been no obvious explanation (Menzel 1983). However, authors such as Cobin (1954), Young (1955), Yee (1982), and Joubert (1970) suggested failure may result from suppression by temperature as lychee is cultivated more often in areas that have temperatures between 5 °C and 14 °C. High rates of premature flowers and, especially, fruit abscission between floral initiation and fruit maturation, low female-to-male sex ratio, lack of pollinating insects, poor pollen transfer, low pollen viability, and failure of pollen tube growth have been linked with poor productivity (Menzel 1983).

Menzel (1984) indicated that floral initiation in lychee is enhanced by dry conditions which inhibit vegetative growth during winter, and fruiting is superior under wet and humid conditions during summer. Menzel (1983) reported that low temperatures and moisture stress restrict vegetative growth and enhance floral initiation. Decrease in vegetative flushing and increase in flowering by both cincturing (ringing) and exogenous auxins with variations in the results have also been reported.

Litchi chinensis Sonn. is the only genus having three subspecies: *L. chinensis* subsp. *chinensis* Forest & Kim Starr, and the commercial form of *L. chinensis* Sonn., which is native to the forests in Chinese provinces of Yunnan, Guangxi, Hainan Island, and Western Guangdong (Huang et al. 2005). *L. chinensis* subsp. *philippinensis* Radlk is native to the Philippines, New Guinea, Malay Peninsula, and Indonesia, and *L. chinensis* subsp. *javensis* Leenh (Fan et al. 2011; Diczbalis 2011) is prevalent in Java (Leenhouts 1994).

The loss of petals in lychee and its ability to grow luxuriantly on drier areas and fruit successfully during winter when temperatures are low are the attributes within the attributes of ecotype. The different varieties of *Litchi chinensis* develop within a narrow range of habitat, occupy a narrow climatic belt, and reveal only slight variations and little adaptability in wider distribution. However, the only contradiction was raised by assigning *L. philippinensis* to this genus, although few attempts to bring it to the botanical gardens and experimental plots continuously failed.

The wild lychee of the Philippines and cultivated lychee of South China are very distinct species of the genus *Litchi*. They are separated taxonomically from other members of the subfamily Nephelieae of the family Sapindaceae. The former is insular and the latter is continental, indicating that both are detached in their origins. Possibly, they developed along parallel lines in their distinctive habitats. By the absence of their petals and less adaptability in distribution, both bear documentary evidence of ecotypic genesis. The latitude in the Philippines and altitude in China produced a climatic cooling effect, as the result of which lychee developed from some of the warmer-loving Nephelieae.

The lychee is distinctively heat and moisture loving because of its Malayan origin, and it enjoys the relatively cool winter temperatures in South China necessary to bring the tree into bearing. The trees flourish in tropical areas in a warmer climate, although they do not bear. The trees are killed to the ground in areas where winter freezes; however, roots remain alive and will sprout again after few years. Temperature below 0 °C may blast the young tip growth in lychee, and long hot summers, relatively high temperatures and yet drop below 4 °C will help in flowering and fruiting. Abundant supply of water and high humidity are necessary for the luxuriant growth and emergence of foliage, respectively. Irrigation is dependent mostly on rainfall as the months of April, May, and June are months of fair rainfall in China. The high humidity is important to prevent leathery skin of the fruits from bursting to expose the fleshy aril to the attack of insects and diseases. The lychee trees can withstand the wind intensity of 75–100 miles per hour as the trunk and root system are highly developed. Unfavorable wind reactions have been noticed to intense sunshine by young plants and light. Lychee grows well in semi-shade light conditions until becoming well established.

The alluvial and laterite soils are usually preferred in lychee cultivation with relatively high humus content. Humus deficiency can be surmounted by the application of liquid fertilizers, mainly night soil and urine, other animal wastes, peanut, animal husks, etc. Correct fertilizer practice is of utmost importance to produce trees of great age with high productivity. The giant stink bug *Tessaratoma papillosa* is the most widespread and destructive species among insects that is supposed to damage lychee in China. The other foes that can damage lychee include negligent husbandmen, frost and snow, unfavorable winds, salt water, bats, and rodents. The trunks, branches, and leaves of lychee are invaded by a number of unidentified species such as fungi and algae. Leaf gall seriously affects foliage, making the leaf thickened and wrinkled with brown, hairy, and velvet-like spots in which galls are embedded. The trunks and branches are further affected by Cerambycidae and other tree borers.

14.4 Nutritional Value

Lychee is sweet in taste and is a source of high energy, protein, and carbohydrates (Table 14.1), and is equipped with different minerals such as calcium, iron, phosphorus, potassium, sodium, zinc, etc. (Table 14.2). It is rich in such vitamins

Table 14.1 Proximity composition of 1 cup (190 g) of lychee

Proximity	Amount	% DV (daily value)
Water	155.34 g	N/D
Energy	125 Kcal	N/D
Energy	524 kJ	N/D
Protein	1.58 g	3.16%
Total fat (lipid)	0.84 g	2.40%
Ash	0.84 g	N/D
Carbohydrate	31.41 g	24.16%
Total dietary fiber	2.5 g	6.58%
Total sugars	28.94 g	N/D

Source:https://ndb.nal.usda.gov/ndb/foods/show/2264

as vitamin E, B_6, niacin, thiamin, folate, and riboflavin (Table 14.3). It also contains fatty acids, viz., myristic acid, palmitic acid, stearic acid, palmitoleic acid, linoleic acid (Table 14.4), and amino acids that include tryptophan, lysine, and methionine (Table 14.5).

14.5 Health Benefits

Lychee is bestowed with different types of compounds such as phenolics, proanthocyanidins, anthocyanins, and flavonols that make it suitable to cure different types of ailments. Leaves, pulp, pericarp, and seeds of lychee have been used to treat different health disorders of people in different regions and in different cultures. The seeds can cure neuralgic disorders, orchitis, hernia, lumbago, ulcers, and digestive disorders (Ahmad and Sharma 2012; Perry and Metzger 1980). Lim (2013) reported that people prepare a tea of the fruit peel to cure smallpox eruptions and diarrhea. Flatulence, heatstroke, and detoxification can be treated by the leaves of lychee, and cough, diarrhea, stomach ulcers, diabetes, dyspepsia, obesity, dysentery, and swelling may be treated by the fruits of lychee (Quisumbing 1951). The pericarp of lychee possesses antitussive, analgesic, antipyretic, hemostatic, and diuretic properties (Liu et al. 2007; Castellain et al. 2014). The seeds are used to cure colds, relieve painful swollen testicles, and reduce premenstrual and postpartum abdominal pains (Yan et al. 1999); the mixture of lychee seeds, cumin, and peel has been used to cure the pain of hernia or testicular swelling in China (Lin et al. 2013). The medicinal tablet made from the lychee nut is suitable for pregnancy diabetes (Shen 1991). The moderate growth inhibition against S aureus is reported by using the seed aqueous extract of lychee. The flesh of lychee helps to overcome tiredness and prevent bronchocele, and its skin is used to treat animal bites.

The compounds present in lychee possess different biological activities that include antiviral, antimicrobial, and antioxidant properties. Compounds such as cinnamtannin B_1 act as inhibitors and may cause cytotoxicity to cure different throat ailments. Root, bark, and flower decoctions of lychee are used (Perry and

Table 14.2 Mineral value of 1 cup (190 g) of lychee

Minerals	Amount	% DV (daily value)
Calcium, Ca	10 mg	1.00%
Iron, Fe	0.59 mg	7.38%
Magnesium, Mg	19 mg	4.52%
Phosphorus, P	59 mg	8.43%
Potassium, K	325 mg	6.91%
Sodium, Na	2 mg	0.13%
Zinc, Zn	0.13 mg	1.18%
Copper, Cu	0.281 mg	31.22%
Manganese, Mn	0.104 mg	4.52%
Selenium, Se	1.1 μg	2.00%

*Source:*https://ndb.nal.usda.gov/ndb/foods/show/2264

Table 14.3 Vitamin composition of 1 cup (190 g) of lychee

Vitamins	Amount	% DV
Water-soluble vitamins		
Vitamin B_1 (thiamin)	0.021 mg	1.75%
Vitamin B_2 (riboflavin)	0.124 mg	9.54%
Vitamin B_3 (niacin)	1.146 mg	7.16%
Vitamin B_6 (pyridoxine)	0.19 mg	14.62%
Vitamin B_9 (folate)	27 μg	6.75%
Folic acid	0 μg	N/D
Folate, food	27 μg	N/D
Folate, DEF	27 μg	N/D
Choline	13.5 mg	2.45%
Vitamin C (ascorbic acid)	135.8 mg	150.89%
Fat-soluble vitamins		
Vitamin E (alpha-tocopherol)	0.13 mg	0.87%
Vitamin K (phylloquinone)	0.8 μg	0.67%

*Source:*https://ndb.nal.usda.gov/ndb/foods/show/2264

Table 14.4 Lipid composition of 1 cup (190 g) of lychee

Lipids	Amount	% DV
Fatty acids, total saturated	0.188 g	N/D
Myristic acid 14:00 (tetradecanoic acid)	0.004 g	N/D
Palmitic acid 16:00 (hexadecanoic acid)	0.133 g	N/D
Stearic acid 18:00 (octadecanoic acid)	0.046 g	N/D
Fatty acids, total monounsaturated	0.228 g	N/D
Palmitoleic acid 16:1 (hexadecenoic acid)	0.002 g	N/D
Oleic acid 18:1 (octadecenoic acid)	0.226 g	N/D
Fatty acids, total polyunsaturated	0.251 g	N/D
Linoleic acid 18:2 (octadecadienoic acid)	0.127 g	N/D
Linolenic acid 18:3 (octadecatrienoic acid)	0.124 g	N/D

*Source:*https://ndb.nal.usda.gov/ndb/foods/show/2264

Table 14.5 Amino acid composition of 1 cup (190 g) of lychee

Amino acids	Amount	% DV[a]
Tryptophan	0.013 g	2.95%
Lysine	0.078 g	2.33%
Methionine	0.017 g	N/D

Source:https://ndb.nal.usda.gov/ndb/foods/show/2264
Calculations are based on an average age of 19 to 50 years and weight of 194 lbs
[a]The aforementioned percent daily values (%DVs) are based on a 2000-calorie diet intake. Mentioned values are recommended by the US Department of Agriculture

Metzger 1980), and ingested lychee can also relieve coughing and gastralgia, tumors, and enlargement of the glands (Cohen and Dubois 2010).

Lychee is also used in traditional medicine and the Ayurveda system of medicine for the treatment of wounds (Wiart 2006), intestinal troubles, neuralgic pain, nerve inflammation (Lim 2013; Miller 2011; Li and Jiang 2007; Perry and Metzger 1980), digestive ulcers, and excretory and reproductive disorders. The seeds of lychee are famous in Chinese and Indian traditional medicine to release stagnant humors, remove chilling, and serve as an analgesic agent (Lin et al. 2013; Xu et al., Wang et al. 2011; Li 2008). The seeds macerated in alcohol are utilized to cure intestinal complaints in India and China (Perry and Metzger 1980). The roots of lychee are used for fever, leaves for poulticing, and the bark as an astringent for tongue diseases in Malaysia (Quisumbing 1951).

The antiinflammatory, analgesic, and antipyretic activities reported by Besra et al. (1996) have similar properties related to ibuprofen and acetyl salicylic acid. Lychee seed is reported to have antidiabetic activity in humans (Zhang and Teng 1986). Oligonol present in lychee helps in the reduction of abdominal circumference and visceral fat volume and improves insulin resistance (Chang et al. 2013). The decrease in serum and hepatic lipids was also observed in high-fat and high-cholesterol diet-fed hamsters. Lychee fruits and seeds inhibit the growth of cancerous cells (Bhat and Al-daihan 2014) and are very effective against breast cancer, being rich in flavonoids.

In addition, the fruit and seed possess many bioactivities such as hypoglycemic, anticancer, antibacterial, antihyperlipidemic, antiplatelet, and antiviral (Xu et al. 2011; Li 2008; Chen et al. 2007) properties, which were also reported by different workers. Oligonol, a flavonoid-rich lychee extract, has numerous health benefits, viz., protection against oxidative stress, treatment of hyperuricemia, and reduction of tiredness (Yamanishi et al. 2014; Kang et al. 2012; Ogasawara et al. 2009; Sakurai et al. 2008). The presence of Flavonoids, tannins, anthocyanins, phenolic acids, triterpenes, and sterols in lychee make it suitable for anticancer, hepatoprotective, antioxidant, antiplatelet, antiviral, antimutagenic, antimicrobial, antihyperlipidemic, antipyretic, and antiinflammatory purposes. A diagrammatic representation of some of the health benefits of lychee is shown in Fig. 14.1.

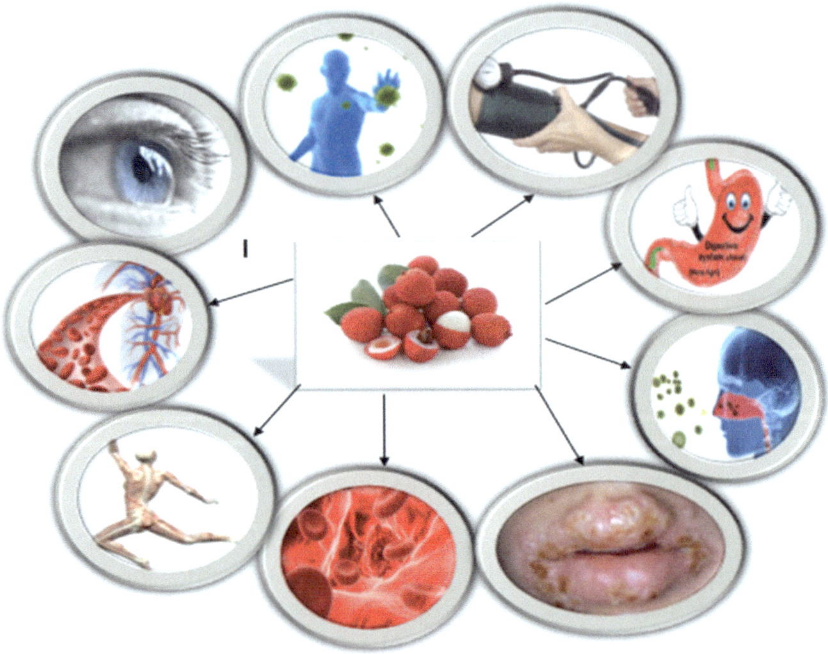

Fig. 14.1 Health benefits of lychee

14.6 Biochemistry

Lychee (*Litchi chinensis* Sonn.) is a tropical and subtropical fruit that has high commercial value partly because of its white and translucent aril and attractive red color (Holcroft and Mitcham 1996). However, the fruit will rapidly lose its bright color and turn brown once harvested (Jaiswal et al. 1987; Jiang 2000; Nip 1988). The lychee fruit pericarp contains a large amount of pigments that are responsible for the red color (Lee and Wicker 1991a, b). Prasad and Jha (1978), using thin-layer chromatography (TLC), first reported that the red color of the lychee fruit was probably the result of a mixture of cyanidin and pelargonidin. The red pigments were later identified as cyanidin-3-rutinoside by TLC and high-performance liquid chromatography (HPLC) (Lee and Wicker 1991a, b). Zhang et al. (2000) reported that cyanindin-3-glucoside and malvidin-3-glucoside may be present in these anthocyanins. Sarni-Manchado et al. (2000) identified the anthocyanins as cyanidin-3-rutinoside, cyanidin glucoside, quercetin-3-rutinoside, and quercetin glucoside, using low-pressure chromatography, HPLC, UV-visible light spectral analysis, mass spectrometry (MS), and nuclear magnetic resonance (NMR). The major anthocyanin of lychee fruit pericarp was cyanidin-3-rutinoside. Differences in the identification of the anthocyanins may be attributed to differences in extraction and purification procedures (Jiang et al. 2004a, b).

Recent research has shown that pigments in postharvest fruits exhibit a strong antioxidant activity (Einbond et al. 2004; Bao et al. 2005). Antioxidative properties of pigments resulted from their high reactivity as hydrogen or electron donors and from their ability to chelate transition metal ions (termination of the Fenton reaction) (Rice-Evans et al. 1997). It is well established that enzymatic browning of postharvest fruits is related to antioxidant activity (Martinez and Whitaker 1995). Unfortunately, little information on the antioxidant activity of skin pigments from harvested lychee fruit is available.

The phenylpropanoid metabolic pathway synthesizes flavonoids in which the amino acid phenylalanine is used to produce 4-cou-maroyl-CoA that combines with malonyl-CoA to yield chalcones, which contain two phenyl rings that result in the three-ring structure of flavones. The metabolic pathway continues through a series of enzymatic modifications to yield flavanones → dihydroflavonols → anthocyanins. Along this pathway, many products can be formed, including the flavonols, flavan-3-ols, proanthocyanidins (tannins), and a host of various other polyphenolics.

Flavonoids are one of the largest nutrient families with more than 6000 identified family members. Some of the best known flavonoids include quercetin, kaempferol, kaempferol-3-*o*-β-D-glucoside, kaempferol-7-*o*-β-D-glucopyranoside, kaempferol-3-*o*-α-rhamnoside, kaempferol-7-*o*-neohesperidoside, quercetin-3-*o*-rutinoside, quercetin-3-*o*-rutinoside-7-*o*-α-L-rhamnoside, tamarixetin 3-*o*-rutinoside, and narcissin from the pericarp seeds and leaves of lychee. The nutrients are famous for their antioxidant and antiinflammatory actions and contribution of vibrant color to foods. Flavonoids are a group of phytonutrients in the chemical category of polyphenols and are most famous for their rich diversity of color-including pigments. Anthocyanins protect against a myriad of human diseases, although they are difficult to study with regard to human health as they interact with other phytochemicals to potentiate biological effects. However, anthocyanin pigments can be of great advantage for health and nutrition research as they can be quickly and easily isolated. Anthocyanins are water-soluble vacuolar pigments with varied colors that are synthesized via the phenylpropanoid pathway. They are odorless but flavorful, contributing to taste as moderately as an astringent sensation. Anthocyanins are present in the lychee pericarp. Plants benefit significantly from environmental adaptations, disease tolerance, and pest tolerance provided by anthocyanins; hence, the biological burden of providing anthocyanin is very high.

Phenols, sometimes called phenolics, are a class of chemical compounds consisting of a hydroxyl group (–OH) bonded directly to an aromatic hydrocarbon group by the shikimate/chorizmate or succinylbenzoate pathway and the acetate/mevalonate pathway. The aromatic amino acid phenylalanine, synthesized in the shikimic acid pathway, is the common precursor of phenol-containing amino acids and phenolic compounds. The phenolics that are usually referred as phenolics flavan-3-ols present in lychee are epicatechin, catechin, gallocatechin, epiafzelechin, epiafzelechin glucoside, epicatechin-3-gallate, and epicatechin-

(7,8-bc)-4β-(4-hydroxyphenyl)-dihroxy-2(3*H*)-pyranone from leaves, pericarps, pulp, and seeds. The proanthocyanidins belong to the family of oxidoreductases, especially those acting on paired donors with O_2 as oxidant and incorporation or reduction of oxygen.

Plant lignans are metabolized into mammalian enterolignans by intestinal bacteria. The lignans present in lychee include schizandriside and isolariciresinol that have been extracted from leaves and pericarp, respectively. Sterols have vital functions in the physiology of eukaryotic organisms as these affect cell membrane fluidity and serve as secondary messengers in developmental signaling. Triterpenes are produced by plants as part of their self-defense mechanism. The sterols and triterpenes present in lychee are β-sitosterol, stigmasterol, lupeol, betulin, betulinic acid, 3-oxotrirucalla-7,24-dien-21-oic acid, and lup-12,20(29) diene-3,27-diol. Methyl dihydrosterculate, 2,5-dihydroxy-hexanoic acid, and litchioside C (3,12-dihyroxy-*cis*-3,4-methylenedodecanoic acid 3-*o*-β-D-glucopyranoside) are the fatty acids present in lychee. Flavonoid or bioflavonoids are secondary metabolites of lychee, with general structure of a 15-carbon skeleton made up of phenyl rings A and B and heterocyclic ring (c). The flavonoids are weakly used to describe non-ketone polyhydroxy polyphenol compounds. They are synthesized by the phenylpropanoid metabolic pathway in which the amino acid phenylamine is used to produce 4-compound cou-maroyl. The different compounds present in leaf, root, stem, fruit, and pericarp are shown in Table 14.6.

14.7 Toxicological Effects

Lychee fruit contains significant amount of profilin and is the major allergen for humans. Profilin is an active binding protein which interacts with some variants of membrane phospholipids that causes sequestration of profilin in an inactive form that can be released by action of enzyme phospholipase. Anaphylactic reactions in patients being sensitized against the plant panallergen profiling may be induced. *L. chinensis* can cause serious inflammation symptoms in people (Zhou et al. 2012). Ingestion of lychee by an undernourished child having low glycogen/glucose stores may cause toxic hypoglycemic syndrome (Spencer et al. 2015). Acute neurological illness affecting young children and characterized by low blood sugar, seizures, and encephalopathy has also been reported (Shrivastava et al. 2015).

14.8 Conclusions

Lychee is famous for its varied biological properties and has received worldwide acceptance for its pharmacological activities against various diseases. The nutritional benefits of lychee have potential for exploitation to promote human health. Almost every part of the plant such as root, shoot, fruit, leaves, and bark is bestowed with key bioactive phytochemicals that might contribute directly or indirectly to the biological properties highlighted in the present chapter. These

Table 14.6 List of the isolated compounds from different parts of *Litchi chinensis*

Compound	Pulp	Pericarp	Seed	Leaves
Phenolics, Flavan-3-ols	(−)-Epicatechin, (þ)-catechin	(−)-Epicatechin, (−)-gallocatechin, epiafzelechin, epicatechin glucoside	(−)-Epicatechin, (−)-gallocatechin, (−)-epicatechin-3-gallate, epicatechin-(7,8-bc)-4β-(4-hydroxyphenyl)-dihydro-2(3*H*)-pyranone	(−)-Epicatechin
Proanthocyanidins	Propelargonidin, procyanidin, prodelphinidin	Procyanidin A2, Epicatechin-(4β-8, 2β-*O*-7)-epicatechin-(4β-8)-epicatechin, proanthocyanidin B2, proanthocyanidin B4	Proanthocyanidin A1, procyanidin A2, proanthocyanidin A6, litchitannin A1 [epicatechin-(2β-*O*-7,4β-6)-epicatechin-(2β-*O*-7,4β-8)-catechin], litchitannin A2 [epicatechin-(2β-*O*-7,4β-6)-epicatechin-(2β-*O*-7,4β-6)-epicatechin], aesculitannin A, epicatechin-(2β-*O*-7,4β-8)-epiafzelechin-(4α-8)-epicatechin, 2α,3α-epoxy-5,7,3′,4′-tetrahydroxyflavan-(4β-8-catechin), 2α,3α-epoxy-5,7,3′,4′-tetrahydroxyflavan-(4β-8-epicatechin), 2β,3β-epoxy-5,7,3′,4′-tetrahydroxyflavan-(4α-8-epicatechin)	Procyanidin A2, proanthocyanidin B2, cinnamtannin B1
Anthocyanins	–	Cyanidin-3-glucoside, cyanidin-3-rutinoside, malvidin-3-glucoside	–	–
Flavonols	Quercetin-3-*O*-rutinoside, quercetin-3-	Kaempferol	Quercetin, kaempferol-7-*O*-β-D-glucopyranoside, kaempferol-7-	Kaempferol-3-*O*-β-D-glucoside, kaempferol-3-

(continued)

Table 14.6 (continued)

Compound	Pulp	Pericarp	Seed	Leaves
	O-rutinoside-7-*O*-α-*L*-rhamnoside		*O*-neohesperidoside, tamarixetin 3-*O*-rutinoside, narcissin	*O*-α-rhamnoside, quercetin-3-*O*-rutinoside
Flavanones	–	–	(2S)-pinocembrin-7-O-(6-O-α-L-rhamnopyranosyl-β-D-glucopyranoside), (2R)-naringenin-7-*O*-(3-*O*-α-L-rhamnopyranosyl-β-D-glucopyranoside), narirutin, naringin, (2R)-pinocembrin-7-neohesperidoside, litchioside D, (−)-pinocembrin-7-*O*-neohesperidoside (onychin), pinocembrin-7-*O*-glucoside, (2S)-pinocembrin-7-*O*-(6″-O-α-L-arabinosyl-β-D-glucopyranoside), pinocembrin-7-O-[(6″-*O*-β-D-glucopyranoside)-β-D-glucopyranoside], pinocembrin-7-*O*-[(2″,6″-di-*O*-α-L-rhamnopyranosyl)-β-D-glucopyranoside]	–
Flavanonols	–	–	Taxifolin-4′-*O*-β-D-glucopyranoside,	–
Dihydrochalcones	–	–	Dihydrocharcone-4′-*O*-β-D-glucopyranoside, phlorizin	–
Phenolic acids	Caffeic acid, chlorogenic acid	Methyl-3,4-dihydroxybenzoate, 2-(2-hydroxy-5-	Protocatechuic acid, coumaric acid, gallic acid, butylated hydroxytoluene	–

		(methoxycarbonyl) phenoxy) benzoic acid		
II-Coumarins	–	–	Scopoletin	–
III-Chromanes	–	–	–	Litchtocotrienols A–G, macrolitchtocotrienol A Cyclolitchtocotrienol A
IV-Lignans	–	Isolariciresinol	–	Schizandriside
V-Sesquiterpenes	–	–	Litchioside A, litchioside B, pumilaside A, funingensin A, pterodontriol-D-6-*O*-β-D-glucopyranoside	–
VI-Fatty acids	–	–	Methyl dihydrosterculate, 2,5-dihydroxy-hexanoic acid, litchioside C (3,12-dihydroxy-*cis*-3,4-methylenedodecanoic acid 3-*O*-β-D-glucopyranoside)	–
VII-Sterols and triterpenes	–	Stigmasterol	3-Oxotrirucalla-7,24-dien-21-oic acid	–
VIII-Miscellaneous	Benzyl alcohol, 5-hydroxymethyl-2-furfuraldehyde, hydrobenzoin	Ethyl shikimate, methyl shikimate	Litchiol A, litchiol B	Secoisolariciresinol-9′-*O*-β-D-xyloside, 4,7,7′,8′,9,9′-hexahydroxy-3,3′-dimethoxy-8,4′-oxyneolignan, ehletianol C, sesquipinsapol B, sesquimarocanol B

Source: Modified from Sabrin R.M. Ibrahim, Gamal A. Mohamed (2015) *Litchi chinensis*: medicinal uses, phytochemistry, and pharmacology. *Journal of Ethnopharmacology* 174:492–513

different compounds present in lychee can be used in the pharmaceutical industry to synthesize different types of medicines. However, in-depth studies are needed to promote research in this regard. The mode of action for different antioxidants present in lychee needs to be explored for further pharmacological studies. Further experiments need to be conducted to cultivate this plant in other regions of the world under field conditions by supplying and keeping environmental conditions favorable.

Acknowledgments We are highly thankful to the Jiwaji University, Gwalior, ITM University Gwalior, and Sant Baba Bhag Singh University, Khiala, Jalandhar, for providing us the facilities for the present work. We are thankful to Javaid Manzoor, Department of Environmental Science, for providing us with support.

References

Ahmad SA, Sharma SCS (2012) Fruit & vegetable juice therapy. Pustak Mahal, New Delhi

Akamine EK (1960) Preventing the darkening of fresh lychees prepared for transport. Technical Programme report. Hawaii Agric Exp Sta Univ Hawaii 127:3–17

Aljadi AM, Kamaruddin MY (2004) Evaluation of the phenolic contents and antioxidant capacities of two Malaysian floral honeys. Food Chem 85:513–518

Baltrušaityte V, Venskutonis PR, Ceksteryte V (2007) Radical scavenging activity of different floral origin honey and beebread phenolic extracts. Food Chem 101(2):502–514

Bao JS, Cai YZ, Wang GY, Corke H (2005) Anthocyanins, flavonols, and free radical scavenging activity of Chinese bayberry (*Myrica rubra*) extracts and their color properties and stability. J Agric Food Chem 53(6):2327–2332

Besra SE, Sharma RM, Gomes A (1996) Anti-inflammatory effect of petroleum ether extract of leaves of *Litchi chinensis* Gaertn. (Sapindaceae). J Ethnopharmacol 54:1–6

Bhat RS, Al-daihan S (2014) Antimicrobial activity of *Litchi chinensis* and *Nephelium lappaceum* aqueous seed extracts against some pathogenic bacterial strains. J King Saud Univ Sci 26:79–82

Brat P, George S, Bellamy A et al (2006) Daily polyphenol intake in France from fruit and vegetables. J Nutr 136(9):2368–2373

Cano MP, Hernández A, De Ancos B (1997) High pressure and temperature effects on enzyme inactivation in strawberry and orange products. J Food Sci 62(1):85–88

Castellain RCL, Gesser M, Tonini F, Schulte RV, Demessiano KZ, Wolff FR, Delle-Monache F, Netz DJA, Cechinel-Filho V, de Freitas RA, de Souza MM, Meyre-Silva C (2014) Chemical composition, antioxidant and antinociceptive properties of *Litchi chinensis* leaves. J Pharm Pharmacol 66:1796–1807

Chang Y, Yang D, Chiu C, Lin Y, Chen J, Chen Y (2013) Antioxidative and anti-inflammatory effects of polyphenol-rich litchi (*Litchi chinensis* Sonn.)-flower water extract on livers of high-fat-diet fed hamsters. J Funct Foods 5:44–52

Chen HB, Huang HB (2001) China litchi industry: development, achievements and problems. Acta Hortic 558:31–39

Chen YB, Wu KS, Gu Y, Chen JZ (2007) Research progress in the chemical constituents and pharmacological effects of lychee seeds. Chin J Infect Tradit Chin Med 14:97–98

Cobin M (1954) Lychee in Florida. Bull Fla Agric Exp Sta 546:1–35

Cohen A, Dubois LJ (2010) Raw food for everyone: essential techniques and 300 simple-to-sophisticated recipes. Avery, New York

Diczbalis Y (2011) Farm and forestry production and marketing profile for lychee (*Litchi chinensis*), specialty crops for Pacific Island agroforestry. Permanent Agriculture Resources (PAR), Holualoa

Ducamp-Collin MN, Ramarson H, Lebrun M, Self G, Reynes M (2008) Effect of citric acid and chitosan on maintaining red colouration of litchi pericarp. Postharvest Biol Technol 49:241–246

Einbond LS, Reynertson KA, Luo XD, Basile MJ, Kennelly EJ (2004) Anthocyanin antioxidants from edible fruits. Food Chem 84:23–28

Fan Q, Chen SF, Zhou RC, Xiang X, Liao WB, Shi SH (2011) Genetic variation of wild litchi (*Litchi chinensis* Sonn. subsp. *chinensis*) revealed by microsatellites. Conserv Genet 12:753–760

Finger RL, Vieira G, Ledsham LR (1997) Maturity standard and pericarp browning of litchi fruit. Rev Bras Fisiol Veg 915–18

Holcroft DM, Mitcham EJ (1996) Postharvest physiology and handling of litchi (*Litchi chinensis* Sonn.) Postharvest Biol Technol 9:265–281. doi:10.1016/S0925-5214(96)00037-3.1

Huang CC, Wang YT (1990) Effects of storage temperature on the colour and quality of litchi fruit. Acta Hortic 269:307–308

Huang X, Subhadrabandhu S, Mitra SK, Ben-Arie R, Stern RA (2005) Origin, history, production and processing. In: Menzel CM, Waite GK (eds) Litchi and longan. Comwell Press, Trowbridge, pp 1–23

Ibrahim SRM, Mohamed GA (2015) Litchi chinensis: medicinal uses, phytochemistry, and pharmacology. J Ethnopharmacol 174:492–513

Jaiswal BP, Sah NL, Prasad US (1987) Regulation of colour break during litchi (*Litchi chinensis* Sonn.) ripening. Indian J Exp Biol 25:66–72

Jiang YM (2000) Role of anthocyanins, polyphenol oxidase and phenols in lychee pericarp browning. J Sci Food Agric 80:305–310

Jiang YM, Duan XW, Joyce DC, Zhang ZQ, Li JR (2004a) Advances in understanding enzymatic browning of harvested litchi fruit. Food Chem 88:443–44446

Jiang YM, Li Y, Li J (2004b) Browning control, shelf extension and quality maintenance of frozen litchi fruit by hydrochloric acid. J Food Eng 63:147–151

Joas J, Caro Y, Ducamp MN, Reynes M (2005) Postharvest control of pericarp browning of litchi fruit (*Litchi chinensis* Sonn. cv. *Kwaiimi*) by treatment with chitosan and organic acids I. Effect of pH and pericarp dehydration. Postharvest Biol Technol 38:128–136

Joubert AJ (1970) The litchi. Bull S Afr Dept Agric Tech Serv 389:1–22

Kalgaonkar S, Nishioka H, Gross HB, Fujii H, Keen CL et al (2010) Bioactivity of a flavanol-rich lychee fruit extract in adipocytes and its effects on oxidant defense and indices of metabolic syndrome in animal models. Phytother Res 24:1223–1228

Kang SW, Hahn S, Kim JK, Yang SM, Park BJ, Chul Lee S (2012)) Oligomerized lychee fruit extract (OLFE) and a mixture of vitamin C and vitamin E for endurance capacity in a double blind randomized controlled trial. J Clin Biochem Nutr 50:106–113

Knight RJ Jr (1980) Origin and world importance of tropical and sub-tropical fruits. In: Nagy S, Shaw PE (eds) Tropical and sub-tropical fruits. Composition, properties and uses. AVI Publishing, Westport, pp 1–20

Kong F, Zhang M, Kuang R, Yu S, Chi J, Wei Z (2010)) Antioxidant activities of different fractions of polysaccharide purified from pulp tissue of litchi (*Litchi chinensis* Sonn.) Carbohydr Polym 81:612–616

Lee HS, Wicker L (1991a) Anthocyanin pigments in the skin of lychee fruit. J Food Sci 56:466–468, 483

Lee HS, Wicker L (1991b) Quantitative changes in anthocyanin pigments of lychee fruit during refrigerated storage. Food Chem 40:263–270

Leenhouts PW (1994) Litchi. In: Adema F, Leenhouts PW, van Welzen PC (eds) Flora Malesiana, Series I, vol 11: Part, vol 3. Rijkssherbarium/Hortus Botanicus/Leiden University, Leiden, pp 653–658

Li JG (2008) The litchi [in Chinese]. China Agricultural Press, Beijing
Li J, Jiang Y (2007) Litchi flavonoids: isolation, identification and biological activity. Molecules 12(745):758
Lim TK (2013) Edible medicinal and non-medicinal plants, vol 6. Fruits. Springer, Dordrecht, pp 45–58
Lin C, Chung Y, Hsu C (2013) Anti-cancer potential of litchi seed extract. World J Exp Med 3 (4):56–61
Liu L, Xie B, Cao S, Yang E, Xu X, Guo S (2007) A-type procyanidins from *Litchi chinensis* pericarp with antioxidant activity. Food Chem 105:1446–1451
Ludikhuyze LR, Hendrickx ME (2001) Effects of high pressure on chemical reactions related to food quality. In: Hendrickx ME, Knorr D (eds) Ultra high pressure treatments of foods. Kluwer Academic/Plenum, New York, pp 167–188
Maity SC, Mitra SK (1990) Litchi. In: Bose TK, Mitra SK (eds) Fruits: tropical and subtropical. Naya Prokask, Calcutta, pp 420–468
Martinez MV, Whitaker JR (1995) The biochemistry and control of enzymatic browning. Trends Food Sci Technol 6:195–200
Menzel CM (1983)) The control of floral initiation in lychee: a review. Sci Hortic 21:201–215
Menzel CM (1984) The pattern and control of reproductive development. Sci Hortic 22:333–345
Menzel CM (2002) The lychee crop in Asia and the Pacific. Food and Agriculture Organization of the United Nations, Regional Office for Asia and the Pacific, Bangkok, pp 8–13
Menzel CM Simpson DR 1994 Lychee. In: Schaffer B, Andersen PC (eds) The handbook of environmental physiology of fruit crops, vol. II. Sub-tropical and tropical. CRC Press, Boca Raton pp 123–1141
Menzel CM, Watson BJ, Simpson DR (1988) The lychee in Australia. Qld Agric J 114(1–2):19–27
Miller L (2011) Ayurvedic curative cuisine for everyone. Lotus Press, Twin Lakes
Nacif SR, Paoli AAS, Salomão LCC (2001) Morphological and anatomical development of the litchi fruit (*Litchi chinensis* Sonn. cv. Brewster). Fruits 56:225–233
Nip WK (1988) Handling and preservation of lychee (*Litchi chinensis* Sonn.) with emphasis on colour retention. Trop Sci 28:5–11
Ogasawara J, Kitadate K, Nishioka H, Fujii H, Sakurai T, Kizaki T, Izawa T, Ishida H, Ohno H (2009) Oligonol, a new lychee fruit-derived low molecular form of polyphenol, enhances lipolysis in primary rat adipocytes through activation of the ERK1/2 pathway. Phytother Res 23:1626–1633
Paull RE, Chen NJ, Deputy J, Huang H, Cheng G, Gao F (1984) Litchi growth and compositional changes during fruit development. J Am Soc Hortic Sci 109:817–820
Perret C (2001) Analyse des Tanins Inhibiteurs de la Stilbène Oxydase Produite par *Botrytis cinerea*. Ph.D. Thesis, Université de Neuchâtel, Faculté des Sciences
Perry LM, Metzger J (1980) Medicinal plants of East and Southeast Asia: attributed properties and uses. MIT Press, Cambridge
Prasad US, Jha OP (1978) Changes in pigmentation patterns during litchi ripening: flavonoid production. Plant Biochem J 5:44–49
Quisumbing E (1951) Medicinal plants of the Philippines, Republic of the Philippines. Philippines (republic) Department of Agriculture and Natural Resources. Technical bulletin, Manila
Rice-Evans CA, Miller NJ, Paganga G (1997 Antioxidant properties of phenolic compounds. Trends Plant Sci 2:152–159
Rivera-López J, Ordorica-Falomir C, Wesche-Ebeling P (1999) Changes in anthocyanin concentration in lychee pericarp during maturation. Food Chem 65:195–200
Sakurai T, Nishioka H, Fujii H, Nakano N, Kizaki T, Radak Z, Izawa T, Haga S, Ohno H (2008) Antioxidative effects of a new lychee fruit-derived polyphenol mixture, Oligonol, converted into a low-molecular form in adipocytes. Biosci Biotechnol Biochem 72:463–476
Salunke DK, Desai B (1984 Postharvest biotechnology of fruits, vol II CRC Press, Boca Raton pp 77–80

Sampson JA (1980) Tropical fruits. Longman, London/New York

Sarni-Manchado P, Le Roux E, Le Guerneve C, Lozano Y, Cheynier V (2000) Phenolic composition of litchi fruit pericarp. J Agric Food Chem 48:5995–6002

Saxena S, Hajare SN, More V, Kumar S, Wadhawan S, Mishra BB, Parte MN (2011) Antioxidant and radioprotective properties of commercially grown litchi (*Litchi chinensis*) from India. Food Chem 126:39–45

Scott KJ, Brown BI, Chaplin GR, Wilcox ME, Bain JM (1982) The control of rotting and browning of litchi fruit by hot benomyl and plastic film. Sci Hortic 16:253–262

Shen MF (1991 Chinese traditional patent. Medicine (Baltim) 11:24–26

Shrivastava A, Srikantiah P, Kumar A, Bhushan G, Goel K, Kumar S, Kumar T et al (2015) Outbreaks of unexplained neurologic illness, Muzaffarpur, India. MMWR Morb Mortal Wkly Rep 64(3):49–53

Sivakumar D, Korsten L (2006) Evaluation of integrated application of two types of modified atmosphere packaging and hot water treatments on the quality and retention of the litchi cultivar McLean's red. J Hortic Sci Biotechnol 81:639–644

Spencer PS, Palmer VS, Mazumder R (2015) Probable toxic cause for suspected lychee-linked viral encephalitis. Emerg Infect Dis 21(5):904–905

Underhill SJR, Coates LM, Saks Y, Mitra LS (eds) (1997) Postharvest physiology and storage of tropical and subtropical fruits. CAB International, Wallingford, pp 191–208

Underhill S, Critchley C (1994) Anthocyanin decolorisation and its role in lychee pericarp browning. Aust J Exp Agric 34:115–122

Wall MM (2006) Ascorbic acid, vitamin A, and mineral composition of banana (*Musa* sp.) and papaya (*Carica papaya*) cultivars grown in Hawai'i. J Food Compos Anal 19:434–445

Wang L, Lou G, Ma Z, Liu X (2011) Chemical constituents with antioxidant activities from litchi (*Litchi chinensis* Sonn.) seeds. Food Chem 126:1081–1087

Weemaes CA, Ludikhuyze LR, Van den Broeck I, Hendrickx ME, Tobback PP (1998) Activity, electrophoretic characteristic and heat inactivation of polyphenol oxidases from apples, avocados, grapes, pears and plums. Lebensmittel-Wissenschaft Und-Technologie-Food Science and Technology 31:44–49

Wenkem NS (1990) Food of Hawaii and the Pacific Basin fruit and fruit products, vol 4. Hawaii Agriculture Experimental Station Research and Extension Services, pp 110–196

Wiart C (2006) Medicinal plants of Asia and the Pacific. CRC Press/Taylor & Francis, Boca Raton

Wu MC, Chen CS (1999) A research note: effect of sugar types and citric acid content on the quality of canned lychee. J Food Qual 22:461–469

Xu X, Xie H, Wang Y, Wei X (2010) A-type proanthocyanidins from lychee seeds and their antioxidant and antiviral activities. J Agric Food Chem 58:11667–11672

Xu X, Xie H, Hao J, Jiang Y, Wei X (2011) Flavonoid glycosides from the seeds of *Litchi chinensis*. J Agric Food Chem 59:1205–1209

Yamanishi R, Yoshigai E, Okuyama T, Mori M, Murase H, Machida T, Okumura T, Nishizawa M (2014) The anti-inflammatory effects of flavanol-rich lychee fruit extract in rat hepatocytes. PLoS One 9:e93818

Yan X, Zhou J, Xie G, Milne GWA (1999) Traditional Chinese medicines: molecular structures, natural sources, and applications. Ashgate, Hampshire

Yee W (1982)The lychee in Hawaii. University of Hawaii Coop, Extension Service, Circular 36622.

Young TW (1955) Branch girdling and root pruning trails on lychees in Florida. Proc Fla Lychee Growers Assoc 3:16–18

Zhang H, Teng Y (1986) Effect of li ren (semen litchi) antidiabetes pills in 45 cases of diabetes mellitus. J Tradit Chin Med 6:277–278

Zhang D, Quantick PC, Grigor JM (2000) Changes in phenolic compounds in litchi (*Litchi chinensis* Sonn.) fruit during postharvest storage. Post Harvest Biol Technol 19:165–172

Zhang Z, Pang X, Xuequan P, Ji Z, Jiang Y (2005) Role of the peroxidase in anthocyanin degradation in litchi fruit pericarp. Food Chem 90:47–52

Zhou Y, Wang H, Yang R, Huang H, Sun Y, Shen Y, Lei H, Gao H (2012) Effects of *Litchi chinensis* fruit isolates on prostaglandin E_2 and nitric oxide production in J774 murine macrophage cells. BMC Compl Altern Med 12:20–28

Litchi Deterioration in India

15

Shiwani Chaudhary, Manoj Kumar, and M.L. Aggarwal

Abstract

The litchi, initially an inhabitant of South China, came to India close to the eighteenth century. It is a significant evergreen soapberry tree 10 to 28 m (33–92 ft) tall and with fleshy pink fruit as long as 5 cm and 4 cm wide, weighing around 20 g. India stands second in the world of litchi production with a total production of 428,900 metric tons. The composition of litchi fruit is juice (60%), rag (8%), seed (19%) and 13% skin, changing with variety and weather. Also, this is an excellent source of vitamins, such as vitamin C, as well as bearing insignificant amounts of protein, fat, pectin and minerals, specially calcium, phosphorus and iron. The fruit is known for high perishability at climate temperatures with 2 to 3 days shelf life. It is cultivated best in regions with short, dry and cool winters and summers with high rainfall (1200 mm). The requirement of high humidity is an essential environmental factor. A mild, cold and dry winter is complementary conditions for litchi flowers. Various types of insects have been reported as fatal attackers of litchi such as *Platypeplus aprobola* Meyer, *Blastobasis* spp., grey weevil, *Indarbela tetraonis*, *Aceria litchi* Keifer, *Planococcus citri*, and snails. It has been observed that frequent disease (fruit rot, brown blight, etc.) is responsible for tree decline and death.

Keywords

Litchi (*Litchi chinensis* Sonn.) • Environmental factors • Major pests and diseases

---

S. Chaudhary • M. Kumar (✉)
Amity Institute of Microbial Technology, Amity University, Noida, Uttar Pradesh-201311, India
e-mail: Shiwani19chaudhary@gmail.com; manojjnu@gmail.com

M.L. Aggarwal
Shriram Institute For Industrial Research, New Delhi, India

M. Kumar et al. (eds.), *Lychee Disease Management*,
DOI 10.1007/978-981-10-4247-8_15

15.1 Introduction

Litchi (*Litchi chinensis* Sonn.), a vital evergreen fruit, is a member of the genus *Litchi* belonging to the family Sapindaceae (Morton 1987). Litchi is famous for its outstanding quality, pleasant flavour and juicy pulp (aril) with an attractive red colour. Although litchi is cherished very much as a table fruit, dried and canned litchis are also popular. The fresh fruit has a "delicate, whitish pulp" with a floral smell and a fragrant, sweet flavour. Litchi was originally an inhabitant of South China; one can still find these wild trees growing in the rainforests of Guangdong Province and on Hainan Island. Informal records in China refer to litchi as far back as 2000 BC (Schaffer and Andersen 1994). In India, litchi was introduced in the eighteenth century through Burma, and from there, it spread to many countries. India stands second in the world in litchi production after China. India and China account for 91% of the world litchi production with a total production of 428,900 metric tonnes from 56,200 hectares of land. Major areas are in north Bihar covering Muzaffarpur, Vaishali, Samastipur, Begusarai, east and west Champaran and Bhagalpur districts. As litchi is exacting in climatic requirements, it is confined to a few states with 74% of production recorded in Bihar. In this state, litchi is the livelihood for millions of people as it provides both on-farm and off-farm employment. Small and marginal farmers obtain additional income from litchi plants in their homesteads. Thus, litchi cultivation is the livelihood security for a large population, especially in the state of Bihar. Because its perfume-like flavour is lost in the process of canning, the fruit is usually eaten fresh (Davidson et al. 2006). These fruits typically have a higher price as its flesh is highly edible (Courtney 2005). The fruit consists of 60% juice, 8% rag, 19% seed and 13% skin, varying with variety and weather. Litchi is also an outstanding source of vitamin C, but it contains insignificant amount of protein, pectin and minerals, particularly calcium, phosphorus and iron. Litchi may vary broadly not only in colour, size, texture and shape but also in bio-input composition and levels (Table 15.1).

The litchi tree litchi is attractive, slow growing, dense and round topped with evergreen, sharply pointed leaves with different colours, from light green to dark green. Flowers are greenish white or yellowish in colour, borne in bunches. The shape of fruits is round or heart with thin, leathery skin and a red, rose or pinkish colour. The aril is the edible portion of the fruit, immediately under the skin. Its flavour differs with the cultivar, which is unique. Seeds are bold, but in some cases seeds are moderately developed from failure of pollination and are denoted as 'chicken-tongue' seed. The small-seeded fruits of litchi are more valued because the portion of pulp is larger. As litchi fruit is highly specific to climatic requirements, litchi cultivation is limited in the world. As predicted before its consumption, the postharvest loss of this fruit may reach 50% (Jiang and Fu 1998).

Litchi is also known as an important crop for export. Presently, as the domestic market is stretched, the export of litchi remains relatively small in India. For export to distant domestic markets, the product is normally packed in 2-kg cartons after pre-cooling and sulphuring. The growth of litchi in different states under different climatic conditions has advantages regarding earliness and extended harvest. Fruits

Table 15.1 Food value (per 100 g) of edible portion of litchi (Neog and Saika 2010)

Bio-inputs	Fresh	Dried
Calories	63–64	277
Moisture	81.9–84.83%	17.9–22.3%
Protein	0.68–1.0 g	2.9–3.8 g
Fat	0.3–0.58 g	0.2–1.2 g
Carbohydrates	13.31–16.4 g	70.7–77.5 g
Fibre	0.23–0.4 g	1.4 g
Ash	0.37–0.5 g	1.5–2.0 g
Calcium	8–10 mg	33 mg
Phosphorus	30–42 mg	–
Iron	0.4 mg	1.7 mg
Sodium	1–3 mg	3 mg
Potassium	170 mg	1100 mg
Thiamine	28 μg	–
Nicotinic acid	0.4 mg	–
Riboflavin	0.05–0.07 mg	0.05 mg
Ascorbic acid	24–60 mg	42 mg

of litchi are ready to use only for 3 to 4 weeks under specific climatic conditions. However, the extent of cultivation over a wide range of climate conditions provides a good option for increasing the period for cropping from the first week of May to the first week of July. Some types of diseases affect mainly three parts of the litchi, the flowers, leaves and fruits, as well as causing death or weakening of the tree. Litchi flowers, fruits, leaves and branches also are affected by numerous groups of insects. Hence, much effort is needed to avoid the death of litchi trees. In a growing market, many chemicals are available to prevent diseases on the flowers and fruit.

The present chapter is focused on the factors responsible for the deterioration of the litchi and also on evaluating the impact of different factors on litchi plants.

15.2 Environmental Factors

In litchi trees, the finest vegetative and reproductive growth are highly sensitive to the climate. The climate determines the optimum growth and development of a particular commodity in a specific area. The central force in the environment is climate, which constitutes different factors with interactions and actions that must be examined in detail to successfully produce litchi in a particular area. Temperature, relative humidity and light are the major climatic elements that vary at different stages of litchi production, as described in detail here.

15.2.1 Temperature

The fruit of litchi is the most temperature-sensitive fruit, growing best in winters (short, dry and cool but fog free) and summers (long and hot). Litchi flowers best with a daytime temperature below 20 °C. Factors such as duration, time of flowering, strength, sex ratio, fruit set, fruit growth, development and ripening time of fruits are governed by temperature. The total period of low temperatures has been considered more important than frequency. Therefore, it is concluded that the winters (cold and dry) are favourable conditions for flowering. Both the length of flushing and the pause between flushes are reduced by high temperature. As well, the rate of reproductive growth with panicles developing is also affected by low temperatures. During early flower development, most female flowers are associated with a temperature of 18 °C and fewer female flowers are seen at temperatures of 23 °C. However, a relationship occurs between the rate of flower opening and the number of flowers per panicle. Below 15 °C or 16 °C, growth is slowed.

Adult trees are killed at −4 °C or −5 °C, even though they may have withstood temperatures at −6 °C for some hours in the resting state with only slight damage, and freezing temperatures, −1 °C or −2 °C, can cause serious damage to tender shoots and young trees, which usually causes death.

Temperature shows strong effects on pollination. A temperature of 19–22 °C is optimal for pollination, with the maximum rate of fertilization obtained after 7 days. At 15 °C, elongation of the pollen tube is strongly inhibited. However, at least 10% of ovules contain pollen tubes at 15–27 °C, indicating that they are fertilized.

15.2.2 Relative Humidity

High relative humidity is usually associated with the months of July to September. Humid conditions increase the risk of disease occurrence. The chances of fruit borer infestation increase with high relative humidity. Sufficient irrigation water with humidity and low rainfall provide growth of young trees with less chance of disease. However, for rapid growth and early high yields, warm and humid are favorable conditions. The cost of production is increased with high humidity as well as reduced fruit cracking.

15.2.3 Light

Light may bound the development of flowers although high sunshine hours are predictable to be correlated with higher temperatures and therefore early anthesis. The photoperiod also affects floral initiation. When it is reduced from 16 to 8 h, there is an increase in flower anthesis (type II). It is generally thought that bright light is essential for commercial production; however, sometimes 9–15 h of sunlight per day has no effect on production.

Cloudy weather or low sunshine hours are responsible for the reduction of fruit development. It has been reported that more than 75% of terminal branches flower grow, even if the plants are shaded for several months before flowering. Heavy shade for 1 week reduces fruit production. Thus, solar radiation is an essential factor for litchi fruit crops.

15.3 Major Pests and Diseases

Litchi leaves, flowers and fruit are affected by a few diseases that cause the death or decline of trees. In China, India and Australia, diseases such as anthracnose (*Colletotrichum gloeosporioides*) are found in the litchi fruits as well as parasitic algae and nematodes affecting some orchards. Such diseases can be prevented with chemicals available in the market.

15.3.1 Insect Pests

(a) Leaf roller (*Platypeplus Aprobola* Meyer)

The new flushes are attacked by this pest throughout the litchi growing area. The green-coloured caterpillar rolls the leaf and feeds on lamina within the roll. Plants belonging to the genus *Hibiscus* act as the main host plant. Therefore; such plants should not be grown near litchi orchards.

(b) Litchi nut borer (*Blastobasis* spp.)

Litchi nut borers feed inside the fruit of litchi by entering through a small pinhead-size hole near the attachment of the peduncle. Fruits are damaged and unfit for consumption after being affected by this insect.

(c) Grey weevil

A small grey weevil feeds on leaf edges, causing U-shaped cuts. The weevil remains active throughout the flushing period. The larvae feeding on roots could be checked by deep hoeing before panicle emergence.

(d) Bark eating-caterpillar (*Indarbela tetraonis*)

This species is a serious pest in neglected litchi orchards. Only larvae after hatching make holes in scaffold crotches; they feed on the surface of the bark. The faeces and brass of wood hangs down at the holes. The girdling caused by bark eating can be injurious to trees.

(e) Erionose mite (*Aceria litchi* keifer)

Both adults and nymphs damage the leaves. Mites feed on the leaf undersurface. The leaf tissues are punctured and lacerate by mites through their sucking cell sap and stout rostrum. Such leaves become thick, curl up forming hollow cylinders and finally dry. Young plants and the nursery are severely attacked. Mites usually attack in March, and maximum activity has been observed in July–August. Subsequent infection occurs when the mites are stimulated around the orchard by direct contact between trees, or transmitted by wind, bees and orchard workers (Waite and McAlpine 1992).

(f) Citrus mealy bug (*Planococcus citri*)

Mealy bugs cause damage during the flowering and fruiting stages, that is, February to April. Large numbers of nymphs crawls up the trees, congregate on the panicles or young shoots, and suck the cell sap. In the month of June the first maturation of nymphs occurs, although old adults still occur on the litchi trees for up to a year, and mostly die in the month of August. The new adults do not mate immediately; they mature over the winter and then mate. Adults and nymphs subsist on terminal, which may be killed, and also cause fall of fruit and flowers. Liu and Lai (1998) declare that up to 30% of fruit in commercial orchards are damaged despite chemical applications.

(g) Snails

Snails do not belong to class Insecta but have become a major pest of litchi orchards, attacking the trunks and scaffolds and sucking cell sap. Snails are very active during the rainy season, July through August.

(h) Bats and birds

Both cause considerable loss of fruit. Bats and birds can be scared off by the beating of drums or firing off crackers or flagging with bright strips (Fig. 15.1).

15.3.2 Diseases

After harvesting and before picking of the fruit, many fruits are infected by diseases. The flowers, fruits and leaves of litchi are infected by organisms that sometimes are responsible for the decline and death of litchi trees. For controlling diseases on the flowers and fruit, chemicals are generally available in the market.

15.3.3 Fruit Rot

Fruit rot at the time of fruit harvest has been observed in some orchards. Rotted fruit cause rotting in other healthy fruits. The marketing of fruits is adversely affected.

Fig. 15.1 Effects of insects and pests on *Litchi chinensis*: leaf roller (**a**), litchi nut borer (**b**), grey weevil (**c**), bark-eating caterpillar (**d**), erionose mite (**e**), citrus mealy bug (**f**), snails (**g**) (Source: Department of Agriculture and Fisheries, ICAR, India)

Fig. 15.2 Some effects on litchi leaves and fruit: shoot drying (**a**) and lichens (**b**) (From Agricultural Info Zone)

15.3.4 Brown Blight

In China, India and Thailand, brown blight (*Peronophythora litchi*) is a most important disease in litchi, although more prevalent in China. During harvest the blight attacks and damages leaves and panicles, as well as fruit. The production and shelf life of litchi are reduced by the infection. The flower panicles are exceptionally susceptible. Immature fruit turn white from mildew growing on the skin to brown after infection. The infection of fungus spores is spread by insects, wind and rain, from soil or from an old infected fruit. Most infection occurs in constant wet weather and temperatures between 22 and 25 °C. After harvesting it is recommended that growers clean their orchard by removing infected dead branches and shade. The use of copper oxychloride during winter and of copper sulphate in spring also help decrease inoculum levels. Copper oxychloride and copper sulphate are replaced by fosetyl Al or metalaxylt for the development of the litchi flowers and fruit (Fig. 15.2).

References

Courtney M (2005) Litchi and longan: botany, production and uses. CABI, Wallingford, p 26. ISBN 0-85199-696-5

Davidson JL, Davidson A, Saberi H, Jaine T (2006) The Oxford companion to food. Oxford University Press, Oxford. ISBN 0-19-280681-5

Jiang Y, Fu J (1998) Inhibition of polyphenol oxidase and the browning control of litchi fruit by glutathione and citric acid. Food Chem 62(1):49–52

Liu XD, Lai CQ (1998) Experiment on control of litchi stink bug by using *Anastatus japonicus* Ashmead. South China Fruits 27:31

Morton JF (1987) Litchi in fruits of warm climates. Center for New Crops & Plant Products. Purdue University, Department of Horticulture and Landscape Architecture, West Lafayette, pp 249–259. Retrieved 30 October 2013

Neog M, Saika L (2010) Control of post-harvest pericarp browning of litchi (*Litchi chinensis* Sonn.) J Food Sci Technol 1:100–104

Schaffer B, Andersen PC (eds) (1994) Handbook of environmental physiology of fruit crops, Temperate crops, vol I. CRC Press, Boca Raton, p 358

Waite GK, McAlpine JD (1992) Honey bees as carriers of litchi erinose mite *Eriophyes litchii* (Acari: Eriophyiidae). Exp Appl Acarol 15:299–302

Integrated Approach for the Management of Differential Patterns of Diseases and Pest Incidence in Lychee

16

Nitika Thakur

Abstract

The chapter deals with the increasing threat of various diseases, insects and pests to litchi growers and their management by an integrated IPM approach. Litchi plants are generally least affected by the action of pathogens as compared to other fruits. The incidence of diseases is more prominent after the harvest stage, but many may be attacked at the time of picking. A number of organisms are associated with infection of leaves, flowers and fruit, resulting in tree decline and death. It has been reported that the fungal pathogens are more responsible for diseases. Enormous losses to litchi have been reported from pests and insects through direct infection on plant parts. Chemical use is an important traditional practice to fight against pests. To upgrade the lychee fruit quality, a more advanced technological intervention is necessary which can be achieved by the integration of conventional methods of practices with biocontrol methods. The new approach is referred as integrated pest management (IPM), which has arisen as an important principle of plant protection. This approach stands as an eco-friendly method for managing pests and disease problems by integrating such methods of disease, pest, and insect control. IPM manages scientific control in such a manner that economic losses are minimized and there is less hazardous chemical effect. IPM packages consist of a well-planned and well-executed management practice including pest monitoring, survey, and field scouting to monitor every developmental stage of a pest and to develop an effective strategy. On the whole these practices not only will minimize the ill effects of chemicals but will also prove a boost for the farmers, researchers and litchi growers to choose eco-friendly methods of management.

N. Thakur (✉)
Department of Biotechnology, Shoolini University, Solan, HP, India
e-mail: nitikathakur45@gmail.com

M. Kumar et al. (eds.), *Lychee Disease Management*,
DOI 10.1007/978-981-10-4247-8_16

Keywords
Lychee • IPM • Biocontrol methods • Quality

16.1 Introduction

Lychee plants are known for having fewer problems with pests and disease as compared to other seed-bearing plants, but an effective management system is needed to prevent these fewer diseases at an initial point for healthy fruits in the future. Litchi specifically is in demand for its fruit superiority and pulp nutrition (Chadha 1968). Pests, insects and postharvest losses affect its production and productivity. The major insect pests affecting the litchi:

1. Leaf mites and leaf miners
2. Borers of fruit and shoots and bark-feeding caterpillars

Phytophthora and Peronospora, for example, contribute significantly to pre-harvest losses (Prasad and Bilgrami 1974; Awasthi et al. 2005).

Surveillance study of pests and insect complexes of litchi showed such pests as leaf roller, fruit borer, midrib borer, litchi bug, and litchi leaf curl mite were the most damaging (GBPUA&T, Pantnagar). The leaf roller and midrib borer caused more than 50% of the damage, whereas litchi bug and leaf curl mite were less damaging. Litchi fruit borer and leaf roller infestations reached 43.50% and 58.90%, respectively, at BCKV, Mohanpur. However, the application of Tricho cards as biocontrol agents, with *Bacillus thuringiensis* (2 g l^{-1}), reduced the infestation as much as 9.67% at RAU, Pusa. The pruning and burning of all affected twigs/leaves just after harvest, followed by spraying of dicofol (0.05%) at the time of new flush, reduced the infestation for litchi leaf curl mite up to 71% at GBPUA&T, Pantnagar. The survey and surveillance of pollinators in litchi during the peak season of flowering showed the presence of *Apis* spp. (79.4%), non-*Apis* species (3.32%) and syrphids (8.26%) at medium levels at GBPUA&T, Pantnagar. However, *Apis dorsata*, *Apis mellifera*, *Trigona* spp., and the flies *Sarcophaga* spp. and *Lucilia* spp. were the major pollinators at BCKV, Mohanpur (Annual Report 2009–2010, Central Institute for Subtropical Horticulture) (Kumar et al. 2011). The losses at the post-harvest stage of fruits and vegetables vary in different zones of the countries (Cappellini and Ceponis 1984; Eckert and Ogawa 1985; Holmes and Eckert 1999; Prusky et al. 1985; Rosenberger and Meyer 1981; Vinas et al. 1991; Gullino and Kuijpers 1994; Ragsdale and Sisler 1994; Wojciech and Korsten 2002; Azizah et al. 2009).

16.1.1 Diseases Encountered in Lychee and Their Effective Management

Some diseases are sampled from any local growing conditions in a lychee orchard. Leaf spots which initiate at the leaf tip are caused by *Botryodiplodia theobromae* and *Colletotrichum gloeosporioides*, whereas lichens and algae are seen to infect the fruits, so a good circulation of air should be provided to keep the fruit dry and cool.

16.1.1.1 Brown Blight

The fungus *Peronophythora litchi* is known to cause blossom blight or downy mildew attacking the fruit (Fig. 16.1). This infection leads to deterioration in self-protection, thus reducing the shelf life. There are cases of infection of *Abacetus blandus* in litchi from Uttar Pradesh (UP) (Singh 1974).

16.1.1.1.1 Management Strategies

1. The lychee orchard should be cleared of any debris.
2. For reducing pathogen inoculum levels, a spray of copper oxychloride in combination with copper sulphate is recommended.
3. Use of metalaxyl spray for reducing disease occurrence is recommended.

16.1.2 Anthracnose

The disease anthracnose (Fig. 16.2) caused by *Botryodiplodia theobromae* is characterized by deep chocolate spots. The infection appears more on the upper side of the leaves (Anonymous 1992; Menzel 1990; Oosthuizen 2001) than on lower surfaces. These spots are characterised by irregularity and brown-coloured spores and fruit (Millian et al. 1994).

16.1.2.1 Management

1. Overcrowding of trees and branches in the orchard leads to increase in infection.
2. Burning and pruning of plants with affected regions is recommended.
3. Integrated management of the disease can reduce the rates of latent infection of the fruits.

A species of *Fusarium* has been identified for its association with necrosis and ultimately plant death.

1. Tree needs to be pruned.
2. Addition of organic manure is needed.
3. Propagation of trees from the diseased area cuttings in the soil should not be done.

Fig. 16.1 Brown blight of litchi. (Source: Miscellaneous plants pest and diseases image gallery)

Fig. 16.2 Anthracnose in litchi. (Source: Miscellaneous plants pest and diseases image gallery)

16.1.3 Red Rust

The highly infected leaves exhibit curling inwards towards the dorsal side. The disease first initiates infection on the young unfolded tender branches, as small dark

Fig. 16.3 Red rust in litchi. (Source: Miscellaneous plants pest and diseases image gallery)

isolated patches which spread very fast, developing into the velvet reddish brown to orange coloured cushion-like growth. The growth of this alga (Fig. 16.3) leads to the distortion of cork tissue in a few leaves, ultimately leading to death.

16.1.3.1 Disease Management

1. Spray with 5:5:50 Bordeaux mixture at 20-day intervals.
2. Ziram (0.25%) spray is known to reduce diseases.

16.1.4 Fruit Infected by Rot

Advancement of the disease depresses the diseased areas because of its deeper penetration into the developing pulp. Litchi stands as a host to a series of post-harvest pathogens, initiating a cascade of infections.

16.1.4.1 Management

1. The most important means of slowing development of rot is storage at low temperature conditions.
2. Health concerns restrict fungicide use.

16.1.5 Decline in Trees by Rot infection

The fungus may survive in the soil for many years, resulting in a decline, thus favouring death.

16.1.6 Important Insect Pests

A total of 42 combinations of pathogens have been recorded.

16.1.6.1 Erinose Mite

The erinose mite (a major litchi pest) leads to severe disease incidence. It has been recorded that the mites remain active (January to October), under the hairy and velvety growth (erineum), as they can change from one stage to another.

Both the nymphs and adult stages lead to leaf and fruit damage. The mites can puncture the sap of the cell and finally cause undernutrition of the leaves, causing deteriorated epidermal growth of the undersurfaces of the leaves. A severe attack leads to failure of leaf buds to bear flower or fruit.

Developing nursery plantlets are very susceptible to infection through wind dispersal from infected leaves.

16.1.6.1.1 Management

Litchi mite control measures must be employed for successive control once the mite is stable, hence depending upon infection severity:

1. Preparation of layers from uninfected plants.
2. Young saplings must be treated with dimethoate (0.05%) when they leave the nursery.
3. Infected portions should be regularly burned or cut off.
4. In the month of October, trees should be treated with dimethoate either alone or in combination with 0.12% dicofol

16.1.6.2 Bark-Eating Caterpillar

The bark-eating caterpillars damage litchi to a considerable extent, attacking trees of all ages, particularly those at the mature stage, degrading their usefulness. Old and neglected orchards are more open to attack by this pest, which infests the entire branch, and the tree ultimately dies.

The female moth lays eggs in clusters in June in bark crevices. Eggs hatch after 8 to 10 days and newly emerged caterpillars appear. The newly emerged caterpillars start coming out at the bark, characterised by the presence of long, thick, blackish or brownish ribbon-like masses of small chips of wood and excreta, which intermix with the help of a material that acts as an adhesive usually secreted by the caterpillar. After an interval of 2 to 3 days, the larvae bore into the trunk or main branches, usually making a downward tunnel with one larva in each hole; there may be 2 to 16 holes in each tree, depending upon the intensity of infection and the tree age. By continuously damaging the tissues, the larvae go through the stem and branches. The caterpillars finally remain within the gradually bored holes during the day and come out at night to feed upon the bark. The attacked trees show the presence of extensive faecal matter which generally disrupts the translocation of sap, and the growth of the tree is arrested and put to a halt, which in turn adversely affects the fruiting capacity of the tree. Severe injury weakens the stem, resulting in branch drying and finally that of the tree itself.

16.1.6.2.1 Management

1. Insertion of an iron spoke into the tunnels can cause killing of the caterpillars.
2. Injection of kerosene oil into the tunnels has also been successfully employed for effective control.

16.1.6.3 Litchi Leaf Roller

High leaf roller incidence is observed during July, extending to February, resulting in higher larval frequency from December to February. The last instar larvae pupate in a larval clip, concealing themselves by bending in a small portion of the leaf on the margin, both anteriorly and posteriorly.

Rolling of the tender leaf shows the symptoms and feeding of larvae, leading to leaf injuries that result in distortion and withering of infested twigs.

16.1.6.3.1 Management

1. Leaf roller damage is tolerable until it is infecting foliage and not the initiation of flowering process.
2. To minimize the damage, it is necessary to apply carbaryl 2 g/l.

16.1.6.4 Litchi Fruit Borer

This borers generally elicits economic losses, as is apparently found infecting the litchi most often.

The infested fruits do not attain normal size, and infection can be traced by the formation of black spots near the pedicel. Although the larvae do not enter much deeper into the pulp, feeding is witnessed below the calyx (15 mm deep). The pests come out and start spinning cocoons and pupate on the leaf surface, causing direct damage to litchi fruits.

The females generally depends on fruits over shoots for oviposition because the survival rate is much higher on shoots when compared to that in fruits.

16.1.6.4.1 Management

1. Exclusion of moths.
2. Fallen fruit should be removed.
3. Fruits must be inspected at weekly intervals.
4. Affected shoots should be removed.

16.1.6.5 Shoot Borer

The caterpillars damage the new shoots by tunnelling from the growing tip downwards. As a result, the shoots finally droops, dries, and withers away. Serious damage from August to October caused by the shoot borer has been seen throughout the country. The female moths preferably lay eggs on tender leaves; after hatching, the larvae first bore into the veins of young leaves and later into shoots with soft

Fig. 16.4 Litchi bug. (Source: Miscellaneous plants pest and diseases image gallery)

stems and inflorescence stalks. When fully grown, these caterpillars come out and invade cracks in the tree bark, dried inflorescences or cracks for pupation.

16.1.6.5.1 Management

1. Capture moths in sufficient darkness.
2. Properly screen and bag moths.
3. Practice regular and proper disposal of all spoiled and attacked fruit.

16.1.6.6 Litchi Bugs

The attack of a litchi bug (*Tessaratoma javanica*) (Fig. 16.4) is the most pathetic invasion as it lays globular pink eggs mostly in branches or on leaves (lower surface).

The soft-bodied and white-coloured nymph feeds on new buds, leaf petioles and fruit stalks, leading to fruit drop.

16.1.6.6.1 Management

1. Eggs are laid in groups by *T. javanica* and so are easily visible and can be removed and destroyed.
2. Natural enemies can be used for biological means of control.

16.1.6.7 Leaf-Eating Weevils

This weevil is a polyphagous pest with adults having a long grey-coloured snout. It attacks leaves, shoots and flowers.

Adult weevils feed on the tender leaves, nibble irregular holes on the leaves and sometimes destroy the entire leaf, leaving the midrib only.

Another weevil recorded recently at NRC Litchi, thus far not properly identified, feeds on the tender leaves. The damage of this weevil is more severe at the time of shoot emergence.

16.1.6.7.1 Management

Spraying of carbaryl at 2 ml/l is recommended for severe damage.

16.1.6.8 Litchi Affected by the Nut Borer

The nut borer is a pest infesting 33 crops. Eggs are scale like, laid singly, infecting fruits.

The damaged mature fruit may affect the other healthy fruits in a cluster, resulting in more damage.

16.1.6.8.1 Management

1. Checking the stage of fruit development weekly for larval entry holes is an effective control.
2. The nut is damaged by borers, leading to economic disturbances among farmers.

16.1.6.9 Scale

Disturbance of the natural balance of predators leads to an increase in scale populations. The infestation of scale insects is generally a slow process because of their immobile nature.

16.1.7 A Shift from Conventional to an Integrated Approach

Insect pests through direct and indirect invasion cause enormous loss to litchi; however, nowadays various side effects have been encountered with the use of chemicals. The successful method for effective management is called integrated pest management (IPM).

16.1.7.1 Strategies for Successful IPM

16.1.7.1.1 Stage 1: Pest Monitoring

(a) Survey
   1. The purpose of the survey is to trace the initial development of pest and insects (Fig. 16.5) in endemic areas.

(b) Field scouting
   1. This step is taken by the agencies and the farmers to trace out the increasing and decreasing patterns of insect pests and the effectiveness of the biocontrol methods employed. This step should be done immediately after the appearance of new flush after the fall of old leaves as succulent tissues are more susceptible to pests at this stage of the crop.

Major Pests

Pests of National Significance	Pests of regional significance
a) Insect Pest	**Insect pest**
1. Fruit and stone borer	Bark eating caterpillar
2. Litchi mite	leaf miner
3. Litchi folder	Shoot borer
4. Mealy bug	White fly
b) Diseases	**Diseases**
1. Fruit rot	Dieback
2. Powdery mildew	

Fig. 16.5 Pests of national and regional significance

(c) Pest monitoring through traps

1. Through yellow sticky traps

   The sticky traps are generally set as one trap for five trees. Locally prepared yellow Palmolive tins coated with grease or Vaseline can also be used.

2. Through pheromone traps

   Some pests require the installation of pheromone traps for initial pest buildup. Sticky pheromone traps of 5 to 7 traps/ha can be used.

(d) Sampling in fruit crops:

1. The fruit crops are perennial in nature, and before starting the surveillance process, an inspector or scout who is going to implement the activity should know about the nature of the crop as well as different crop stages and its growth stages.

(e) For insect pests:

1. For mites, mealy bug, leaf roller, shoot borer, leaf miner and whitefly: randomly record the number of both nymphs and adults present on five apical twigs
2. For fruit borer, fruit-sucking moth, nut borer and fruit fly: record the total number of fruits, fruits damaged by fruit borer and larval population

(f) Disease stage:

1. In the scouting stage one should be aware that symptoms of plant disease problems may be caused by any biotic factors such as fungal and bacterial pathogens or abiotic factors such as weather, fertilizers, nutrient deficiencies, pesticides and soil problems. For correct and close examination, general laboratory culture diagnosis is required. Fungal diseases cause clear symptoms with irregularity in growth and colour pattern (except viruses).

2. Sampling (root): Observations should be recorded on the prevailing signs of the causal organism (fungal growth or ooze). It is necessary to wash the roots with water to examine them properly. If the roots are well developed, examine the roots by cutting portions to observe internal infections. Finally, count the total number of pseudo-stems damaged by rot.

(g) Sampling (leaf): Examine and record lesions on all leaves on each tree generally during the seedling and flowering stages. Observations should be recorded on a number of infected leaves (leaf area diameter).

16.1.7.1.2 Stage 2: IPM Strategies

(a) Cultural practices
   1. This practice involves repeatedly ploughing and wrapping alkathene bands around the tree during November–December to prevent destruction by mealy bugs. Repeated practices such as neem application, weeding and hoeing are important.

(b) Mechanical control practices
   1. After the mud plastering process, gauge alkathene should be used for wrapping the trunk portion and protecting it from juvenile stages.

(c) Biological control practices
   1. A number of parasites, predators and pathogens are effective against various pests of litchi.

(d) Botanical
   1. NSKE at 5% helps reduce pest populations.

(e) Chemical control practices
   1. Many foliar sprays are recommended such as a spray of Endosulphan at 2.0 ml/l water just before the flowering phase to prevent fruit and stone borers.

16.1.8 Pest Management at Integrated Level (IPM) Using AESA (Agro-Ecosystem Analysis)

The IPM has been evolving strategies to explain the disastrous impacts of chemical pesticides on the environment that ultimate affecting the growers' interest and economy.

16.1.9 Understanding Intricate Interactions in an Ecosystem

Understanding of the important interactions in an ecosystem is important in pest management.

Decision making in pest management requires a rigorous analysis of the agro-ecosystem. Farmers have to learn how to keep a check, analyse the situation and observe the crop to make proper decisions for crop management.

16.1.10 Strategies of AESA

The participants in AESA make a drawing to record all the observations on a large piece of paper (60 × 80 cm), requiring the participants/growers to observe closely and intensively the focal point of analysis and discussions that can be kept in a record.

AESA is an approach that can be employed by growers to analyse orchard situations with regard to pests.

16.1.11 Basic Components of AESA

(a) Population dynamics of pest and disease
(b) Conditions of soil
(c) Factors affecting climate
(d) Growers' past experience

16.2 Conclusion

Insect pests have been a major concern affecting the production and productivity of litchi. The most important pest complex consists mainly of *Conopomorpha* spp. (Lepidoptera, Cossidae). The consumers safety and security, as well as environment issues, are highly dependent on chemical pest management. It has been seen that most stakeholders depend upon economic threshold levels (ETL). Future work and studies are ongoing for development made by resource personnel for upgrading field technologies. AESA-based IPM packages focus on stakeholders and the farming community to promote strategies which are sustainable and eco-friendly.

References

Anonymous (1992) The cultivation of litchis, Bulletin 425 of the Agricultural Research Council–ITSC. Agricultural Research Council, Nelspruit

Awasthi DP, Sarkar S, Mishra NK, Kaiser SAKM (2005) Disease situation of some major fruit crops in new alluvial plains of West Bengal. Environ Ecol 23:497

Azizah O, Saari N, Saleh R, Bakar J, Zainal ND, Yacob M (2009) Postharvest handling practices on selected local fruits and vegetables at different levels of the distribution chain. J Agribus Market 2:39–53

Cappellini RA, Ceponis MJ (1984) Postharvest losses in fresh fruits and vegetables. In: Moline HE (ed) Postharvest pathology of fruits and vegetables: postharvest losses in perishable crops. Agriculture Experiment Station, Berkeley, pp 24–30

Chadha KL (1968) Litchi cultivation in India. Indian Hortic 12:13–16

Eckert JW, Ogawa JM (1985) The chemical control of postharvest diseases: subtropical and tropical fruits. Annu Rev Phytopathol 23:421–454

Gullino ML, Kuijpers L (1994) Social and political implications of managing plant diseases with restricted fungicides in Europe. Annu Rev Phytopathol 32:559–579

Holmes GJ, Eckert JW (1999) Sensitivity of *Penicillium digitatum* and *P. italicum* to postharvest citrus fungicides in California. Phytopathology 89:716–721

Kumar V, Kumar A, Nath V (2011) Emerging pests and diseases of litchi (*Litchi chinensis* Sonn.) Nat Res Pest Manag Hortic Ecosyst 17(1):11–13

Menzel CM (1990) The lychee (*Litchi chinensis* Sonn.) in South Africa, Reunion and Mauritius. Queensland Department of Primary Industries, Nambour, p 18

Oosthuizen JH (2001) Lychee cultivation in South Africa. Yearb Australian Lychee Growers Assoc 1:51–55

Prasad SS, Bilgrami RS (1974) Investigations on diseases of litchi: VI. Post harvest diseases of fruit in India. Plant Dis Rep 58(12):1134–1136

Prusky D, Bazak M, Ben Arie R (1985) Development, persistence, survival and strategies for control of thiabendazole resistant strains of *Penicillium expansum* on pome fruits. Phytopathology 75:877–882

Ragsdale NN, Sisler HD (1994) Social and political implications of managing plant diseases with decreased availability of fungicides in the United States. Annu Rev Phytopathol 32:545–557

Rosenberger DA, Meyer FW (1981) Postharvest fungicides for apples: development of resistance to benomyl, vinclozolin, and iprodione. Plant Dis 65:1010–1013

Singh H (1974) *Apoderus blandus* Faust (Coleoptera: Curculionidae): a new pest of litchi (*Litchi chinensis* Sonn). Indian J Entomol 36:239–240

Vinas I, Usall J, Sanchis V (1991) Tolerance of *Penicillium expansum* to postharvest fungicide treatments in apple packing houses in Lerida (Spain). Mycopathologia 113:15–18

Wojciech JJ, Korsten L (2002) Biological control of postharvest diseases of fruits. Annu Rev Phytopathol 40:411–441

MIX
Papier aus verantwortungsvollen Quellen
Paper from responsible sources
FSC® C105338

If you have any concerns about our products,
you can contact us on
ProductSafety@springernature.com

In case Publisher is established outside the EU,
the EU authorized representative is:
Springer Nature Customer Service Center GmbH
Europaplatz 3, 69115 Heidelberg, Germany

Printed by Libri Plureos GmbH
in Hamburg, Germany